Communication Systems for the Mobile Information Society

Communication Systems for the Mobile Information Society

Martin Sauter
Nortel Networks, Germany

John Wiley & Sons, Ltd

Other Wiley Editorial Offices

John Wiley & Sons Inc., 111 River Street, Hoboken, NJ 07030, USA

Jossey-Bass, 989 Market Street, San Francisco, CA 94103-1741, USA

Wiley-VCH Verlag GmbH, Boschstr. 12, D-69469 Weinheim, Germany

John Wiley & Sons Australia Ltd, 42 McDougall Street, Milton, Queensland 4064, Australia

John Wiley & Sons (Asia) Pte Ltd, 2 Clementi Loop #02-01, Jin Xing Distripark, Singapore 129809

John Wiley & Sons Canada Ltd, 22 Worcester Road, Etobicoke, Ontario, Canada M9W 1L1

Wiley also publishes its books in a variety of electronic formats. Some content that appears in print may not be available in electronic books.

British Library Cataloguing in Publication Data

A catalogue record for this book is available from the British Library

ISBN-13 978-0-470-02676-2 (HB)
ISBN-10 0-470-02676-6 (HB)

Typeset in 10/12pt Times by Integra Software Services Pvt. Ltd, Pondicherry, India

FSC
Mixed Sources
Product group from well-managed
forests and other controlled sources

Cert no. SGS-COC-2953
www.fsc.org
© 1996 Forest Stewardship Council

Contents

Preface

Wireless technologies such as GSM/UMTS, wireless LAN, 802.16 (WiMAX), and Bluetooth have revolutionized the way we communicate and exchange data by making services like telephony and Internet access available at anytime and from almost anywhere. Today, a great variety of technical publications offer background information about these technologies but they all fall short in one way or another. Books covering these technologies usually describe only one of the systems in detail and are generally too complex as a first introduction. The Internet is also a good source, but the articles one finds are usually too short and superficial or only deal with a specific mechanism of one of the systems. Because of this, it was difficult for me to recommend a single publication to students in my telecommunication classes, which I've been teaching in addition to my chosen profession as a wireless systems consultant. This book aims to change this.

All wireless technologies discussed in the book continue to evolve, with increasing transmission speeds being the driving goal. This book covers some of the evolutions such as HSDPA and HSUPA enhancements, which deliver increased transmission speeds in UMTS networks, and EDGE, which does the same thing for GPRS. As WiMAX already offers high transmission speeds for stationary users (802.16d), the evolution path of this system introduces mobility. Therefore, the mobility extension of WiMAX (802.16e) is also discussed.

Beyond speed and mobility improvements, research is being performed into how future multi-mode wireless devices can offer anytime, anywhere connectivity. The challenge of this approach is determining how to offer a seamless transition from one radio technology to another for users roaming out of the coverage area of a network. As this book describes the similarities and differences between the major radio technologies, which will form the basis of such 4G networks, it also provides a wealth of information for readers involved in this area of research.

Each of the six chapters in this book gives a detailed introduction and overview of one of the wireless systems mentioned above. Special emphasis has also been put into explaining the thoughts and reasoning behind the development of each system. Not only the 'how', but also the 'why' is of central importance in each chapter. Furthermore, comparisons are made between the different technologies to show the differences and commonalities of the systems. For some applications, several technologies compete directly with each other, while in other cases only a combination of different wireless technologies creates a practical application for the end user. For readers who want to test their understanding of a system, each chapter concludes with a list of questions. For further investigation, all chapters contain references to the relevant standards and other documents. These provide an ideal additional source to find out more about a specific system or topic.

While working on the book, I've tremendously benefited from the wireless technologies that are already available today. Whether at home or while traveling, wireless LAN, Bluetooth, UMTS, and EDGE have provided reliable connectivity for my research and have allowed me to communicate with friends and loved ones at anytime, from anywhere. In a way, the book is a child of the technologies it describes.

The decision to write books about wireless systems in my free time came to me quite suddenly. While browsing a Paris bookshop, I discovered a book by Pierre Lescuyer, an author whom I did not know at this time but was working for the same company as myself. After I got in contact with him, he explained to me, over an extended lunch, how he went from an idea to his first finished book. I would like to thank Pierre for his invaluable advice, which has helped me many times since then.

Furthermore, my sincere thanks go to Berenice, who has stood by me during this project with her love, friendship, and good advice.

Also, I would like to thank Prashant John, Timothy Longman, Tim Smith, Peter van den Broek, Prem Jayaraj, Kevin Wriston, and Gregg Beyer for revising the different chapters in their free time and for their invaluable suggestions on content, style, and grammar.

Martin Sauter
Paris, France

List of Figures

List of Tables

List of Abbreviations

3GPP	3rd Generation Partnership Project
8DPSK	eight phase differential phase shift keying
A2DP	advanced audio distribution profile
AAA	authentication, authorization, and accounting
AAS	adaptive antenna system
ACL	asynchronous connection-less
ACM	address complete message
AES	advanced encryption standard
AFH	adaptive frequency hopping
AG	audio gateway
AGCH	access grant channel
AK	authentication key
AMC	adaptive modulation and coding
AMR	adaptive multi rate
ANM	answer message
APDU	application protocol data unit
APN	access point name
ARFCN	absolute radio frequency channel number
ARQ	automatic retransmission request
ASN	access service network
ASN-GW	access service network gateway
ATM	asynchronous transfer mode
AVCTP	audio/video control transport protocol
AVDTP	audio/video distribution transfer protocol
AVRCP	audio/video remote control profile
BCCH	broadcast common control channel
BCSM	basic call state model
BEP	bit error probability
BICN	bearer independent core network
BNEP	Bluetooth network encapsulation protocol
BS	best effort service
BSC	base station controller
BSN	block sequence number

BSS	base station subsystem
BSS	basic service set
BSSMAP	base station subsystem mobile application part
BTS	base transceiver station
CA	certification authority
CAMEL	customized applications for mobile enhanced logic
CCK	complementary code keying
CDMA	code division multiple access
CDR	call detail record
CID	connection-ID
COA	care-of IP address
CQI	channel quality index
CS	convergence sublayer
CSCF	call session control function
CSMA/CA	carrier sense multiple access/collision avoidance
CSMA/CD	carrier sense multiple access/collision detect
CTS	clear to send
DBPSK	differential binary phase shift keying
DCD	downlink channel description
DCF	distributed coordination function
DES	data encryption standard
DHCP	dynamic host configuration protocol
DLP	direct link protocol
DNS	domain name service
DQPSK	differential quadrature phase shift keying
DS0	digital signal level 0
DSP	digital signal processor
DSSS	direct sequence spread spectrum
DTAP	direct transfer application part
DTM	dual transfer mode
DTMF	dual tone multi frequency
DUN	dial-up network
EDGE	enhanced data rates for GSM evolution
EDR	enhanced data rates
EMLPP	enhanced multi level precedence and preemption
ESS	extended service set
ETSI	European Telecommunication Standards Institute
FACCH	fast associated control channel
FBSS	fast base station switching
FCCH	frequency correction channel
FCH	frame control header
FDD	frequency division duplex
FDMA	frequency division multiple access
FEC	forward error correction
FHS	frequency hopping synchronization
FHSS	frequency hopping spread spectrum

FTDMA	frequency and time division multiple access
FTP	file transfer profile
FTP	file transfer protocol
GAP	generic access profile
GFSK	Gaussian frequency shift keying
GGSN	gateway GPRS support node
GIF	graphics interchange format
GMM	GPRS mobility management
GMM/SM	GPRS mobility management and session management
GMSK	Gaussian minimum shift keying
GN	group ad-hoc network
GOEP	general object exchange profile
GPRS	General Packet Radio Service
GSM	Global System for Mobile Communications
GTP	GPRS tunneling protocol
HARQ	hybrid automatic retransmission request
HCI	host controller interface
HEC	header error check
HLR	home location register
HSDPA	high speed downlink packet access
HSUPA	high speed uplink packet access
HTML	hypertext markup language
HTTP	hypertext transfer protocol
IAM	initial address message
IAPP	inter access point protocol
IE	information element
IEEE	Institute of Electrical and Electronics Engineers
IMS	IP multimedia subsystem
IMSI	international mobile subscriber identity
IN	intelligent network subsystem
IP	Internet Protocol
IRAU	inter SGSN routing area update
ISDN	Integrated Services Digital Network
ISM	industrial, scientific, and medial
ISUP	ISDN user part
ITU	International Telecommunication Union
KEK	key encryption key
L2CAP	logical link control and adaptation protocol
LAC	location area code
LAN	local area network
LAP	LAN access profile
LAPD	link access protocol
LLC	logical link control
LMP	link manager protocol
LOS	line of sight
MAC	medium access control protocol

MAN	metropolitan area network
MAP	mobile application part
MCC	mobile country code
MDHO	macro diversity handover
MGCF	media gateway control function
MIME	multipart Internet mail extension
MIMO	multiple input–multiple output
MM	mobility management
MMS	multimedia messaging service
MNC	mobile network code
MS	mobile station
MSC	mobile switching center
MSIN	mobile subscriber identification number
MSRN	mobile station roaming number
MTP-1	message transfer part 1
NACC	network assisted cell change
NAP	network access point
NAV	network allocation vector
NLOS	non-line of sight
NOM	network operation mode
NrtPS	non real-time polling service
NSAPI	network subsystem access point identifier
NSS	network subsystem
OFDM	orthogonal frequency division multiplexing
OFDMA	orthogonal frequency division multiple access
OMA	Open Mobile Alliance
OVSF	orthogonal variable spreading factor
PACCH	packet associated control channel
PAGCH	packet access grant channel
PAN	personal area network
PCCCH	packet common control channel
PCH	paging channel
PCM	pulse code modulated
PCU	packet control unit
PDP	packet data protocol
PDTCH	packet data traffic channel
PDU	packet data unit
PHS	packet header suppression
PLCP	physical layer convergence procedure
PLMN	public land mobile network
PPCH	packet paging channel
PPP	point-to-point protocol
PRACH	packet random access channel
PSM	protocol service multiplexer
PSTN	public standard telephone network
PTCCH	packet timing advance control channel

PTT	push to talk
QAM	quadrature amplitude modulation
QoS	quality of service
RAB	radio access bearer
RACH	random access channel
RADIUS	remote authentication dial-in server
RANAP	radio access network application part
RAU	routing area update
REL	release
RISC	reduced instruction set
RLC	radio link control
RLC/MAC	radio link control/medium access control
RNC	radio network controller
RNSAP	radio network subsystem application part
RRC	radio resource control
RSSI	received signal strength indication
RTD	round-trip delay
RTG	receive/transmit transition gap
RtPS	real-time polling service
RTS	ready to send
SACCH	slow associated control channel
SAFER+	secure and fast encryption routine
SAP	service access point
SBC	sub-band codec
SCCP	signaling connection and control part
SCH	synchronization channel
SCO	synchronous connection-oriented
SCP	service control point
SDCCH	standalone dedicated control channel
SDP	service discovery protocol
SGSN	serving GPRS support node
SIB	system information block
SIFS	short interframe space
SIP	session initiation protocol
SM	session management
SMIL	synchronized multimedia integration language
SMS	short messaging service
SMSC	short messaging service center
SMTP	simple mail transfer protocol
SNDCP	subnetwork dependent convergence protocol
SPP	serial port profile
SRES	signed response
SRNS	serving radio network subsystem
SSL	secure socket layer
SSN	subsystem number

SSP	service switching point
STM	synchronous transfer mode
STP	signaling transfer point
TBF	temporary block flow
TCAP	transaction capability application part
TCH	traffic channel
TCP	transfer control protocol
TDD	time division duplex
TDMA	time division multiple access
TEK	traffic encryption key
TFI	temporary flow identity
TFS	transport format set
TFTP	trivial file transfer protocol
TID	tunnel identifier
TIM	traffic indication map
TMSI	temporary mobile subscriber identity
TRAU	transcoding and rate adaptation unit
TS	technical specification
TTG	transmit/receive transition gap
TTI	transmission time interval
UART	universal asynchronous receiver and transmitter
UDP	user datagram protocol
UE	user equipment
UGS	unsolicited grant service
UMTS	Universal Mobile Telecommunications System
URL	universal resource locator
USB	universal serial bus
USF	uplink state flag
USSD	unstructured supplementary service data
UTRAN	UMTS terrestrial radio access network
UUID	universally unique ID
VAD	voice activity detection
VBS	voice broadcast service
VGCS	voice group call service
VLR	visitor location register
VoIP	voice over IP
WAP	wireless application protocol
WBMP	wireless application bitmap protocol
WCDMA	wideband code division multiple access
WEP	wired equivalent privacy
WiMAX	worldwide interoperability for microwave access
WML	wireless markup language
WPA	wireless protected access
WSP	wireless session protocol
WTP	wireless transaction protocol

1

Global System for Mobile Communications (GSM)

At the beginning of the 1990s, GSM, the Global System for Mobile Communications triggered an unprecedented change in the way people communicate with each other. While earlier analog wireless systems were used by only a few people, GSM was used by over 1.5 billion subscribers worldwide at the end of 2005. This has mostly been achieved by the steady improvements in all areas of telecommunication technology and due to the steady price reductions for both infrastructure equipment and mobile phones. The first chapter of this book discusses the architecture of this system, which also forms the basis for the packet-switched extension called GPRS, discussed in Chapter 2, and for the Universal Mobile Telecommunications System (UMTS), which is described in Chapter 3. While the first designs of GSM date back to the middle of the 1980s, GSM is still the most widely used wireless technology worldwide and it is not expected to change any time soon. Despite its age and the evolution towards UMTS, GSM itself continues to be developed. As will be shown in this Chapter, GSM has been enhanced with many new features in recent years. Therefore, many operators continue to invest in their GSM networks in addition to their UMTS activities to introduce new functionality and to lower their operational cost.

1.1 Circuit-Switched Data Transmission

The GSM mobile telecommunication network has been designed as a circuit-switched network in a similar way to fixed-line phone networks. At the beginning of a call, the network establishes a direct connection between two parties, which is then used exclusively for this conversation. As shown in Figure 1.1, the switching center uses a switching matrix to connect any originating party to any destination party. Once the connection has been established, the conversation is then transparently transmitted via the switching matrix between the two parties. The switching center only becomes active again to clear the connection in the switching matrix if one of the parties wants to end the call. This approach is identical in both mobile and fixed-line networks. Early fixed-line telecommunication networks were only

Communication Systems for the Mobile Information Society   Martin Sauter
© 2006 John Wiley & Sons, Ltd

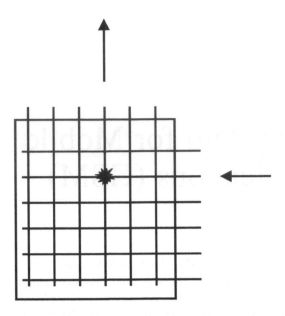

Figure 1.1 Switching matrix in a switching center

designed for voice communication for which an analog connection between the parties was established. In the mid-1980s, analog technology was superseded by digital technology in the switching center. This means that today, calls are no longer sent over an analog line from originator to terminator. Instead, the switching center digitizes the analog signal it receives from the subscribers, which are directly attached to it, and forwards the digitized signal to the terminating switching center. There, the digital signal is again converted back to an analog signal which is then sent over the copper cable to the terminating party. In some countries ISDN (Integrated Services Digital Network) lines are quite popular. With this system, the transmission is fully digital and the conversion back into an analog audio signal is done directly in the phone.

GSM reuses much of the fixed-line technology that was already available at the time the standards were created. Thus, existing technologies such as switching centers and long-distance communication equipment were used. The main development for GSM was the means to wirelessly connect the subscribers to the network. In fixed-line networks, subscriber connectivity is very simple as only two dedicated wires are necessary per user. In a GSM network, however, the subscribers are mobile and can change their location at any time. Thus, it is not always possible to use the same input and output in the switching matrix for a user as in fixed-line networks.

As a mobile network consists of many switching centers, with each covering a certain geographical area, it is not even possible to predict in advance which switching center a call should be forwarded to for a certain subscriber. This means that the software for subscriber management and routing of calls of fixed-line networks cannot be used for GSM. Instead of a static call-routing mechanism, a flexible mobility management architecture is necessary in the core network, which needs to be aware of the current location of the subscriber and is thus able to route calls to the subscribers at any time.

It is also necessary to be able to flexibly change the routing of an ongoing call as a subscriber can roam freely and thus might leave the coverage area of the radio transmitter

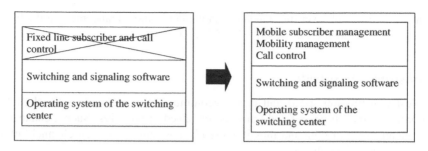

Figure 1.2 Necessary software changes to adapt a fixed-line switching center for a wireless network

of the network over which the call was established. While there is a big difference in the software of a fixed and a mobile switching center, the hardware as well as the lower layers of the software which are responsible for example for the handling of the switching matrix are mostly identical. Therefore, most telecommunication equipment vendors like Siemens, Nortel, Ericsson, Nokia, or Alcatel offer their switching center hardware both for fixed-line as well as for mobile networks. Only the software in the switching center decides if the hardware is used in a fixed or mobile network (see Figure 1.2).

1.2 Standards

As many telecom companies compete globally for orders of telecommunication network operators, standardization of interfaces and procedures is necessary. Without standards, which are defined by the International Telecommunication Union (ITU), it would not be possible to make phone calls internationally and network operators would be bound to the supplier they initially select for the delivery of their network components. One of the most important ITU standards discussed in Section 1.4 is the signaling system number 7 (SS-7), which is used for call routing. Many ITU standards, however, only represent the smallest common denominator as most countries have specified their own national extensions. In practice, this incurs a high cost for software development for each country as a different set of extensions needs to be implemented in order for a vendor to be able sell its equipment. Furthermore, the interconnection of networks of different countries is complicated by this.

GSM, for the first time, set a common standard for Europe for wireless networks, which has also been adopted by many countries outside Europe. This is the main reason why subscribers can roam in GSM networks across the world that have roaming agreements with each other. The common standard also substantially reduces research and development costs as hardware and software can now be sold worldwide with only minor adaptations for the local market. The European Telecommunication Standards Institute (ETSI), which is also responsible for a number of other standards, was the main body responsible for the creation of the GSM standard. The ETSI GSM standards are composed of a substantial number of standards documents each called a technical specification (TS), which describe a particular part of the system. In the following chapters, many of those specifications will be referenced and can thus be used for further information about a specific topic. All standards are freely available on the Internet at http://www.etsi.org [1] or at http://www.3gpp.org [2], which is the organization that took over the standards maintenance and enhancement at the beginning of the UMTS standardization as described in Chapter 3.

1.3 Transmission Speeds

The smallest transmission speed unit in a telecommunication network is the digital signal level 0 (DS0) channel. It has a fixed transmission speed of 64 kbit/s. Such a channel can be used to transfer voice or data and thus it is usually not called a speech channel but simply referred to as a user data channel.

The reference unit of a telecommunication network is an E-1 connection in Europe and a T-1 connection in the United States, which use either a twisted pair or coaxial copper cable. The gross data rate of an E-1 connection is 2.048 Mbit/s and 1.544 Mbit/s for a T-1. An E-1 is divided into 32 timeslots of 64 kbit/s each while a T-1 is divided into 24 timeslots of 64 kbit/s each. One of the timeslots is used for synchronization which means that 31 timeslots for an E-1 or 23 timeslots for a T-1 respectively can be used to transfer data. In practice, only 29 or 30 timeslots are used for user data transmission while the rest (usually one or two) are used for SS-7 signaling data (see Figure 1.3). More about SS-7 can be found in Section 1.4.

Most of the time a single E-1 connection with 31 DS0s is not enough to connect two switching centers with each other. In this case E-3 connections can be used, which are also carried over twisted pair or coaxial cables. An E-3 connection is defined at a speed of 34.368 Mbit/s, which corresponds to 512 DS0s.

For higher transmission speeds and for long distances, optical systems are used which use the synchronous transfer mode (STM) standard. Table 1.1 shows some data rates and the number of 64 kbit/s DS0 channels which are transmitted per pair of fiber.

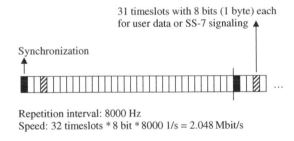

Figure 1.3 Timeslot architecture of an E-1 connection

Table 1.1 STM transmission speeds and number of DS0s

STM level	Speed	Approx. number of DS0 connections
STM-1	155.52 Mbit/s	2300
STM-4	622.08 Mbit/s	9500
STM-16	2488.32 Mbit/s	37,000
STM-64	9953.28 Mbit/s	148,279

1.4 The Signaling System Number 7

For establishing, maintaining, and clearing a connection, signaling information needs to be exchanged between the end user and network devices. In the fixed-line network, analog phones signal their connection request when the receiver is lifted off the hook and by dialing a phone number which is sent to the network either via pulses (pulse dialing) or via tone dialing which is called dual tone multi frequency (DTMF) dialing. With fixed-line ISDN phones and GSM mobile phones the signaling is done via a dedicated signaling channel, and information such as the destination phone number is sent via messages.

If several components in the network are involved in the call establishment, for example if originating and terminating parties are not connected to the same switching center, it is also necessary that the different nodes in the network exchange information with each other. This signaling is transparent for the user and a protocol called the signaling system number 7 (SS-7) is used for this purpose. SS-7 is also used in GSM networks and the standard has been enhanced by ETSI in order to be able to fulfill the special requirements of mobile networks, for example subscriber mobility management.

The SS-7 standard defines three basic types of network nodes:

- Service switching points (SSPs) are switching centers that are more generally referred to as network elements which are able to establish, transport, or forward voice and data connections.
- Service control points (SCPs) are databases and application software that can influence the establishment of a connection. In a GSM network, SCPs can be used for example for storing the current location of a subscriber. During call establishment to a mobile subscriber the switching centers query the database for the current location of the subscriber in order to be able to forward the call. More about this procedure can be found in Section 1.6.3 about the home location register.
- Signaling transfer points (STPs) are responsible for the forwarding of signaling messages between SSPs and SCPs as not all network nodes have a dedicated link to all other nodes of the network. The principal functionality of an STP can be compared to an IP router in the Internet, which also forwards packets to different branches of the network. Unlike IP routers however, STPs only forward signaling messages which are necessary for the establishing, maintaining, and clearing of a call. The calls themselves are directly carried on dedicated links between the SSPs.

Figure 1.4 shows the general structure of an SS-7 circuit-switched telecommunication network and how the nodes described above are interconnected with each other.

1.4.1 The SS-7 Protocol Stack

SS-7 comprises a number of protocols and layers. A well-known model for describing telecommunication protocols and different layers is the OSI 7 layer model which is used in Figure 1.5 to show the layers on which the different SS-7 protocols reside.

The message transfer part 1 (MTP-1) protocol describes the physical properties of the transmission medium on layer 1 of the OSI model. Thus, this layer is also called the physical layer. Properties that are standardized in MTP-1 are for example the definition of the different kinds of cables that can be used to carry the signal, signal levels, and transmission speeds.

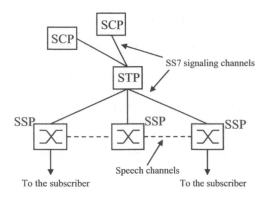

Figure 1.4   An SS-7 network with an STP, two SCP databases, and three switching centers

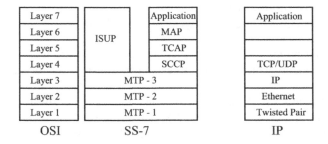

Figure 1.5   Comparison of the SS-7, OSI, and TCP/IP protocol stacks

On layer 2, the data link layer, messages are framed into packets and a start and stop identification at the beginning and end of each packet is inserted into the data stream so the receiver is able to detect where a message ends and a new message begins.

Layer 3 of the OSI model, which is called the network layer, is responsible for packet routing. In order to enable network nodes to forward incoming packets to other nodes, each packet gets a source and destination address on this layer. This is done by the MTP-3 protocol of the SS-7 stack. For readers who are already familiar with the TCP/IP protocol stack it may be noted at this point that the MTP-3 protocol fulfills the same tasks as the IP protocol. Instead of IP addresses, however, the MTP-3 protocol uses so-called point codes to identify the source and the destination of a message.

A number of different protocols are used on layers 4 to 7 depending on the application. If a message needs to be sent for the establishment or clearing of a call the ISDN user part (ISUP) protocol is used. Figure 1.6 shows how a call is established between two parties by using ISUP messages. In the example, party A is a mobile subscriber while party B is a fixed-line subscriber. Thus, A is connected to the network via a mobile switching center (MSC) while B is connected via a fixed-line switching center.

In order to call B, the phone number of B is sent by A to the MSC. The MSC then analyzes the national destination code of the phone number, which usually comprises the first two to four digits of the number, and detects that the number belongs to a subscriber in the fixed-line network. In the example shown in Figure 1.6, the MSC and the fixed-line

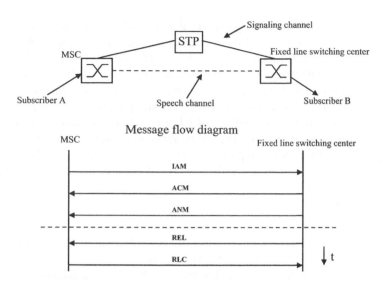

Figure 1.6 Establishment of a voice call between two switching centers

switching center are directly connected with each other. Therefore, the call can be directly forwarded to the terminating switching center. This is quite a realistic scenario as direct connections are often used if for example a mobile subscriber calls a fixed-line phone in the same city.

As B is a fixed-line subscriber, the next step for the MSC is to establish a voice channel to the fixed-line switching center. This is done by sending an ISUP initial address message (IAM). The message contains among other data the phone number of B and informs the fixed-line switching center and the channel which the MSC would like to use for the voice path. In the example, the IAM message is not sent directly to the fixed-line switching center. Instead, an STP is used to forward the message.

On the other end, the fixed-line switching center receives the message, analyzes the phone number, and establishes a connection via its switching matrix to subscriber B. Once the connection is established via the switching matrix, the switch applies a periodic current to the line of the fixed-line subscriber so the fixed-line phone can generate an alerting tone. To indicate to the originating subscriber that the phone number is complete and the destination party was found, the fixed-line switch sends back an address complete message (ACM). The MSC then knows that the number is complete and that the terminating party is being alerted of the incoming call.

If B answers the call, the fixed-line switching center sends an answer message (ANM) to the MSC and conversation can start.

When B ends the call, the fixed-line switching center resets the connection in the switching matrix and sends a release (REL) message to the MSC. The MSC confirms the termination of the connection by sending back a release complete (RLC) message. If A had terminated the call the messages would have been identical with only the direction of the REL and RLC reversed.

For the communication between the switching centers (SSPs) and the databases (SCPs), the signaling connection and control part (SCCP) is used on layer 4. SCCP is very similar to TCP and UDP in the IP world. Protocols on layer 4 of the protocol stack enable the distinction of different applications on a single system. TCP and UDP use ports to do this. If a PC for example is used as a web server and FTP server at the same time, both applications would be accessed over the network via the same IP address. However, while the web server can be reached via port 80, the FTP server waits for incoming data on port 21. Therefore, it is quite easy for the network protocol stack to decide which application to forward incoming data packets. In the SS-7 world, the task of forwarding incoming messages to the right application is done by SCCP. Instead of port numbers, SCCP uses subsystem numbers (SSNs).

For database access, the transaction capability application part (TCAP) protocol has been designed as part of the SS-7 family of protocols. TCAP defines a number of different modules and messages that can be used to query all kinds of different databases in a uniform way.

1.4.2 SS-7 Protocols for GSM

Apart from the fixed-line network SS-7 protocols, the following additional protocols were defined to address the special needs of a GSM network.

The mobile application part (MAP): this protocol has been standardized in 3GPP TS 29.002 [3] and is used for the communication between an MSC and the home location register (HLR) which maintains subscriber information. The HLR is queried for example if the MSC wants to establish a connection to a mobile subscriber. In this case, the HLR returns the information about the current location of the subscriber. The MSC is then able to forward the call to the responsible switching center for the mobile subscriber by establishing a voice channel between itself and the next hop by using the ISUP message flow that has been shown in Figure 1.6. MAP is also used between two MSCs if the subscriber moves into the coverage area of a different MSC while a call is ongoing. As shown in Figure 1.7, the MAP protocol uses the TCAP, SCCP, and MTP protocols on lower layers.

The base station subsystem mobile application part (BSSMAP): this protocol is used for the communication between the MSC and the radio network. Here, the additional protocol is necessary for example to establish a dedicated radio channel for a new connection to a mobile subscriber. As BSSMAP is not a database query language like the MAP protocol, BSSMAP is based on SCCP directly instead of using TCAP in between.

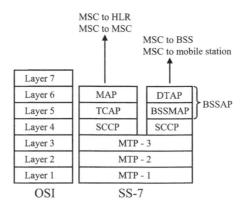

Figure 1.7 Enhancement of the SS-7 protocol stack for GSM

The direct transfer application part (DTAP): this protocol is used between the user's mobile phone, which is also called mobile station (MS), to communicate transparently with the MSC. In order to establish a voice call the MS sends a setup message to the MSC. As in the example in Section 1.4.1, this message contains among other things the phone number of the called subscriber. As it is only the MSC's task to forward calls, all network nodes between the MS and the MSC forward the message transparently and thus need not understand the DTAP protocol.

1.5 The GSM Subsystems

A GSM network is split into three subsystems which are described in more detail below:

- The base station subsystem (BSS), which is also called 'radio network', contains all nodes and functionalities that are necessary to wirelessly connect mobile subscribers over the radio interface to the network. The radio interface is usually also referred to as the 'air interface'.
- The network subsystem (NSS), which is also called 'core network', contains all nodes and functionalities that are necessary for switching of calls, for subscriber management and mobility management.
- The intelligent network subsystem (IN) comprises SCP databases which add optional functionality to the network. One of the most important optional IN functionality of a mobile network is the prepaid service, which allows subscribers to first fund an account with a certain amount of money which can then be used for network services like phone calls, SMS messages, and of course data services via GPRS and UMTS as described in Chapters 2 and 3. When a prepaid subscriber uses a service of the network, the responsible IN node is contacted and the amount the network operator charges for a service is deducted from the account in real time.

1.6 The Network Subsystem

The most important responsibilities of the NSS are call establishment, call control, and routing of calls between different fixed and mobile switching centers and other networks. Other networks are, for example, the national fixed-line network which is also called the public standard telephone network (PSTN), international fixed-line networks, other national and international mobile networks, and voice over IP (VoIP) networks. Furthermore, the NSS is responsible for subscriber management. The nodes necessary for these tasks are shown in Figure 1.8 and are further described in the next sections.

1.6.1 The Mobile Switching Center (MSC)

The mobile switching center (MSC) is the central element of a mobile telecommunication network, which is also called a public land mobile network (PLMN) in the standards. All connections between subscribers are managed by the MSC and are always routed over the switching matrix even if two subscribers that have established a connection communicate over the same radio cell. The management activities to establish and maintain a connection

are part of the call control (CC) protocol, which is generally responsible for the following
tasks:

- Registration of mobile subscribers: when the mobile station (MS) is switched on, it
  registers to the network and is then reachable by all other subscribers of the network.
- Call establishment and call routing between two subscribers.
- Forwarding of SMS (short messaging service) messages.

As subscribers can roam freely in the network, the MSC is also responsible for the mobility
management (MM) of subscribers. This activity is comprises the following tasks:

- Authentication of subscribers at connection establishment: this is necessary because a
  subscriber cannot be identified as in the fixed network by the pair of copper cables over
  which the signaling arrives. Authentication of subscribers and the authentication center
  are further discussed in Section 1.6.4.
- If no active connection exists between the network and the mobile station, the MSC has
  to report a change of location to the network in order to be reachable for incoming calls
  and SMS messages. This procedure is called location update and is further described in
  Section 1.8.1.
- If the subscriber changes its location while a connection is established with the network,
  the MSC is part of the process that ensures that the connection is not interrupted and is
  rerouted to the next cell. This procedure is called handover and is described in more detail
  in Section 1.8.3.

In order to enable the MSC to communicate with other nodes of the network, it is connected
to them via standardized interfaces as shown in Figure 1.8. This allows network operators
to buy different components for the network from different network vendors.

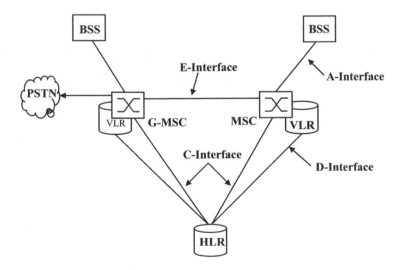

Figure 1.8 Interfaces and nodes in the NSS

The base station subsystem (BSS), which connects all subscribers to the core network, is connected to the MSCs via a number of 2 Mbit/s E-1 connections. This interface is called the A-interface. As has been shown in Section 1.4 the BSSMAP and DTAP protocols are used over the A-interface for communication between the MSC, the BSS, and the mobile stations. As an E-1 connection can only carry 31 channels, many E-1 connections are necessary to connect an MSC to the BSS. In practice, this means that many E-1s are bundled and sent over optical connections such as STM-1 to the BSS. Another reason to use an optical connection is that electrical signals can only be carried over long distances with great effort and it is not unusual that an MSC is over 100 kilometers away from the next BSS node.

As an MSC only has a limited switching capacity and processing power, a PLMN is usually composed of dozens or even hundreds of independent MSCs. Each MSC thus covers only a certain area of the network. In order to ensure connectivity beyond the immediate coverage area of an MSC, E-1s, which are again bundled into optical connections, are used to interconnect the different MSCs of a network. As a subscriber can roam into the area that is controlled by a different MSC while a connection is active, it is necessary to change the route of an active connection to the new MSC (handover). The necessary signaling connection is called the E-interface. ISUP is used for the establishment of the speech path between different MSCs and the MAP protocol is used for the handover signaling between the MSCs. Further information about the handover process can be found in Section 1.8.3.

The C-interface is used to connect the MSCs of a network with the home location register (HLR) of the mobile network. While the A-and E-interface, described previously, always consist of signaling and speech path links, the C-interface is a pure signaling link. Speech channels are not necessary for the C-interface as the HLR is a pure database which cannott accept or forward calls. Despite being only a signaling interface, E-1 connections are used for this interface. All timeslots are used for signaling purposes or are unused.

As has been shown in Section 1.3, a voice connection is carried on a 64 kbit/s E-1 timeslot in a circuit-switched fixed line or mobile network. Before the voice signal can be forwarded, it needs to be digitized. For an analog fixed-line connection this is done in the switching center, while an ISDN fixed-line phone or a GSM mobile phone digitizes the voice signal themselves.

An analog voice signal is digitized in three steps: in the first step, the bandwidth of the input signal is limited to 300–3400 Hz in order to be able to carry the signal with the limited bandwidth of a 64 kbit/s timeslot. Afterwards, the signal is sampled at a rate of 8000 times a second. The next processing step is the quantization of the samples, which means that the analog samples are converted into eight-bit digital values that can each have a value from 0 to 255. See Figure 1.9.

The higher the volume of the input signal, the higher the amplitude of the sampled value and its digital representation. In order to be able to also transmit low-volume conversations, the quantization is not linear over the whole input range but only in certain areas. For small amplitudes of the input signal a much higher range of digital values is used than for high amplitude values. The resulting digital data stream is called a pulse code modulated (PCM) signal. Which volume is represented by which digital eight-bit value is described in the A-law standard for European networks and in the μ-law standard in North America.

The use of different standards unfortunately complicates voice calls between networks that use different standards. Therefore, it is necessary for example to convert a voice signal for a connection between France and the United States.

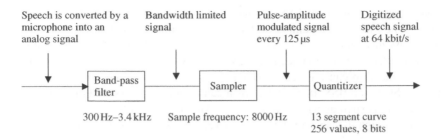

Figure 1.9 Digitization of an analog voice signal

As the MSC controls all connections, it is also responsible for billing. This is done by creating a billing record for each call which is later transferred to a billing server. The billing record contains information like the number of caller and calling party, cell ID of the cell from which the call was originated, time of call origination, the duration of the call, etc. Calls for prepaid subscribers are treated differently as the charging is already done while the call is running. The prepaid billing service is usually implemented on an IN system and not on the MSC as is further described in Section 1.11.

1.6.2 The Visitor Location Register (VLR)

Each MSC has an associated visitor location register (VLR), which holds a record of each subscriber that is currently served by the MSC (Figure 1.10). These records are only a copy of the original records, which are stored in the HLR (see Section 1.6.3). The VLR is mainly used to reduce the signaling between the MSC and the HLR. If a subscriber roams into the area of an MSC, the data is copied to the VLR of the MSC and is thus locally available for every connection establishment. The verification of the subscriber's record at every connection establishment is necessary, as the record contains information about which

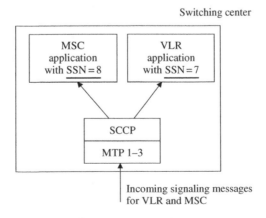

Figure 1.10 Mobile switching center (MSC) with integrated visitor location register (VLR)

services are active and from which services the subscriber is barred. Thus, it is possible, for example, to bar outgoing calls while allowing incoming calls to prevent abuse of the system. While the standards allow implementing the VLR as an independent hardware component, all vendors have implemented the VLR simply as a software component in the MSC. This is possible because MSC and VLR use different SCCP subsystem numbers (see Section 1.4.1) and can thus run on a single physical node.

When a subscriber leaves the coverage area of an MSC, the subscriber's record is copied from the HLR to the VLR of the new MSC, and is then removed from the VLR of the previous MSC. The communication with the HLR is standardized in the D-interface specification which is shown together with other MSC interfaces in Figure 1.8.

1.6.3 The Home Location Register (HLR)

The HLR is the subscriber database of a GSM network. It contains a record for each subscriber, which contains information about the individually available services.

The international mobile subscriber identity (IMSI) is an internationally unique number that identifies a subscriber and used for most subscriber-related signaling in the network (Figure 1.11). The IMSI is stored in the subscriber's SIM card and in the HLR and is thus the key to all information about the subscriber. The IMSI consists of the following parts:

- The mobile country code (MCC): the MCC identifies the subscriber's home country. Table 1.2 shows a number of MCC examples.
- The mobile network code (MNC): this part of the IMSI is the national part of a subscriber's home network identification. A national identification is necessary because there are usually several independent mobile networks in a single country. In the UK for example the following MNCs are used: 10 for O2, 15 for Vodafone, 30 for T-Mobile, 33 for Orange, 20 for Hutchison 3G, etc.
- The mobile subscriber identification number (MSIN): the remaining digits of the IMSI form the MSIN, which uniquely identifies a subscriber within the home network.

As an IMSI is internationally unique, it enables a subscriber to use his phone abroad if a GSM network is available that has a roaming agreement with his home operator. When the mobile phone is switched on, the IMSI is retrieved from the SIM card and sent to the MSC. There, the MCC and MNC of the IMSI are analyzed and the MSC is able to request the subscriber's record from the HLR of the subscriber's home network.

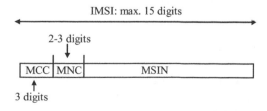

Figure 1.11 The international mobile subscriber identity (IMSI)

Table 1.2 Mobile country codes

MCC	Country
234	United Kingdom
310	United States
228	Switzerland
208	France
262	Germany
604	Morocco
505	Australia

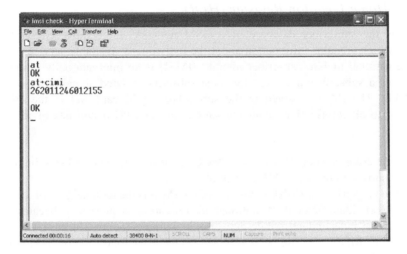

Figure 1.12 A terminal program can be used to retrieve the IMSI from the SIM card

For information purposes, the IMSI can also be retrieved from the SIM card with a PC and a serial cable that connects to the mobile phone. By using a terminal program such as HyperTerminal, the mobile can be instructed to return the IMSI by using the 'at+cimi' command, which is standardized in 3GPP TS 27.007 [4]. Figure 1.12 shows how the IMSI is returned by the mobile phone.

The phone number of the user, which is called the mobile subscriber ISDN number (MSISDN) in the GSM standards, has a length of up to 15 digits and consists of the following parts:

- The country code is the international code of the subscriber's home country. The country code has one to three digits such as +44 for the UK, +1 for the US, +353 for Ireland.
- The national destination code (NDC) usually represents the code with which the network operator can be reached. It is normally three digits in length. It should to be noted that mobile networks in the US use the same NDCs as fixed-line networks. Thus, it is not possible for a user to distinguish if he is calling a fixed line or a mobile phone. This

impacts both billing and routing, as the originating network cannot deduct which tariff to apply from the NDC.

* The remainder of the MSISDN is the subscriber number, which is unique in the network.

There is usually a 1:1 or 1:N relationship in the HLR between the IMSI and the MSISDN. Furthermore, a mobile subscriber is normally assigned only a single MSISDN. However, as the IMSI is the unique identifier of a subscriber in the mobile network, it is also possible to assign several numbers to a single subscriber.

Another advantage of using the IMSI as the key to all subscriber information instead of the MSISDN is that the phone number of the subscriber can be changed without replacing the user's SIM card or changing any information on it. In order to change the MSISDN, only the HLR record of the subscriber needs to be changed. In effect, this means that the mobile station is not aware of its own phone number. This is not necessary because the MSC automatically adds the user's MSISDN to the message flow for a mobile-originated call establishment so it can be presented to the called party.

Many countries have introduced a functionality called mobile number portability (MNP), which allows a subscriber to keep his MSISDN if he wants to change his mobile network operator. This is a great advantage for the subscribers and for competition between the mobile operators, but also implies that it is no longer possible to discern the mobile network to which the call will be routed from the NDC. Furthermore, the introduction of MNP also increased the complexity of call routing and billing in both fixed-line and mobile networks, because it is no longer possible to use the NDC to decide which tariff to apply to a call. Instead of a simple call-routing scheme based on the NDC, the networks now have to query a mobile number portability database for every call to a mobile subscriber to find out if the call can be routed inside the network or if it has to be forwarded to a different national mobile network.

Apart from the IMSI and MSISDN, the HLR contains a variety of information about each subscriber, such as which services he is allowed to use. Table 1.3 shows a number of 'basic services' that can be activated on a per subscriber basis:

In addition to the basic services described above, the GSM network offers a number of other services that can also be activated on a per subscriber basis. These services are called supplementary services and are shown in Table 1.4.

Table 1.3   Basic services of a GSM network

Basic service	Description
Telephony	If this basic service is activated, a subscriber can use the voice telephony services of the network. This can be partly restricted by other supplementary services which are described below
Short messaging service (SMS)	If activated, a subscriber is allowed to use the SMS
Data service	Different circuit-switched data services can be activated for a subscriber with speeds of 2.4, 4.8, 9.6, and 14.4 kbit/s data calls
FAX	Allows or denies a subscriber the use of the FAX service that can be used to exchange FAX messages with fixed-line or mobile terminals

Table 1.4 Supplementary services of a GSM network

Supplementary service	Description
Call forward unconditional (CFU)	If this service is configured, a number can be configured to which all incoming calls are forwarded immediately [5]. This means that the mobile phone will not even be notified of the incoming call even if it is switched on
Call forward busy (CFB)	This service allows a subscriber to define a number to which calls are forwarded if he is already engaged in a call when a second call comes in
Call forward no reply (CFNRY)	If this service is activated, it is possible to forward the call to a user-defined number if the subscriber does not answer the call within a certain time. The subscriber can change the number to which to forward the call to as well as the timeout value (e.g. 25 seconds)
Call forward not reachable (CFNR)	This service forwards the call if the mobile phone is attached to the network but is not reachable momentarily (e.g. temporary loss of network coverage)
Barring of all outgoing calls (BAOC)	This functionality can be activated by the network operator if, for example, the subscriber has not paid his monthly invoice in time. It is also possible for the network operator to allow the subscriber to change the state of this feature together with a PIN (personal identification number) so the subscriber can lend the phone to another person for incoming calls only [6]
Barring of all incoming calls (BAIC)	Same functionality as provided by BAOC for incoming calls [6]
Call waiting (CW)	This feature allows signaling an incoming call to a subscriber while he is already engaged on another call [7]. The first call can then be put on hold to accept the incoming call. The feature can be activated or barred by the operator and switched on or off by the subscriber
Call hold (HOLD)	This functionality is used to accept an incoming call during an already active call or to start a second call [7]
Calling line identification presentation (CLIP)	If activated by the operator for a subscriber, the functionality allows the switching center to forward the number of the caller
Calling line identification restriction (CLIR)	If allowed by the network, the caller can instruct the network not to show his phone number to the called party
Connected line presentation (COLP)	Shows the calling party the MSISDN to which a call is forwarded, if call forwarding is active at the called party side
Connected line presentation restriction (COLR)	If COLR is activated at the called party, the calling party will not be notified of the MSISDN the call is forwarded to
Multi party (MPTY)	Allows subscribes to establish conference bridges with up to six subscribers [8]

Most supplementary services can be activated by the network operator on a per subscriber basis and allow the operator to charge an additional monthly fee for some services if desired. Other services, like multi party, can be charged on a per use basis. Most services can be configured by the subscriber via a menu on the mobile phone. The menu, however, is just a graphical front end for the user and the mobile phone translates the user's commands into numerical strings which start with a '*' character. These strings are then sent to the network by using an unstructured supplementary service data (USSD) message. The codes are standardized in 3GPP TS 22.030 [9] and are thus identical in all networks. As the menu is only a front end for the USSD service, the user can also input the USSD strings himself via the keypad. After pressing the 'send' button, which is usually the button that is also used to start a phone call after typing in a phone number, the mobile phone sends the string to the HLR via the MSC, where the string is analyzed and the requested operation is performed. For example, call forwarding to another phone (e.g. 0782 192 8355), while a user is already engaged in another call (CFB), is activated with the following string: **67*07821928355# + call button.

1.6.4 The Authentication Center

Another important part of the HLR is the authentication center (AC). The AC contains an individual key per subscriber (Ki) which is a copy of the Ki in the SIM card of the subscriber. As the Ki is secret, it is stored in the AC and especially on the SIM card in a way that prevents it being read directly.

For many operations in the network, for instance during the establishment of a call, the subscriber is identified by using this key. Thus it can be ensured that the subscriber's identity is not misused by a third party. Figures 1.13 and 1.14 show how the authentication process is performed.

The authentication process is initiated when a subscriber establishes a signaling connection with the network before the actual request (e.g. call establishment request) is sent. In the first step of the process, the MSC requests an authentication triplet from the HLR/authentication center. The AC retrieves the Ki of the subscriber and the authentication algorithm (A3 algorithm) based on the IMSI of the subscriber that is part of the message from the MSC. The Ki is then used together with the A3 algorithm and a random number to generate the authentication triplet which contains the following values:

* RAND: a 128-bit random number.
* SRES: the signed response (SRES) is generated by using Ki, RAND, and the authentication A3 algorithm, and has a length of 32 bits.
* Kc: the ciphering key, Kc, is also generated by using Ki and RAND. It is used for the ciphering of the connection once the authentication has been performed successfully. Further information on this topic can be found in Section 1.7.5.

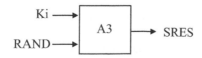

Figure 1.13   Creation of a signed response (SRES)

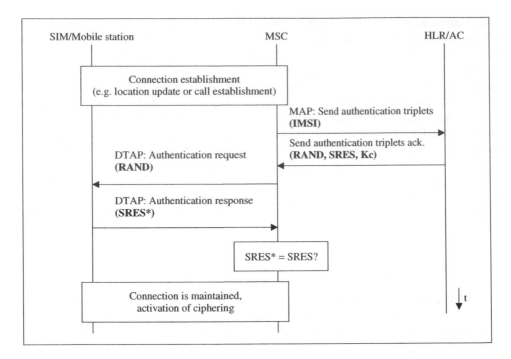

Figure 1.14 Message flow during the authentication of a subscriber

RAND, SRES, and Kc are then returned to the MSC, which then performs the authentication of the subscriber. It is important to note that the secret Ki key never leaves the authentication center.

In order to speed up subsequent connection establishments the AC usually returns several authentication triplets per request. These are buffered by the MSC/VLR and are used during the next connection establishments.

In the next step, the MSC sends the RAND inside an authentication request message to the mobile station. The terminal forwards the RAND to the SIM card which then uses the Ki and the authentication A3 algorithm to generate a signed response (SRES*). The SRES* is returned to the mobile station and then sent back to the MSC inside an authentication response message. The MSC then compares SRES and SRES* and if they are equal the subscriber is authenticated and allowed to proceed with the communication.

As the secret key, Ki, is not transmitted over any interface that could be eavesdropped on, it is not possible for a third party to correctly calculate an SRES. As a fresh random number is used for the next authentication, it is also pointless to intercept the SRES* and use it for another authentication. A detailed description of the authentication procedure and many other procedures between the mobile station and the core network can be found in [10].

Figure 1.15 shows some parts of an authentication request and an authentication response message. Apart from the format of RAND and SRES, it is also interesting to note the different protocols which are used to encapsulate the message (see Section 1.4.2).

Extract of a decoded Authentication Request message
```
SCCP MSG: Data Form 1
DEST. REF ID: 0B 02 00
DTAP MSG   LENGTH: 19
PROTOCOL DISC.: Mobility Management
DTAP MM MSG: Auth. Request
Ciphering Key Seq.: 0
RAND in hex: 12 27 33 49 11 00 98 45
             87 49 12 51 22 89 18 81 (16 byte = 128 bit)
```

Extract of a decoded Authentication Response message
```
SCCP MSG: Data Form 1
DEST. REF ID: 00 25 FE
DTAP MSG   LENGTH: 6
PROTOCOL DISC.: Mobility Management
DTAP MM MSG: Auth. Response
SRES in hex: 37 21 77 61 (4 byte = 32 bit)
```

Figure 1.15 Authentication between network and mobile station

1.6.5 The Short Messaging Service Center (SMSC)

Another important network element is the short message service center (SMSC) which is used to store and forward short messages. The short messaging service was only introduced about four years after the first GSM networks went into operation as add on and has been specified in 3GPP TS 23.040 [11]. Most industry observers were quite skeptical at the time as the general opinion was that if it is needed to convey some information, it is done by calling someone rather than to cumbersomely type in a text message on the small keypad. However, they were proven wrong and today most GSM operators generate over 15% of their revenue from the short messaging service alone with a total number of over 25 billion SMS messages exchanged annually in the United Kingdom.

The short messaging service can be used for person-to-person messaging as well as for notification purposes of received email messages or a new call forwarded to the voice mail system. The transfer method for both cases is identical.

The sender of an SMS prepares the text for the message and then sends the SMS via a signaling channel to the MSC. As a signaling channel is used, an SMS is just an ordinary DTAP SS-7 message and thus, apart from the content, very similar to other DTAP messages, such as a location update message or a setup message to establish a voice call. Apart from the text, the SMS message also contains the MSISDN of the destination party and the address of the SMSC which the mobile station has retrieved from the SIM card. When the MSC receives an SMS from a subscriber it transparently forwards the SMS to the SMSC. As the message from the mobile station contains the address of the subscriber's SMSC, international roaming is possible and the foreign MSC can forward the SMS to the home SMSC without the need for an international SMSC database. See Figure 1.16.

In order to deliver a message, the SMSC analyses the MSISDN of the recipient and retrieves its current location (the responsible MSC) from the HLR. The SMS is then forwarded to the responsible MSC. If the subscriber is currently attached, the MSC tries to contact the mobile station and if an answer is received, the SMS is forwarded. Once the mobile station

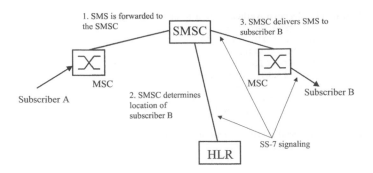

Figure 1.16 SMS delivery principle

has confirmed the proper reception of the SMS, the MSC notifies the SMSC as well and the SMS is deleted from the SMSC's data storage.

If the subscriber is not reachable because the battery of the mobile station is empty, the network coverage has been lost temporarily, or if the device is simply switched off, it is not possible to deliver the SMS. In this case, the message waiting flag is set in the VLR and the SMSC is stored in the SMSC. Once the subscriber communicates with the MSC, the MSC notifies the SMSC to reattempt delivery.

As the message waiting flag is also set in the HLR, the SMS also reaches a subscriber that has switched off the mobile station in London for example and switches it on again after a flight to Los Angeles. When the mobile station is switched on in Los Angeles, the visited MSC reports the location to the subscriber's home HLR (location update). The HLR then sends a copy of the user's subscription information to the MSC/VLR in Los Angeles including the message waiting flag and thus the SMSC can also be notified that the user is reachable again.

The SMS delivery mechanism does not unfortunately include a delivery reporting functionality for the sender of the SMS. The sender is only notified that the SMS has been correctly received by the SMSC. If and when the SMS is also correctly delivered to the recipient, however, is not signalled to the originator of the message. Most SMSC vendors have therefore implemented their own proprietary solutions. Some vendors use a code for this purpose that the user has to include in the text message. With some operators for example, '*N#' or '*T#' can be put into the text message at the beginning to indicate to the SMSC that the sender wishes a delivery notification. The SMSC then removes the three-character code and returns an SMS to the originator once the SMS was successfully delivered to the recipient.

1.7 The Base Station Subsystem (BSS)

While most functionality required in the NSS for GSM could be added via additional software, the BSS had to be developed from scratch. This was mainly necessary because earlier generation systems were based on analog transmission over the air interface and thus had not much in common with the GSM BSS.

1.7.1 Frequency Bands

In Europe, GSM was initially only specified for operation in the 900 MHz band between 890–915 MHz in the uplink direction and between 935–960 MHz in the downlink direction (Figure 1.17). 'Uplink' refers to the transmission from the mobile station to the network and 'downlink' to the transmission from the network to the mobile station. The bandwidth of 25 MHz is split into 125 channels with a bandwidth of 200 kHz each.

It soon became apparent that the number of available channels was not sufficient to cope with the growing demand in many European countries. Therefore, the regulating bodies assigned an additional frequency range for GSM which uses the frequency band from 1710–1785 MHz for the uplink and 1805–1880 for the downlink. Instead of a total bandwidth of 25 MHz as in the 900 MHz range, the 1800 MHz band offers 75 MHz of bandwidth which corresponds to 375 additional channels. The functionality of GSM is identical on both frequency bands, with the channel numbers, also referred to as the absolute radio frequency channel numbers (ARFCNs), being the only difference. See Table 1.5.

While GSM was originally intended only as a European standard, the system soon spread to countries in other parts of the globe. In countries outside Europe, GSM sometimes competes with other technologies, such as CDMA. Today, only a few countries, like Japan and South Korea, are not covered by GSM systems. However, some of the operators in these countries operate W-CDMA UMTS networks (see Chapter 3). Therefore, GSM/UMTS subscribers with dual-mode phones can also roam in these countries.

In North America, analog mobile networks continued to be used for some time before second-generation networks, with GSM being one of the technologies used, were introduced. Unfortunately, however, the 900 MHz as well as the 1800 MHz band were already in use by other systems and thus the North American regulating body chose to open frequency bands for the new systems in the 1900 MHz band and later on in the 850 MHz band. The disadvantage of this approach is that many US GSM mobile phones cannot be used in Europe

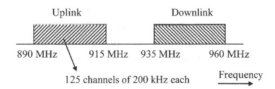

Figure 1.17   GSM uplink and downlink in the 900 MHz frequency band

Table 1.5   GSM frequency bands

Band	ARFCN	Uplink (MHz)	Downlink (MHz)
GSM 900 (Primary)	0–124	890–915	935–960
GSM 900 (Extended)	975–1023, 0–124	880–915	925–960
GSM 1800	512–885	1710–1785	1805–1880
GSM 1900 (North America)	512–810	1850–1910	1930–1990
GSM 850 (North America)	128–251	824–849	869–894
GSM-R	0–124, 955–1023	876–915	921–960

and vice versa. Fortunately, many new GSM and UMTS phones support the US frequency bands as well as the European frequency bands, which are also used in most countries in other parts of the world. These tri-band or quad-band phones thus enable a user to truly roam globally.

The GSM standard is also used by railway communication networks in Europe and other parts of the world. For this purpose, GSM was enhanced to support a number of private mobile radio and railway specific functionalities and is known as GSM-R. The additional functionalities include:

- The voice group call service (VGCS): this service offers a circuit-switched walkie-talkie functionality to allow subscribers that have registered to a VGCS group to communicate with all other subscribers in the area who have also subscribed to the group. In order to talk, the user has to press a push to talk button. If no other subscriber holds the uplink, the network grants the request and blocks the uplink for all other subscribers while the push to talk button is pressed. The VGCS service is very efficient especially if many subscribers participate in a group call, as all mobile stations that participate in the group call listen to the same timeslot in downlink direction. Further information about this service can be found in 3GPP TS 43.068 [12].
- The voice broadcast service (VBS): same as VGCS with the restriction that only the originator of the call is allowed to speak. Further information about this service can be found in 3GPP TS 43.069 [13].
- Enhanced multi level precedence and preemption (eMLPP): this functionality, which is specified in 3GPP TS 23.067 [14], is used to attach a priority to a point-to-point, VBS, or VGCS call. This enables the network and the mobile stations to automatically preempt ongoing calls for higher priority calls to ensure that emergency calls (e.g. a person has fallen on the track) is not blocked by lower priority calls and a lack of resources (e.g. because no timeslots are available).

As GSM-R networks are private networks, it has been decided to assign a private frequency band in Europe for this purpose which is just below the public 900 MHz GSM band. To use GSM-R, mobile phones need to be slightly modified to be able to send and receive in this frequency range. This requires only minor software and hardware modifications. In order to be also able to use the additional functionalities described above, further extensions of the mobile station software are necessary. More about GSM-R can be found at http://gsm-r.uic.asso.fr [15].

1.7.2 The Base Transceiver Station (BTS)

Base stations, which are also called base transceiver stations (BTSs), are the most visible network elements of a GSM system (Figure 1.18). Compared to fixed-line networks, the base stations replace the wired connection to the subscriber with a wireless connection which is also referred to as the air interface. The base stations are also the most numerous components of a mobile network as according to press reports each wireless operator in the UK for example has well over 10,000 base stations.

In theory, a base station can cover an area with a radius of up to 35 km. This area is also called a cell. As a base station can only serve a limited number of simultaneous users,

Figure 1.18 A typical antenna of a GSM base station. The optional microwave directional antenna (round antenna at the bottom of the mast) connects the base station with the GSM network

cells are much smaller in practice especially in dense urban environments. There, cells cover areas with a radius between 3 and 4 km in residential and business areas, and down to only several 100 m and minimal transmission power in heavily frequented areas like shopping centers and downtown streets. Even in rural areas, a cell's coverage area is usually less then 15 km as the transmission power of the mobile station of one or two watts is the limiting factor in this case.

As the emissions of different base stations of the network must not interfere with each other, all neighboring cells have to send on different frequencies. As can be seen in Figure 1.19, a singe base station usually has quite a number of neighboring sites. Therefore, only a limited number of different frequencies can be used per base station in order to increase capacity.

To increase the capacity of a base station, the coverage area is usually split into two or three sectors which are then covered on different frequencies by a dedicated transmitter.

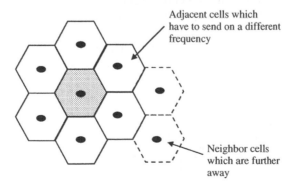

Figure 1.19 Cellular structure of a GSM network

Figure 1.20 Sectorized cell configurations

This allows the reuse of frequencies in two-dimensional space better than if only a single frequency was used for the whole base station. Each sector of the base station therefore forms its own independent cell (Figure 1.20).

1.7.3 The GSM Air Interface

The transmission path between the BTS and the mobile terminal is referred to in the GSM specifications as the air interface or the Um interface. To allow the base station to communicate with several subscribers simultaneously, two methods are used. The first method is frequency division multiple access (FDMA) which means that users communicate with the base station on different frequencies. The second method used is time division multiple access (TDMA). See Figure 1.21. GSM uses carrier frequencies with a bandwidth of 200 kHz over which up to eight subscribers can communicate with the base station simultaneously.

Subscribers are time multiplexed by dividing the carrier into frames with durations of 4.615 ms. Each frame contains eight physically independent timeslots, each for communication with a different subscriber. The timeframe of a timeslot is called a burst and the burst duration is 577 microseconds. If a mobile station is allocated timeslot number two for a voice call for example, the mobile station will send and receive only during this burst. Afterwards, it has to wait until the next frame before it is allowed to send again.

By combining the two multiple access schemes it is possible to approximately calculate the total capacity of a base station. For the following example it is assumed that the base station is split into three sectors and each sector is covered by an independent cell. Each cell is equipped with two transmitters and receivers, a configuration that is used quite often. In each sector, $2 \times 8 = 16$ timeslots are thus available. Two timeslots are usually assigned for signaling purposes which leaves 14 timeslots per sector for user channels. Let us further assume that four timeslots or more are used for the packet-switched GPRS service (see Chapter 2). Therefore, 10 timeslots are left for voice calls per sector, which amounts to 30

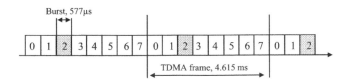

Figure 1.21 A GSM TDMA frame

channels for all sectors of the base station. In other words this means that 30 subscribers can communicate simultaneously per base station.

A single BTS, however, provides service for a much higher number of subscribers, as they do not all communicate at the same time. Mobile operators, therefore, base their network dimensioning on a theoretical call profile model in which the number of minutes a subscriber statistically uses the system per hour is one of the most important parameters. A commonly used value for the number of minutes a subscriber uses the system per hour is one minute. This means that a base station is able to provide service for 60 times the number of active subscribers. In this example a base station with 30 channels is therefore able to provide service for about 1800 subscribers.

This number is quite realistic as the following calculation shows: Vodafone Germany had a subscriber base of about 25 million in 2005. If this value is divided by the number of subscribers per cell, the total number of base stations required to serve such a large subscriber base can be determined. With our estimation above, the number of base stations required for the network would be about 14,000. This value is quite accurate and in line with numbers published by the operator.

Each burst of a TDMA frame is divided into a number of different sections as shown in Figure 1.22. Each burst is encapsulated by a guard time in which no data is sent. This is necessary because the distance of the different subscribers relative to the base station can change while they are active. As airwaves 'only' propagate through space at the speed of light, the signal of a far away subscriber takes a longer time to reach the base station compared to a subscriber that is closer to the base station. In order to prevent any overlap, guard times were introduced. These parts of the burst are very short, as the network actively controls the timing advance of the mobile station. More about this topic can be found below.

The training sequence in the middle of the burst always contains the same bit pattern. It is used to compensate for interference caused for example by reflection, absorption, and multi-path propagation. On the receiver side these effects are countered by comparing the received signal to the training sequence and thus adapting the analog filter parameters for the signal. The filter parameters calculated this way can then be used to modify the rest of the signal and thus to better recreate the original signal.

At the beginning and end of each burst, another well-known bit pattern is sent to enable the receiver to detect the beginning and end of a burst correctly. These fields are called 'tails'. The actual user data of the burst, i.e. the digitized voice signal, is sent in the two user data fields with a length of 57 bits each. This means, that a 577-microsecond burst transports 114 bits of user data. Finally, each frame contains two bits to the left and right of the training sequence which are called 'stealing bits'. These bits indicate if the data fields

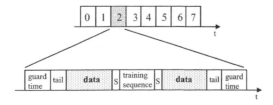

Figure 1.22 A GSM burst

contain user data or are used ('stolen') for urgent signaling information. User data of bursts which carry urgent signaling information, however, is lost. As shown below, the speech decoder is able to cope with short interruptions of the data stream quite well and thus are not normally audible to the user.

For the transmission of user or signaling data, the timeslots are arranged into logical channels. A user data channel for the transmission of digitized voice data for example is a logical channel. On the first carrier frequency of a cell the first two timeslots are usually used for common logical signaling channels while the remaining six independent timeslots are used for user data channels or GPRS. As there are more logical channels then physical channels (timeslots) for signaling, 3GPP TS 45.002 [16] describes how 51 frames are grouped into a multiframe to be able to carry a number of different signaling channels over the same timeslot. In such a multiframe, which is infinitely repeated, it is specified in which bursts on timeslots 0 and 1 which logical channels are transmitted. For user data timeslots (e.g. voice) the same principle is used. Instead of 51 frames, these timeslots are grouped into a 26-multiframe pattern. In order to visualize this principle, Figure 1.23 shows how the eight timeslots of a frame are grouped into a two-dimensional table. Figure 1.24 then uses this principle to show how the logical channels are assigned to physical timeslots in the multiframe.

Logical channels are arranged into two groups. If data on a logical channel is dedicated to a single user, the channel is called a dedicated channel. If the channel is used for data that needs to be distributed to several users, the channel is called a common channel.

Let us take a look at the dedicated channels first:

- The traffic channel (TCH) is a user data channel. It can be used to transmit a digitized voice signal or circuit-switched data services of up to 14.4 kbit/s.
- The fast associated control channel (FACCH) is transmitted on the same timeslot as a TCH. It is used to send urgent signaling messages like a handover command. As these messages do not have to be sent very often, no dedicated physical bursts are allocated to the FACCH. Instead, user data is removed from a TCH burst. In order to inform the

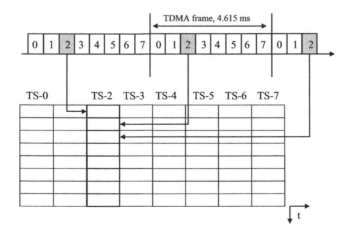

Figure 1.23 Arrangement of bursts of a frame for the visualization of logical channels in Figure 1.24

FN	TS-0	TS-1		FN	TS-2	...	TS-7
0	FCCH	SDCCH/0		0	TCH		TCH
1	SCH	SDCCH/0		1	TCH		TCH
2	BCCH	SDCCH/0		2	TCH		TCH
3	BCCH	SDCCH/0		3	TCH		TCH
4	BCCH	SDCCH/1		4	TCH		TCH
5	BCCH	SDCCH/1		5	TCH		TCH
6	AGCH/PCH	SDCCH/1		6	TCH		TCH
7	AGCH/PCH	SDCCH/1		7	TCH		TCH
8	AGCH/PCH	SDCCH/2		8	TCH		TCH
9	AGCH/PCH	SDCCH/2		9	TCH		TCH
10	FCCH	SDCCH/2		10	TCH		TCH
11	SCH	SDCCH/2		11	TCH		TCH
12	AGCH/PCH	SDCCH/3		12	SACCH		SACCH
13	AGCH/PCH	SDCCH/3		13	TCH		TCH
14	AGCH/PCH	SDCCH/3		14	TCH		TCH
15	AGCH/PCH	SDCCH/3		15	TCH		TCH
16	AGCH/PCH	SDCCH/4		16	TCH		TCH
17	AGCH/PCH	SDCCH/4		17	TCH		TCH
18	AGCH/PCH	SDCCH/4		18	TCH		TCH
19	AGCH/PCH	SDCCH/4		19	TCH		TCH
20	FCCH	SDCCH/5		20	TCH		TCH
21	SCH	SDCCH/5		21	TCH		TCH
22	SDCCH/0	SDCCH/5		22	TCH		TCH
23	SDCCH/0	SDCCH/5		23	TCH		TCH
24	SDCCH/0	SDCCH/6		24	TCH		TCH
25	SDCCH/0	SDCCH/6		25	free		free
26	SDCCH/1	SDCCH/6		0	TCH		TCH
27	SDCCH/1	SDCCH/6		1	TCH		TCH
28	SDCCH/1	SDCCH/7		2	TCH		TCH
29	SDCCH/1	SDCCH/7		3	TCH		TCH
30	FCCH	SDCCH/7		4	TCH		TCH
31	SCH	SDCCH/7		5	TCH		TCH
32	SDCCH/2	SACCH/0		6	TCH		TCH
33	SDCCH/2	SACCH/0		7	TCH		TCH
34	SDCCH/2	SACCH/0		8	TCH		TCH
35	SDCCH/2	SACCH/0		9	TCH		TCH
36	SDCCH/3	SACCH/1		10	TCH		TCH
37	SDCCH/3	SACCH/1		11	TCH		TCH
38	SDCCH/3	SACCH/1		12	SACCH		SACCH
39	SDCCH/3	SACCH/1		13	TCH		TCH
40	FCCH	SACCH/2		14	TCH		TCH
41	SCH	SACCH/2		15	TCH		TCH
42	SACCH/0	SACCH/2		16	TCH		TCH
43	SACCH/0	SACCH/2		17	TCH		TCH
44	SACCH/0	SACCH/3		18	TCH		TCH
45	SACCH/0	SACCH/3		19	TCH		TCH
46	SACCH/1	SACCH/3		20	TCH		TCH
47	SACCH/1	SACCH/3		21	TCH		TCH
48	SACCH/1	free		22	TCH		TCH
49	SACCH/1	free		23	TCH		TCH
50	free	free		24	TCH		TCH
				25	free		free

Figure 1.24 Use of timeslots in downlink direction as per 3GPP TS 45.002 [16]

mobile station, the stealing bits to the left and right of the training sequence, as shown in Figure 1.22, are used. This is the reason why the FACCH is not shown in Figure 1.24.

- The slow associated control channel (SACCH) is also assigned to a dedicated connection. It is used in the uplink direction to report signal quality measurements of the serving cell and neighboring cells to the network. The network then uses these values for handover decisions and power control. In the downlink direction, the SACCH is used to send power control commands to the mobile station. Furthermore, the SACCH is used for timing advance control which is described in Section 1.7.4 and Figure 1.29. As these messages are only of low priority and the necessary bandwidth is very small, only a few bursts are used on a 26 multiframe at fixed intervals.
- The standalone dedicated control channel (SDCCH) is a pure signaling channel which is used during call establishment when a subscriber has not yet been assigned a traffic channel. Furthermore, the channel is used for signaling which is not related to call establishment such as for the location update procedure or for sending or receiving a text message (SMS).

Besides the dedicated channels, which are always assigned to a single user, there are a number of common channels that are monitored by all subscribers in a cell:

- The synchronization channel (SCH) is used by mobile stations during network and cell searches.
- The frequency correction channel (FCCH) is used by the mobile stations to calibrate their transceiver units und is also used to detect the beginning of a multiframe.
- The broadcast common control channel (BCCH) is the main information channel of a cell and broadcasts SYS_INFO messages that contain a variety of information about the network. The channel is monitored by all mobile stations, which are switched on but currently not engaged in a call or signaling connection (idle mode), and broadcasts among many other things the following information:

  - the MCC and MNC of the cell;
  - the identification of the cell which consists of the location area code (LAC) and the cell ID;
  - to simplify the search for neighboring cells for a mobile station, the BCCH also contains information about the frequencies used by neighboring cells. Thus, the mobile station does not have to search the complete frequency band for neighboring cells.

- The paging channel (PCH) is used to inform idle subscribers of incoming calls or SMS messages. As the network is only aware of the location area the subscriber is roaming in, the paging message is broadcast in all cells belonging to the location area. The most important information element of the message is the IMSI of the subscriber or a temporary identification called the temporary mobile subscriber identity (TMSI). A TMSI is assigned to a mobile station during the network attach procedure and can be changed by the network every time the mobile station contacts the network once encryption has been activated. Thus, the subscriber has to be identified with the IMSI only once and is then addressed with a constantly changing temporary number when encryption is not yet activated for the communication. This increases anonymity in the network and prevents eavesdroppers from creating movement profiles of subscribers.

- The random access channel (RACH) is the only common channel in the uplink direction. If the mobile station receives a message via the PCH that the network is requesting a connection establishment or if the user wants to establish a call or send an SMS, the RACH is used for the initial communication with the network. This is done by sending a channel request message. Requesting a channel has to be done via a 'random' channel because subscribers in a cell are not synchronized with each other. Thus, it cannot be ensured that two devices do not try to establish a connection at the same time. Only once a dedicated channel (SDCCH) has been assigned to the mobile station by the network can there no longer be any collision between different subscribers of a cell. If a collision occurs during the first network access, the colliding messages are lost and the mobile stations do not receive an answer from the network. Thus, they have to repeat their channel request messages after expiry of a timer which is set to an initial random value. This way, it is not very likely that the mobile stations will interfere with each other again during their next connection establishment attempts because they are performed at different times.
- The access grant channel (AGCH): if a subscriber sends a channel request message on the RACH, the network allocates an SDCCH or in exceptional cases a TCH and notifies the subscriber on the AGCH via an immediate assignment message. The message contains information about which SDCCH or TCH the subscriber is allowed to use.

Figure 1.25 shows how PCH, AGCH, and SDCCH are used during the establishment of a signaling link between the mobile station and the network. The BSC, which is responsible for assigning SDCCH and TCH channels of a base station, is further described in Section 1.7.4.

As can also be seen in Figure 1.24, not all bursts on timeslots 2 to 7 are used for traffic channels. Every twelfth burst of a timeslot it used for the SACCH. Furthermore, the 25th

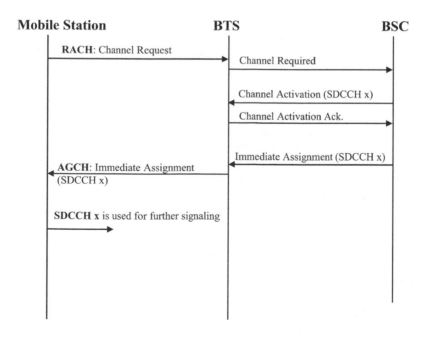

Figure 1.25   Establishment of a signaling connection

burst is also not used for carrying user data. This gap is used to enable the mobile station to perform signal strength measurements of neighboring cells on other frequencies. This is necessary so that the network can redirect the connection into a different cell (handover) to maintain the call while the user is moving.

The GSM standard offers two possibilities to use the available frequencies. The simplest case, which has been described so far, is the use of a constant carrier frequency (ARFCN) for each channel. In order to improve the transmission quality it is also possible to use alternating frequencies for a single channel of a cell. This concept is known as frequency hopping and changes the carrier frequency for every burst during a transmission. This increases the probability that only few bits are lost if one carrier frequency experiences a lot of interference from other sources like neighboring cells. In the worst case only a single burst is affected because the next burst is already sent on a different frequency. Up to 64 different frequencies can be used per base station for frequency hopping. In order to inform the mobile of the use of frequency hopping, the immediate assignment message used during the establishment of a signaling link contains all the information about which frequencies are used and which hopping pattern is applied to the connection.

For carriers that transport the SCH, FCCH, and BCCH channels, frequency hopping must not be used. This restriction is necessary because it would be very difficult for mobile stations to find neighboring cells.

In practice, network operators use static frequencies as well as frequency hopping in their networks.

The interface which connects the base station to the network and which is used to carry the information for all logical channels is called the A-bis interface. An E-1 connection is usually used for the A-bis interface and due to its 64 kbit/s timeslot architecture the logical channels are transmitted in a different way than on the air interface. All common channels as well as the information sent and received on the SDCCH and SACCH channels are sent over one or more common 64 kbit/s E-1 timeslots. This is possible because these channels are only used for signaling data which is not time critical. On the A-bis interface these signaling messages are sent by using the link access protocol (LAPD). This protocol was initially designed for the ISDN D-channel of fixed-line networks and has been reused for GSM with only minor modifications.

For traffic channels that use a bandwidth of 13 kbit/s on the A-bis interface, only one-quarter of an E-1 timeslot is used. This means that all eight timeslots of an air interface frame can be carried on only two timeslots of the E-1 interface. A base station composed of three sectors which uses two carriers each thus requires 12 timeslots on the A-bis interface plus an additional timeslot for the LAPD signaling. The remaining timeslots of the E-1 connection can be used for the communication between the network and other base stations. For this purpose, several cells are usually daisy chained via a single E-1 connection. See Figure 1.26.

1.7.4 The Base Station Controller (BSC)

While the base station is the interface element that connects the mobile stations with the network, the base station controller (BSC) is responsible for the establishment, release, and maintenance of all connections of cells which are connected to it.

If a subscriber wants to establish a voice call, send an SMS, etc., the mobile station sends a channel request message to the BSC as shown in Figure 1.25. The BSC then checks if

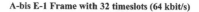

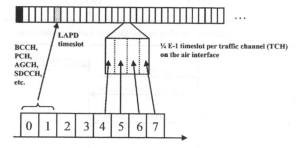

A-bis E-1 Frame with 32 timeslots (64 kbit/s)

BCCH, PCH, AGCH, SDCCH, etc.

LAPD timeslot

¼ E-1 timeslot per traffic channel (TCH) on the air interface

One carrier with 8 timeslots on the air interface (Um)

Figure 1.26 Mapping of E-1 timeslots to air interface timeslots

an SDCCH is available and activates the channel in the BTS. Afterwards, the BSC sends an immediate assignment message to the mobile station on the AGCH which includes the number of the assigned SDCCH. The mobile station then uses the SDCCH to send DTAP messages which the BSC forwards to the MSC.

The BSC is also responsible for establishing signaling channels for incoming calls or SMS messages. In this case, the BSC receives a paging message from the MSC which contains the IMSI and TMSI of the subscriber, as well as the location area ID in which the subscriber is currently located. The BSC in turn has a location area database which it uses to identify all cells in which the subscriber needs to be paged. When the mobile station receives the paging message, it responds to the network in the same way as in the example above by sending a channel request message.

The establishment of a traffic channel for voice calls is always requested by the MSC for both mobile-originated and mobile-terminated calls. Once the mobile station and the MSC have exchanged all necessary information for the establishment of a voice call via an SDCCH, the MSC sends an assignment request for a voice channel to the BSC as shown in Figure 1.27.

The BSC then verifies if a TCH is available in the requested cell and if so, activates the channel in the BTS. Afterwards, the mobile station is informed via the SDCCH that a TCH is now available for the call. The mobile station then changes to the TCH and FACCH. To inform the BTS that it has switched to the new channel, the mobile station sends a message to the BTS on the FACCH which is acknowledged by the BTS. In this way, the mobile also has a confirmation that its signal can be decoded correctly by the BTS. Finally, the mobile station sends an assignment complete message to the BSC which in turn informs the MSC of the successful establishment of the traffic channel.

Apart from the establishment and release of a connection, another important task of the BSC is the maintenance of the connection. As subscribers can roam freely through the network while a call is ongoing it can happen that the subscriber roams out of the coverage area of the cell in which the call was initially established. In this case, the BSC has to redirect the call to the appropriate cell. This procedure is called handover. In order to be able to perform a handover into another cell, the BSC requires signal quality measurements for the air interface. The results of the downlink signal quality measurements are reported to the

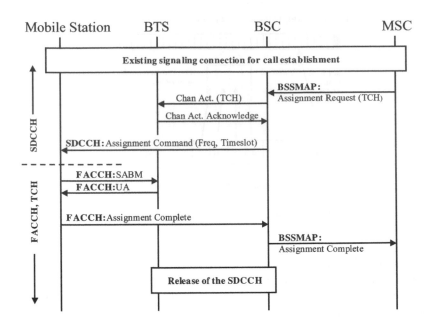

Figure 1.27 Establishment of a traffic channel (TCH)

BSC by the mobile station, which continuously performs signal quality measurements which it reports via the SACCH to the network. The uplink signal quality is constantly measured by the BTS and also reported to the BSC. Apart from the signal quality of the user's current cell, it is also important that the mobile station reports the quality of signals it receives from other cells. To enable the mobile station to perform these measurements, the network sends the frequencies of neighbouring cells via the SACCH during an ongoing call. The mobile station then uses this information to perform the neighbouring cell measurements while the network communicates with other subscribers and reports the result via measurement report messages in the uplink SACCH.

The network receives these measurement values and is thus able to periodically evaluate if a handover of an ongoing call to a different cell is necessary. Once the BSC decides to perform a handover, a TCH is activated in the new cell as shown in Figure 1.28. Afterwards, the BSC informs the mobile station via the old cell with a handover command message that is sent over the FACCH. Important information elements of the message are the new frequency and timeslot number of the new TCH. The mobile station then changes its transmit and receive frequency, synchronizes to the new cell if necessary, and sends a handover access message in four consecutive bursts. In the fifth burst, an SABM message is sent which is acknowledged by the BTS to signal to the mobile station that the signal can be received. At the same time, the BTS informs the BSC of the successful reception of the mobile station's signal with an establish indication message. The BSC then immediately redirects the speech path into the new cell.

From the mobile's point of view the handover is now finished. The BSC, however, has to release the TCH in the old cell and has to inform the MSC of the performed handover

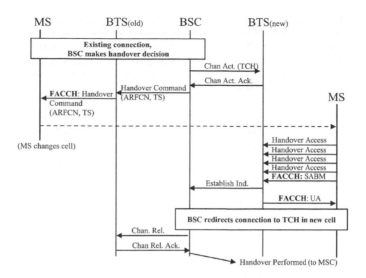

Figure 1.28   Message flow during a handover procedure

before the handover is finished from the network's point of view. The message to the MSC is only informative and has no impact on the continuation of the call.

In order to reduce interference, the BSC is also in charge of controlling the transmission power for every air interface connection. For the mobile station an active power control has the advantage that the transmission power can be reduced under favorable reception conditions. The control of the mobile station's transmission power is done using the signal quality measurements of the BTS for the connection. If the mobile station's transmission power has to be increased or decreased, the BSC sends a power control message to the BTS. The BTS in turn forwards the message to the mobile station and repeats the message on the SACCH in every frame. In practice, it can be observed that power control and adaptation is performed every one to two seconds. During call establishment, the mobile station always uses the highest allowed power output level which is then reduced or increased again by the network step by step. Table 1.6 gives an overview of the mobile station power classes. A distinction is made for the 900 MHz versus the 1800 MHz band. While mobile stations operating on the 900 MHz band are allowed to use up to 2 watts, connections on the 1800 MHz band are limited to 1 watt. For stationary devices or car phones with external antennas, power values for up to 8 watts are allowed. The power values in the table represent the power output when the transmitter is active in the assigned timeslot. As the mobile station only sends on one of the eight timeslots of a frame, the average power output of the mobile station is only one-eighth of this value. The average power output of a mobile station which sends with a power output of 2 watts is thus only 250 milliwatts.

The BSC is also able to control the power output of the base station. This is done by evaluating the signal measurements of the mobile stations in the current cell. It is important to note that power control can only be performed for downlink carriers which do not broadcast the common channels like FCH, SCH, and BCCH of a cell. On such carriers the power output has to be constant in order to allow mobile stations, which are currently located in other cells of the network, to perform their neighbouring cell measurements. This would not

Table 1.6 GSM power levels and corresponding power output

GSM 900 Power level	GSM 900 Power output	GSM 1800 Power level	GSM 1800 Power output
(0–2)	(8 W)		
5	2 W	0	1 W
6	1.26 W	1	631 mW
7	794 mW	2	398 mW
8	501 mW	3	251 mW
9	316 mW	4	158 mW
10	200 mW	5	100 mW
11	126 mW	6	63 mW
12	79 mW	7	40 mW
13	50 mW	8	25 mW
14	32 mW	9	16 mW
15	20 mW	10	10 mW
16	13 mW	11	6.3 mW
17	8 mW	12	4 mW
18	5 mW	13	2.5 mW
19	3.2 mW	14	1.6 mW
		15	1.0 mW

be possible if the signal amplitude would varies over time as the mobile stations can only listen to the carrier signal of neighbouring cells for a short time.

Due to the limited speed of radio waves, a time shift of the arrival of the signal can be observed when a subscriber moves away from a base station during an ongoing call. If no countermeasures are taken, this would mean that at some point the signal of a subscriber would overlap with the next timeslot despite the guard time of each burst which is shown in Figure 1.22. Thus, the signal of each subscriber has to be carefully monitored and the timing of the transmission of the subscriber has to be adapted. This procedure is called timing advance control (Figure 1.29).

The timing advance can be controlled in 64 steps (0 to 63) of 550 m. The maximum distance between a base station and a mobile subscriber is in theory $64 \times 550 \, \text{m} = 35.2 \, \text{km}$. In practice, such a distance is not reached very often as base stations usually cover a much smaller area due to capacity reasons. Furthermore, the transmission power of the terminal is also not sufficient to bridge such a distance under non-line-of-sight conditions to the base

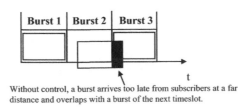

Without control, a burst arrives too late from subscribers at a far distance and overlaps with a burst of the next timeslot.

Figure 1.29 Time shift of bursts of distant subscribers without timing advance control

station. Therefore, one of the few scenarios where such a distance has to be overcome is in costal areas from ships at sea.

The control of the timing advance already starts with the first network access on the RACH with a channel request message. This message is encoded into a very short burst that can only transport a few bits in exchange for large guard periods at the beginning and end of the burst. This is necessary because the mobile phone is unaware of the distance between itself and the base station when it attempts to contact the network. Thus, the mobile station is unable to select an appropriate timing advance value. When the base station receives the burst it measures the delay and forwards the request including a timing advance value required for this mobile station to the BSC. As has been shown in Figure 1.25, the BSC reacts to the connection request by returning an immediate assignment message to the mobile station on the AGCH. Apart from the number of the assigned SDCCH, the message also contains a first timing advance value to be used for the subsequent communication on the SDCCH. Once the connection has been successfully established, the BTS continually monitors the delay experienced for this channel and reports any changes to the BSC. The BSC in turn instructs the mobile station to change its timing advance by sending a message on the SACCH.

For special applications, like coastal communication, the GSM standard offers an additional timeslot configuration in order to increase the maximum distance to the base station to up to 120 km. This is achieved by only using every second timeslot per carrier which allows a burst to overlap into the following (empty) timeslot. While this dramatically increases the range of a cell, the number of available communication channels is cut in half. Another issue is that mobile phones that are limited to a transmission power of 1 watt (1800 MHz band) or 2 watts (900 MHz band) may be able to receive the BCCH of such a cell at a great distance but are unable to communicate with the cell in the uplink. Thus, such an extended range configuration mostly makes sense with permanently installed mobile phones with external antennas that can transmit with a power level of up to 8 watts.

1.7.5 The TRAU for Voice Data Transmission

For the transmission of voice data, a TCH is used in GSM as described in Section 1.7.3. A TCH uses all but two bursts of a 26-burst multiframe with one being reserved for the SACCH as shown in Figure 1.24, and one which remains empty to allow the mobile station to perform neighbouring cell measurements. As has been shown in the preceding section, a burst which is sent to or from the mobile every 4.615 ms can carry exactly 114 bits of user data. When taking the two bursts into account, which are not used for user data of a 26-burst multiframe, this results in a raw data rate of 22.8 kbit/s. As will be shown in the remainder of this section, a substantial part of the bandwidth of a burst is required for error detection and correction bits. The resulting data rate for the actual user data is thus around 13 kbit/s.

The narrow bandwidth of a TCH stands in contrast to how a voice signal is transported in the core network. Here, the PCM algorithm is used (see Section 1.6.1) to digitize the voice signal, which makes full use of the available 64 kbit/s bandwidth of an E-1 timeslot to encode the voice signal. See Figure 1.30

A simple solution for the air interface would have been to define air interface channels that can also carry 64 kbit/s PCM-coded voice channels. This has not been done because the scarce resources on the air interface have to be used as efficiently as possible. The decision to compress the speech signal was taken during the first standardization phase in the 1980s

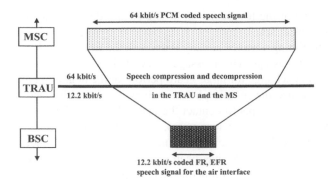

Figure 1.30 GSM speech compression

because it was foreseeable that advances in hardware and software processing capabilities would allow compression of a voice data stream in real time.

In the mobile network, the compression and decompression of the voice data stream is performed in the transcoding and rate adaptation unit (TRAU) which is located between the MSC and a BSC and controlled by the BSC. During an ongoing call, the MSC sends the 64 kbit/s PCM-encoded voice signal towards the radio network and the TRAU converts the voice stream in real time into a 13 kbit/s compressed data stream which is transmitted over the air interface. In the other direction, the BSC sends a continuous stream of compressed voice data towards the core network and the TRAU converts the stream into a 64 kbit/s coded PCM signal. In the mobile station, the same algorithms are implemented as in the TRAU to compress and decompress the speech signal. See Figure 1.31.

While the TRAU is a logical component of the BSS, it is most often installed next to an MSC in practice. This has the advantage that four compressed voice channels can be transmitted in a single E-1 timeslot. After compression, each voice channel uses a 16 kbit/s sub-timeslot. Thus, only one-quarter of the transmission capacity between an MSC and BSC is needed in comparison to an uncompressed transmission. As the BSCs of a network are usually located in the field and not close to an MSC, this helps to reduce transmission costs for the network operator substantially.

The TRAU offers a number of different algorithms for speech compression. These algorithms are called speech codecs or simply codecs. The first codec that was standardized for GSM is the full-rate (FR) codec which reduces the 64 kbit/s voice stream to about 13 kbit/s.

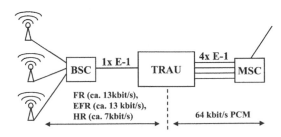

Figure 1.31 Speech compression with a 4:1 compression ratio in the TRAU

At the end of the 1990s, the enhanced full-rate (EFR) codec was introduced and is still the most widely used codec in operational GSM networks today. The EFR codec not only compresses the speech signal to about 13 kbit/s but also offers a superior voice quality compared to the FR codec. The disadvantage of the EFR codec is the higher complexity of the compression algorithm which requires more processing power. However, the processing power available in mobile phones has increased substantially in recent years and thus modern GSM phones easily cope with the additional complexity.

Besides those two codecs, a half-rate (HR) codec has been defined for GSM which only requires a bandwidth of 7 kbit/s. While there is almost no audible difference between the EFR codec compared to a PCM-coded speech signal, the voice quality of the HR codec is noticeably inferior. The advantage for the network operator of the HR codec is that the number of simultaneous voice connections per carrier can be doubled. With HR codec, a single timeslot, which is used for a single EFR voice channel, can carry two TCH (HR). In practice, however, operators do not use the HR codec very often. Even during big events like fairs, operators still assign a TCH (FR) or TCH (EFR) to the subscriber for a voice call.

The latest speech codec development is the adaptive multi rate (AMR) algorithm [17]. Instead of using a single codec, which is selected at the beginning of the call, AMR allows a change to the codec during a call. The considerable advantage of this approach is the ability to switch to a speech codec with a higher compression rate during bad radio signal conditions in order to increase the number of error detection and correction bits. If signal conditions permit, a lower rate codec can be used which only uses every second burst of a frame for the call. This in effect doubles the capacity of the cell as a single timeslot can be shared by two calls similarly to the HR codec. Unlike the HR codec, however, the AMR codecs, which only use every second burst and which are thus called HR AMR codecs, still have a voice quality which is comparable to the EFR codec. While AMR is optional for GSM, it has been chosen for the UMTS system as a mandatory feature. In the United States, AMR is used by some network operators to increase the capacity of their network, especially in very dense traffic areas like New York, where it has become very difficult to increase the capacity of the network any further with over half a dozen carrier frequencies per sector already used. In Europe, however, it is not certain that AMR will be widely deployed as most operators invested heavily in the deployments of their UMTS networks which offer ample capacity for both voice and data communication, even in high density traffic areas. Further information about AMR can also be found in Chapter 3.

While the PCM algorithm digitizes analog volume levels by statically mapping them to digital values, the GSM speech digitization is much more complex to reach the desired compression rate. In the case of the FR codec, which is specified in 3GPP TS 46.010 [18], the compression is achieved by emulating the human vocal system. This is done by using a source-filter model (Figure 1.32). In the human vocal system, the speech is created in the larynx and by the vocal cords. This is emulated in the mathematical model in the signal creation part while the filters represent the signal forming which is done in the human throat and mouth.

On a mathematical level, the speech forming is simulated by using two time-invariant filters. The period filter creates the periodic vibrations of the human voice while the vocal tract filter simulates the envelope. The filter parameters are generated from the human voice, which is the input signal into the system. In order to digitize and compress the human voice, the model is used in reverse direction as shown in Figure 1.32. As time variant filters are

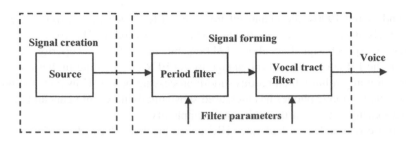

Figure 1.32   Source-filter model of the GSM FR codec

hard to model, the system is simplified by generating a pair of filter parameters for an interval of 20 milliseconds. As an input to the algorithm, a speech signal is used that has previously been converted into an 8- or 13-bit PCM codec. As the PCM algorithm delivers 8000 values per second, the FR codec requires 160 values for a 20 ms interval to calculate the filter parameters. As eight bits are used per value, 8 bits × 160 values = 1280 input bits are used per 20 ms interval. For the period filter, the input bits are used to generate a filter parameter with a length of 36 bits. Afterwards, the filter is applied to the original input signal. The resulting signal is then used to calculate another filter parameter with a length of 36 bits for the vocal tract filter. Afterwards, the signal is again sent through the vocal tract filter with the filter parameter applied. The signal, which is thus created, is called the 'rest signal' and coded into 188 bits. See Figure 1.33.

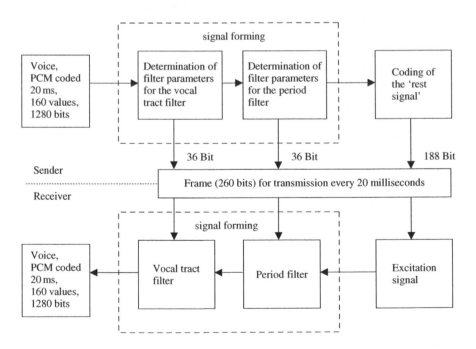

Figure 1.33   Complete transmission chain with transmitter and receiver of the GSM FR codec

Once all parameters have been calculated, the two 36-bit filter parameters and the rest signal, which is coded into 188 bits, are sent to the receiver. Thus, the original information, which was coded in 1280 bits, has been reduced to 260 bits. In the receiver, the filter procedure is applied in reverse order on the rest signal and thus the original signal is recreated. As the procedure uses a lossy compression algorithm, the original signal and the recreated signal at the other end are no longer exactly identical. For the human ear, however, the differences are almost inaudible.

Before a 260-bit data frame is transmitted over the air interface every 20 ms, it traverses a number of additional functional blocks which are not implemented in the TRAU but in the base station. These additional functional blocks are shown in Figure 1.34.

In a first step, the voice frames are processed in the channel coder unit, which adds error detection and correction information to the data stream. This step is very important as the transmission over the air interface is prone to frequent transmission errors due to the constantly changing radio environment. Furthermore, the compressed voice information is very sensitive and even a few bits that might be changed while the frame is transmitted over the air interface create an audible distortion. In order to prevent this, the channel coder separates the 260 bits of a voice data frame into three different classes as shown in Figure 1.35.

Fifty of the 260 bits of a speech frame are class Ia bits and extremely important for the overall reproduction of the voice signal at the receiver side. Such bits are for example the higher order bits of the filter parameters. In order to enable the receiver to verify the correct transmission of those bits, a three-bit CRC checksum is calculated and added to the data stream. If the receiver later on cannot recreate the checksum with the received bits, the frame is discarded.

Another 132 bits of the frame are also quite important and are thus put into class Ib. However, no checksum is calculated for them. In order to generate the exact amount of bits that are necessary to fill a GSM burst, four filler bits are inserted. Afterwards, the class Ia

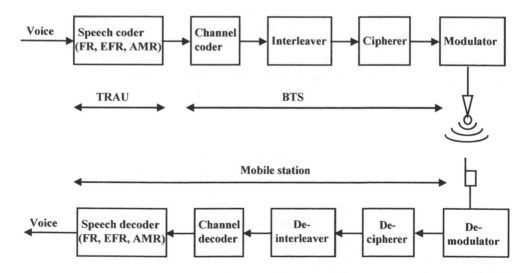

Figure 1.34 Transmission path in the downlink direction between network and mobile station

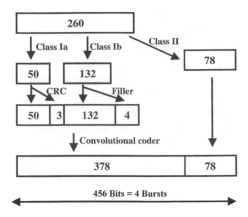

Figure 1.35 GSM channel coder for full-rate speech frames

bits, checksum, class Ib bits, and the four filler bits are treated by a convolutional coder which adds redundancy to the data stream. For each input bit, the convolutional decoder calculates two output bits. For the computation of the output bits the coder uses not only the current bit but also uses information about the values of the previous bits. For each input bit, two output bits are calculated. This mathematical algorithm is also called a half-rate convolutional coder.

The remaining 78 bits of the original 260-bit data frame belong to the third class which is called class II. These are not protected by a checksum and no redundancy is added for them. Errors which occur during the transmission of these bits can neither be detected nor corrected.

As has been shown, the channel coder uses the 260-bit input frame to generate 456 bits on the output side. As a burst on the air interface can carry exactly 114 bits, four bursts are necessary to carry the frame. As the bursts of a traffic channel are transmitted every 4.6152 ms, the time it takes to transmit the frame over the air interface is about 20 ms. In order to get to exactly 20 ms, the empty burst and the burst used for the SACCH per 26-burst multiframe has to be included in the calculation.

Due to the redundancy added by the channel coder, it is possible to correct a high number of faulty bits per frame. The convolutional decoder, however, has one weak point: if several consecutive bits are changed during the transmission over the air interface, the convolutional decoder on the receiver side is not able to correctly reconstruct the original frame. This effect is often observed as air interface disturbances usually affect several bits in a row.

In order to decrease this effect, the interleaver changes the bit order of a 456-bit data frame in a specified pattern over eight bursts (Figure 1.36). Consecutive frames are thus interlocked with each other. On the receiver side, the frames are put through the de-interleaver, which puts the bits again into the correct order. If several consecutive bits are changed due to air interface signal distortion, this operation disperses the faulty bits in the frame and the convolutional decoder can thus correctly restore the original bits. A disadvantage of the interleaver, however, is an increased delay of the voice signal. In addition to the delay of 20 ms generated by the full-rate coder, the interleaver adds another 40 ms as a speech frame is spread over eight bursts instead of being transmitted consecutively in four bursts. Compared

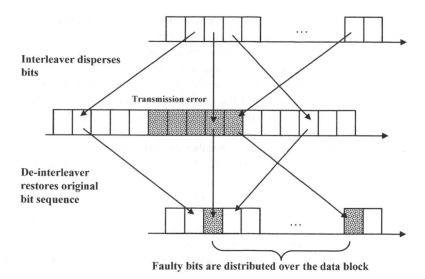

Figure 1.36 Frame interleaving

to a voice call in a fixed-line network, a mobile network thus introduces a delay of at least 60 ms. If the call is established between two mobile phones, the delay is at least 120 ms as the transmission chain is traversed twice.

The next module of the transmission chain is the cipherer (Figure 1.37), which encrypts the data frames it receives from the interleaver. GSM uses, like most communication systems,

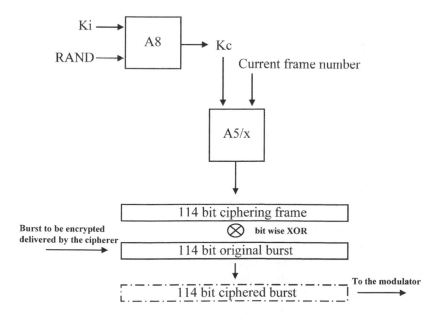

Figure 1.37 Ciphering of an air interface burst

a stream cipher algorithm. In order to encrypt the data stream, a ciphering key (Kc) is calculated in the authentication center and on the SIM card by using a random number (RAND) and the secret key (Ki) as input parameters for the A8 algorithm. Together with the GSM frame number, which is increased for every air interface frame, Kc is used as input parameter for the A5 ciphering algorithm. The A5 algorithm computes a 114-bit sequence which is XOR combined with the bits of the original data stream. As the frame number is different for every burst, it is ensured that the 114-bit ciphering sequence also changes for every burst which further enhances security.

In order to be as flexible as possible, a number of different ciphering algorithms have been specified for GSM. These are called A5/1, A5/2, A5/3, and so on. Thus, it is possible to export GSM network equipment to countries where export restrictions prevent the sale of some ciphering algorithms and technologies. Furthermore, it is possible to introduce new ciphering algorithms into already existing networks in order to react to security issues if a flaw is detected in one of the currently used algorithms. The selection of the ciphering algorithm also depends on the capabilities of the mobile station. During the establishment of a connection, the mobile station informs the network which ciphering algorithms it supports. The network can then choose an algorithm which is supported by the network and the terminal.

When the mobile station establishes a new connection to the network, its identity is verified before the mobile station is allowed to proceed with the call setup. This procedure has already been described in Section 1.6.4. Once the terminal and subscriber have been authenticated, it is also possible for the MSC to start encryption by sending a ciphering command to the mobile station. The ciphering command message contains, among other information elements, the ciphering key, Kc, which is used by the base station for the ciphering of the connection on the air interface. Before the BSC forwards the message to the mobile station, however, the ciphering key is removed from the message because this information must not be sent over the air interface. The mobile station, however, does not need to receive the ciphering key from the network as the SIM card calculates the Kc on its own and forwards the key to the mobile station together with the SRES during the authentication procedure. Figure 1.40 shows how ciphering is activated during a location update procedure.

Unfortunately, there are a number of weak spots in the overall GSM encryption architecture. One serious problem is that encryption has only been specified as an optional feature. Thus, encryption can be easily switched on or off by the network operator. Some mobile phones like the Siemens S series for example show a '*!*' symbol on the display if ciphering is disabled. So far, however, the author of this book has only seen this symbol in a laboratory environment where encryption was deactivated on purpose. Thus, it can be assumed that public networks, in the majority of cases, only very rarely deactivate this feature. Another weakness in the overall security architecture is the fact that a connection is only ciphered between the BTS and the mobile station. All other interfaces between components of the network like the connection between the base station and the BSC or the connection between the TRAU and the MSC are not protected. As many network operators use microwave links between base stations and BSCs, it is possible to intercept calls with suitable microwave equipment without having physical access to any component of the network.

At the end of the transmission chain, the modulator maps the digital data onto an analog carrier, which uses a bandwidth of 200 kHz. This mapping is done by encoding the bits into changes of the carrier frequency. As the frequency change takes a finite amount of time, a method called Gaussian minimum shift keying (GMSK) is used, which smoothes the flanks created by the frequency changes. GMSK has been selected for GSM as its modulation and demodulation properties are easy to handle and implement into hardware and due to the fact that it interferes only slightly with neighboring channels.

In order to reduce the interference on the air interface and to increase the operating time of the mobile station, data bursts are only sent if a speech signal is detected. This method is called discontinuous transmission (DTX) and can be activated independently in the uplink and downlink directions (Figure 1.38). Since only one person is speaking at a time during a conversation, one of the two speech channels can usually be deactivated. In the downlink direction, this is managed by the voice activity detection (VAD) algorithm in the TRAU while in the uplink direction the VAD is implemented in the mobile station.

Simply deactivating a speech channel, however, creates a very undesirable side effect. As no speech signal is transmitted anymore, the receiver no longer hears the background noise of the other side. This can be very irritating especially for high-volume background noise levels such as if a person is driving in a car or sitting in a train. Therefore, it is necessary to generate artificial noise, called comfort noise, which simulates the background noise of the other party to the listener. As the background noise can change over time, the mobile phone or the network respectively analyze the background noise of the channel and calculate an approximation for the current situation. This approximation is then exchanged between the mobile phone and the TRAU every 480 ms. Additional benefits for the network and mobile phone are the ability to perform periodic signal quality measurements of the channel and the ability to use these frames to get an estimation on the current signal timing in order to adapt the timing advance for the call if necessary. How well this method performs is clearly audible as this procedure is used in all mobile phone calls today and the simulation of the background noise in most cases cannot be differentiated from the original signal.

Despite using sophisticated methods for error correction, it is still possible that parts of a frame are destroyed beyond repair during the transmission on the air interface. In these cases, the complete 20 ms voice frame is discarded by the receiver and the previous data block is used instead to generate an output signal. Most errors that are repaired this way remain undetected by the listener. This trick, however, cannot be used indefinitely. If after 320 ms still no valid data block has been received, the channel is muted and the decoder keeps trying to decode the subsequent frames. If, during the following seconds, no valid data frame is received, the connection is terminated and the call drops.

Many of the previously mentioned procedures have specifically been developed for the transmission of voice frames. For circuit-switched data connections, however, a number of modifications are necessary. While it is possible to tolerate a number of faulty bits for voice frames or discarding frames if a CRC error is detected, this is not possible for data calls. If even a single bit is faulty, a retransmission of at least a single frame has to be performed as

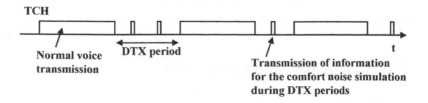

Figure 1.38 Discontinuous transmission (DTX)

most applications cannot tolerate a faulty data stream. In order to increase the likelihood to correctly reconstruct the initial data stream, the interleaver spreads the bits of a frame over a much larger number of bursts than the eight bursts used for voice frames. Furthermore, the channel coder, which separates the bits of a frame into different classes based on their importance, had to be adapted for data calls as well, as all bits are equally important. Thus, the convolutional decoder has to be used for all bits of a frame. Finally, it is also not possible to use a lossy data compression scheme for data calls. Therefore, the TRAU operates in a transparent mode for data calls. If the data stream can be compressed this has to be performed by higher layers or by the data application itself.

With a radio receiver or an amplifier of a stereo set, the different states of a GSM connection can be made audible. This is possible due to the fact that the activation and deactivation of the transmitter of the mobile station induce an audible sound in the amplifier part of audio devices. If the GSM mobile station is held close enough to an activated radio or an amplifier during the establishment of a call, the typical noise pattern can be heard, which is generated by the exchange of messages on the signaling channel (SDCCH). At some point during the signaling phase, a TCH is assigned to the mobile station at which point the noise pattern changes. As a TCH burst is transmitted every 4.615 ms, the transmitter of the mobile station is switched on and off with a frequency of 217 Hertz. If the background noise is low enough or the mute button of the telephone is pressed, the mobile station changes into the discontinuous transmission mode for the uplink part of the channel. This can be heard as well, as the constant 217 Hz hum is replaced by single short bursts every 0.5 s.

For incoming calls, this method can also be used to detect that a mobile phone starts communicating with the network on the SDDCH already one to two seconds before it starts ringing. This delay is due to the fact that the mobile station first needs to go through the authentication phase and the activation of the ciphering for the channel. Only afterwards can the network forward further information to the mobile station as to why the channel was established. This is also the reason why it takes a much longer time for the alerting tone to be heard when calling a mobile phone compared to calling a fixed-line phone.

Some mobile phones possess a number of interesting network monitoring functionalities which are hidden in the mobile phone software and are usually not directly accessible via the phone's menu. These network monitors allow the visualization of many procedures and parameters that have been discussed in this chapter such as the timing advance, channel allocation, power control, the cell-id, neighboring cell information, handover, cell reselection, etc. On the Internet, various web pages can be found that explain how these monitors can be activated, depending on the type and model of the phone. As the activation procedures are different for every phone, it is not possible to give a general recommendation. However, by using the manufacturer and model of the phone in combination with terms like 'GSM network monitor', 'GSM netmonitor' or 'GSM monitoring mode', it is relatively easy to discover if and how the monitoring mode can be activated for a specific phone.

1.8 Mobility Management and Call Control

As all components of a GSM mobile network have now been introduced, the following section gives an overview of the three processes that allow a subscriber to roam throughout the network.

1.8.1 Location Area and Location Area Update

As the network needs to be able to forward an incoming call, the subscriber's location must be known. After the mobile phone is switched on, its first action is to register with the network. Therefore the network becomes aware of the current location of the user, which can change at any time due to the mobility of the user. If the user roams into the area of a new cell it may need to inform the network of this change. In order to reduce the signaling load in the radio network, several cells are grouped into a location area. The network informs the mobile station via the BCCH of a cell not only of the cell-ID but also of the location area that the new cell belongs to. The mobile station thus only has to report its new location if the new cell belongs to a new location area. Grouping several cells into location areas not only reduces the signaling load in the network but also reduces the power consumption of the mobile. A disadvantage of this method is that the network operator is only aware of the current location area of the subscriber but not of the exact cell. Therefore, the network has to search for the mobile station in all cells of a location area for an incoming call or SMS. This procedure is called paging. The size of a location area can be set by the operator depending on his particular needs. In operational networks, usually 20 to 30 cells are grouped into a location area (Figure 1.39).

Figure 1.40 shows how a location area update procedure is performed. After a signaling connection has been established, the mobile station sends a location update request message to the MSC, which is transparently forwarded by the radio network. Before the message can be sent, however, the mobile station needs to authenticate itself first and ciphering is usually activated before as well.

Once the connection is secured against eavesdropping, the mobile station is usually assigned a new TMSI by the network, which it will use for the next connection establishment

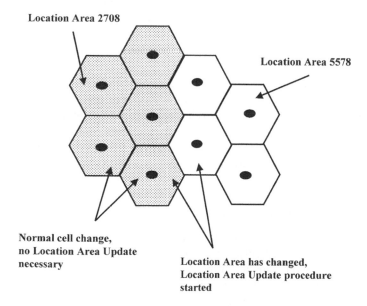

Figure 1.39 Cells in different location areas

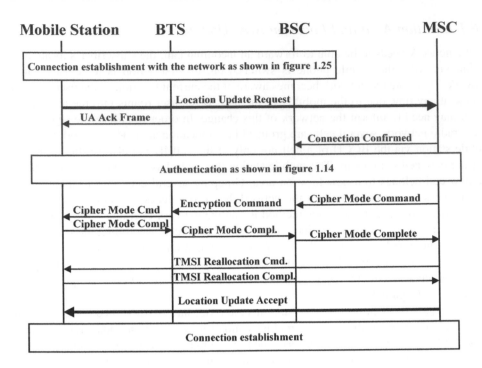

Figure 1.40   Message flow for a location update procedure

to identify itself instead of the IMSI. By using a constantly changing temporary ID, the identity of a subscriber is not revealed to listeners during the first phase of the call which is not ciphered. Once TMSI reallocation has been performed, the location area update message is sent to the network which acknowledges the correct reception. After receiving the acknowledgment, the connection is terminated and the mobile station returns to idle state.

If the old and new location areas are under the administration of two different MSC/VLRs, a number of additional steps are necessary. In this case, the new MSC/VLR has to inform the HLR that the subscriber has roamed into its area of responsibility. The HLR then deletes the record of the subscriber in the old MSC/VLR. This procedure is called an Inter-MSC location update. From the mobile point of view, however, there is no difference to a standard location update as the additional messages are only exchanged in the core network.

1.8.2 The Mobile Terminated Call

An incoming call for a mobile subscriber is called a mobile terminated call by the GSM standards. The main difference between a mobile network and a fixed-line PSTN network is the fact that the telephone number of the subscriber does not hold any information about where the subscriber is located. In the mobile network it is thus necessary to query the HLR for the current location of the subscriber before the call can be forwarded to the correct switching center.

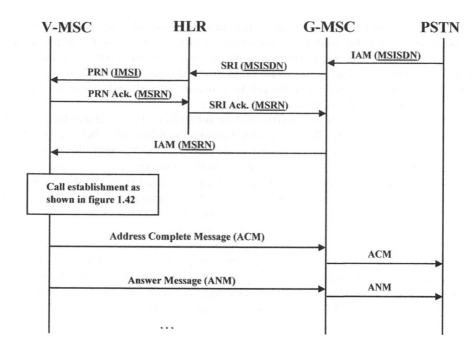

Figure 1.41 Mobile terminated call establishment, part 1

Figure 1.41 shows the first part of the message flow for a mobile terminated call initiated from a fixed-line subscriber. From the fixed-line network, the gateway-MSC (G-MSC) receives the telephone number (MSISDN) of the called party via an ISUP IAM message. The subsequent message flow on this interface is as shown in Figure 1.6 and the fixed-line network does not have to be aware that the called party is a mobile subscriber. The G-MSC in this example is simply a normal MSC with additional connections to other networks. When the G-MSC receives the IAM message, it sends a send routing information message (SRI) to the HLR in order to locate the subscriber in the network. The MSC currently responsible for the subscriber is also called the subscriber's visited MSC (V-MSC).

The HLR then determines the subscriber's IMSI by using the MSISDN to search through its database and thus is able to locate the subscriber's current V-MSC. The HLR then sends a provide roaming number (PRN) message to the V-MSC/VLR to inform the switching center of the incoming call. In the V-MSC/VLR, the IMSI of the subscriber, which is part of the PRN message, is associated with a temporary mobile station roaming number (MSRN) which is returned to the HLR. The HLR then transparently returns the MSRN to the Gateway-MSC.

The G-MSC uses the MSRN to forward the call to the V-MSC. This is possible as the MSRN not only temporarily identifies the subscriber in the V-MSC/VLR but also uniquely identifies the V-MSC to external switches. To forward the call from the G-MSC to the V-MSC, an IAM message is used again, which instead of the MSISDN contains the MSRN to identify the subscriber. This has been done as it is possible, and even likely, that there are transit switching centers between the G-MSC and V-MSC, which are thus able to forward the call without querying the HLR themselves.

As the MSRN is internationally unique instead of only in the subscriber's home network, this procedure can still be used if the subscriber is roaming in a foreign network. The presented procedure therefore works for both national and international roaming. As the MSRN is saved in the billing record for the connection, it is also possible to invoice the terminating subscriber for forwarding the call to a foreign network and to transfer a certain amount of the revenue to the foreign network operator.

In the V-MSC/VLR, the MSRN is used to find the subscriber's IMSI and thus the complete subscriber record in the VLR. This is possible because the relationship between the IMSI and MSRN was saved when the HLR first requested the MSRN. After the subscriber's record has been found in the VLR database, the V-MSC continues the process and searches the subscriber in the last reported location area, which was saved in the VLR record of the subscriber. The MSC then sends a paging message to the responsible BSC. The BSC in turn sends a paging message via each cell of the location area on the PCH. If no answer is received the message is repeated after a number of seconds.

After the mobile station has answered the paging message, an authentication and ciphering procedure has to be executed to secure the connection in a similar way as previously presented for a location update. Only afterwards is the mobile station informed about the details of the incoming call with a setup message. The setup message contains, for example, the telephone number of the caller if the CLIP supplementary service is active for this subscriber and not suppressed by the CLIR option which can be set by the caller (see Table 1.4).

If the mobile station confirms the incoming call with a call confirmed message, the MSC requests the establishment of a TCH for the voice path from the BSC. See Figure 1.42.

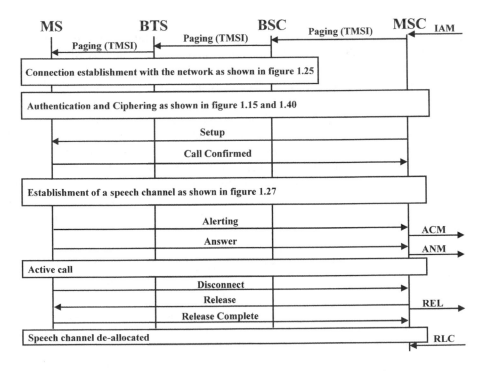

Figure 1.42 Mobile terminated call establishment, part 2

After successful establishment of the speech path, the mobile station returns an alerting message and thus informs the MSC that the subscriber is informed of the incoming call (the phone starts ringing). The V-MSC then forwards this information via the address complete message (ACM) to the G-MSC. The G-MSC then also forwards the alerting indication to the fixed-line switch via its own ACM message.

Once the mobile subscriber accepts the call by pressing the answer button, the mobile station returns an answer message to the V-MSC. Here, an ISUP answer (ANM) message is generated and returned to the G-MSC. The G-MSC forwards this information again via an ANM message back to the fixed-line switching center.

While the conversation is ongoing, the network continues to exchange messages between different components in order to ensure that the connection is maintained. Most of the messages are measurement report messages, which are exchanged between the mobile station, the base station, and the BSC. If necessary, the BSC can thus trigger a handover to a different cell. More about the handover process can be found in Section 1.8.3.

If the mobile subscriber wants to end the call, the mobile station sends a disconnect message to the network. After releasing the traffic channel with the mobile station and after sending an ISUP release (REL) message to the other party, all resources in the network are freed and the call ends.

In this example, it has been assumed that the mobile subscriber is not in the area that is covered by the G-MSC. Such a scenario, however, is quite likely if a call is initiated by a fixed-line subscriber to a mobile subscriber which currently roams in the same region. As the fixed-line network usually forwards the call to the closest MSC to save costs, the G-MSC will in many cases also be the V-MSC for the connection. The G-MSC recognizes such a scenario if the MSRN returned by the HLR in the SRI acknowledge message contains a number, which is from the MSRN pool of the G-MSC. In this case, the call is treated in the G-MSC right away and the ISUP signaling inside the mobile network (IAM, ACM, ANM) is left out. More details about call establishment procedures in GSM networks can be found in [19].

1.8.3 Handover Scenarios

If reception conditions deteriorate during a call due to a change in the location of the subscriber, the BSC has to initiate a handover procedure. The basic procedure and the necessary messages have already been shown in Figure 1.28. Depending on which parts of the network are involved in the handover, one of the following handover scenarios is used to ensure that the connection remains established:

- Intra-BSC handover: in this scenario, the current cell and new cell are connected to the same BSC. This scenario is shown in Figure 1.28.
- Inter-BSC handover: if a handover has to be performed into a cell which is connected to a second BSC, the current BSC is not able to control the handover itself as no direct signaling connection exists between the BSCs of a network. Thus, the current BSC requests that the MSC initiates a handover to the other cell by sending a handover request message. Important parameters of the message are the cell-ID and the location area code (LAC) of the new cell. As the MSC administers a list of all LACs and cells under its control, it can find the correct BSC and request the establishment of a traffic channel for the handover

in a next step. Once the new BSC has prepared the speech channel (TCH) in the new cell, the MSC returns a handover command to the mobile station via the still existing connection over the current BSC. The mobile station then performs the handover to the new cell. Once the new cell and BSC have detected the successful handover, the MSC can switch over the speech path and inform the old BSC that the traffic channel for this connection can be released.

• Inter-MSC handover: if the current and new cells for a handover procedure are not connected to the same MSC, the handover procedure is even more complicated. As in the example before, the BSC detects that the new cell is not in its area of responsibility and thus forwards the handover request to the MSC. The MSC also detects that the LAC of the new cell is not part of its coverage area. Therefore, the MSC looks into another table which lists all LACs of the neighboring MSCs. As the MSC in the next step contacts a second MSC, the following terminology is introduced to unambiguously identify the two MSCs: the MSC which has assigned a MSRN at the beginning of the call is called the anchor-MSC (A-MSC) of the connection. The MSC that receives the call during a handover is called the relay-MSC (R-MSC). See Figure 1.43.

In order to perform the handover, the A-MSC sends a MAP (mobile application part, see Section 1.4.2) handover message to the R-MSC. The R-MSC then asks the responsible BSC to establish a traffic channel in the requested cell and reports back to the A-MSC. The A-MSC then instructs the mobile station via the still existing connection over the current cell to perform the handover. Once the handover has been performed successfully, the R-MSC reports the successful handover to the A-MSC. The A-MSC can then switch the voice path towards the R-MSC. Afterwards, the resources in the old BSC and cell are released.

If the subscriber yet again changes to another cell during the call, which is controlled by yet another MSC, a subsequent inter-MSC handover has to be performed (Figure 1.44).

For this scenario, the current relay-MSC (R-MSC 1) reports to the A-MSC that a subsequent inter-MSC handover to R-MSC 2 is required in order to maintain the call. The A-MSC then instructs R-MSC 2 to establish a channel in the requested cell. Once the speech channel is ready in the new cell, the A-MSC sends the handover command message via R-MSC 1.

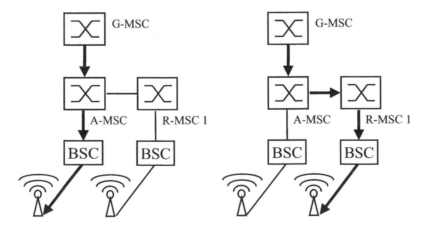

Figure 1.43 Inter-MSC handover

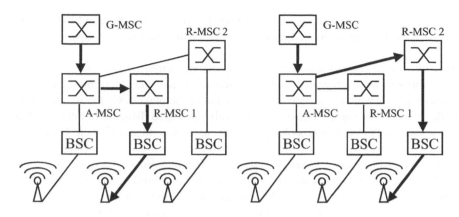

Figure 1.44 Subsequent inter-MSC handover

The mobile station then performs the handover to R-MSC 2 and reports the successful execution to the A-MSC. The A-MSC can then redirect the speech path to R-MSC 2 and instruct R-MSC 1 to release the resources. By having the A-MSC in command in all the different scenarios, it is assured that during the lifetime of a call only the G-MSC, the A-MSC, and at most one R-MSC are part of a call. Additionally, tandem switches might be necessary to route the call through the network or to a roaming network. However, these switches purely forward the call and are thus transparent in this procedure.

Finally, there is also a handover case in which the subscriber, who is served by an R-MSC, returns to a cell which is connected to the A-MSC. Once this handover is performed, no R-MSC is part of the call. Therefore, this scenario is called a subsequent handback.

From the mobile station point of view, all handover variants are performed in the same way, as the handover messages are identical for all scenarios. In order to perform a handover as quickly as possible, however, GSM can send synchronization information for the new cell inside the handover message. This allows the mobile station to immediately switch to the allocated timeslot instead of having to synchronize first. This can only be done, however, if current and new cell are synchronized with each other which is not possible for example if they are controlled by different BSCs. As two cells which are controlled by the same BSC may not necessarily be synchronized, synchronization information is by no means an indication of what kind of handover is being performed in the radio and core network.

1.9 The Mobile Station

Due to the progress of miniaturization of electronic components during the mid-1980s, it was possible to integrate all components of a mobile phone into a single portable device. Only a few years later, mobile phones have shrunk to such a small size that the limiting factor in future miniaturization is no longer the size of the electronic components. Instead, the space required for user interface components like display and keypad limit a further reduction. Due to the continuous improvement and miniaturization of electronic components, it is possible to integrate more and more functionalities into a mobile phone and to improve the ease of use. While mobile phones were at first only used for voice calls, the trend today is a move towards devices 'with an integrated mobile phone' for different user groups:

- PDA with mobile phone for voice and data communication.
- Game consoles with integrated mobile phone for voice and data communication (e.g. multi-user games with a real-time interconnection of the players via the wireless Internet).
- Mobile phones for voice communication with integrated Bluetooth interface that lets devices such as PDAs or notebooks use the phone as a connection to the Internet.

Independent of the size and variety of different functionalities, the basic architecture of all mobile phones, which is shown in Figure 1.45, is very similar. The core of the mobile phone is the base band processor which contains a RISC (reduced instruction set) CPU and a digital signal processor (DSP). The RISC processor is responsible for the following tasks:

- Handling of information that is received via the different signaling channels (BCCH, PCH, AGCH, PCH, etc.).
- Call establishment (DTAP).
- GPRS management and GPRS data flow.
- Parts of the transmission chain: channel coder, interleaver, cipherer (dedicated hardware component in some designs).
- Mobility management (network search, cell reselection, location update, handover, timing advance, etc.).
- Connections via external interfaces like Bluetooth, RS-232, IrDA, USB.
- User interface (keypad, display, graphical user interface).

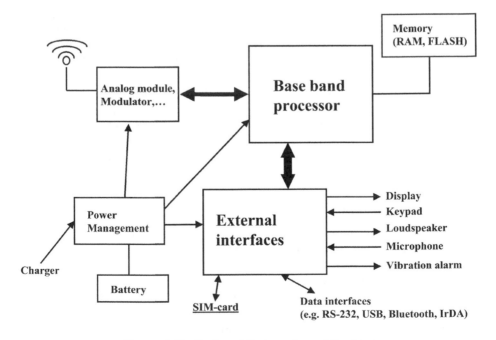

Figure 1.45   Basic architecture of a mobile phone

As many of these tasks have to be performed in parallel, a multitasking embedded real-time operating system is used on the RISC processor. The real-time component of the operating system is especially important as the processor has to be able to provide data for transmission over the air interface according to the GSM frame structure and timing. All other tasks like keypad handling, display update and the graphical user interface, in general, have a lower priority. This can be observed with many mobile phones during a GPRS data session. Here, the RISC CPU is not only used for signaling, but also for treating incoming and outgoing data and forwarding the data stream between the network and an external device like a notebook or PDA. Especially during times of high volume data transfers, it can be observed that the mobile phone reacts slowly to user input, because treating the incoming and outgoing data flow has a higher priority.

The processor capacity of the RISC processor is the main factor when deciding which applications and features to implement in a mobile phone. For applications like recording and displaying digital pictures or videos for example, fast processing capabilities are required. One of the RISC architectures that is used for high-end GSM and UMTS mobile phones is the ARM-9 architecture. This processor architecture allows CPU speeds of over 200 MHz and provides sufficient computing power for calculation intensive applications like those mentioned before. The downside of fast processors, however, is higher power consumption, which forces designers to increase battery capacity while trying at the same time to maintain the physical dimensions of a small mobile phone. Therefore, intelligent power-saving mechanisms are required in order be able to reduce power consumption during times of inactivity.

The DSP is another important component of a GSM and UMTS chipset. Its main task is FR, EFR, HR, or AMR speech compression. Furthermore, the DSP is used in the receiver chain to help decode the incoming signal. This is done by the DSP analyzing the training sequence of a burst (see Section 1.7.3). As the DSP is aware of the composition of the training sequence of a frame, the DSP can calculate a filter which is then used to decode the data part of the burst. This increases the probability that the data can be correctly reconstructed. The DSP 56600 architecture with a processor speed of 104 MHz is often used for these tasks.

Figure 1.46 shows which tasks are performed by the RISC processor and the DSP processor, respectively. If the transmission chain for a voice signal is compared between

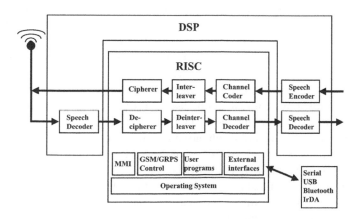

Figure 1.46 Overview of RISC and DSP functionalities

the mobile phone and the network, it can be seen that the TRAU mostly performs the task the DSP unit is responsible for in the mobile phone. All other tasks such as channel coding are performed by the BTS which is thus the counterpart of the RISC CPU of the mobile phone.

As millions of mobile phones are sold every year, there is a great variety of chipsets available on the market. The chipset is in many cases not designed by the manufacturer of the mobile phone. While Motorola design its own chipsets, Nokia relies on chipsets of STMicroelectronics and Texas Instruments. Other GSM chipset developers include Infineon, Analog Devices, and Philips, as well as many Asian companies.

Furthermore, mobile phone manufacturers are also outsourcing parts of the mobile phone software development. BenQ/Siemens for example uses the WAP browser of OpenWave, which the company has also sold to other mobile phone manufacturers. This demonstrates that many companies are involved in the development and production of a mobile phone. It can also be observed that most GSM and UMTS phones today are shipped with a device-independent Java runtime environment, which is called the Java 2 Micro Edition (J2ME) [20]. This allows third-party companies and individuals to develop programs which can be ported with no or only minor effort to other mobile phones as well. Most games for example, which are available for GSM and UMTS mobile phones today, are based on J2ME and many other applications like email and other office software is available via the mobile network operator or directly via the Internet.

1.10 The SIM Card

Despite its small size, the SIM card is one of the most important parts of a GSM network because it contains all the subscription information of a subscriber. Since it is standardized, a subscriber can use any GSM or UMTS phone by simply inserting the SIM card. Exceptions are phones that contain a 'SIM lock' and thus only work with a single SIM card or only with the SIM card of a certain operator. However, this is not a GSM restriction. It was introduced by mobile phone operators to ensure that a subsidized phone is only used with SIM cards of their network.

The most important parameters on the SIM card are the IMSI and the secret key (Ki), which is used for authentication and the generation of ciphering keys (Kc). With a number of tools, which are generally available on the Internet free of charge, it is possible to read out most parameters from the SIM card, except for sensitive parameters that are read protected. Figure 1.47 shows such a tool. Protected parameters can only be accessed with a special unlock code that is not available to the end user.

Astonishingly, a SIM card is much more than just a simple memory card as it contains a complete microcontroller system that can be used for a number of additional purposes. The typical properties of a SIM card are shown in Table 1.7.

As shown in Figure 1.48, the mobile phone cannot access the information on the EEPROM directly, but has to request the information from the SIM's CPU. Therefore, direct access to sensitive information is prohibited. The CPU is also used to generate the SRES during the network authentication procedure based on the RAND which is supplied by the authentication center (see Section 1.6.4). It is imperative that the calculation of the SRES is done on the SIM card itself and not in the mobile phone in order to protect the secret Ki key. If the

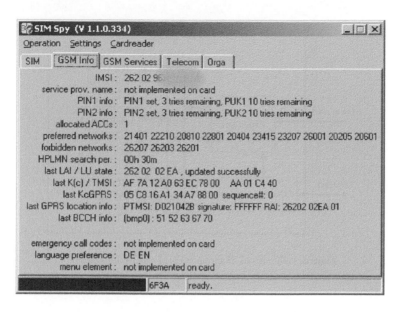

Figure 1.47   Example of a tool to visualize the data contained on a SIM card

Table 1.7   SIM card properties

CPU	8- or 16-bit CPU
ROM	40–100 kbyte
RAM	1–3 kbyte
EEPROM	16–64 kbyte
Clock rate	10 MHz, generated from clock supplied by mobile phone
Operating voltage	3 V or 5 V

calculation was done in the mobile phone itself, this would mean that the SIM card would have to hand over the Ki to the mobile phone or any other device upon request. This would seriously undermine security as tools like the one shown in Figure 1.47 would be able to read the Ki which then could be used to make a copy of the SIM card.

Furthermore, the microcontroller system on the SIM can also execute programs which the network operator may have installed on the SIM card. This is done via the SIM application toolkit (SAT) interface, which is specified in 3GPP TS 31.111 [21]. With the SAT interface, programs on the SIM card can access functionalities of the mobile phone such as waiting for user input, or can be used to show text messages and menu entries on the display. Many mobile network operators use this functionality to put an operator-specific menu item into the overall menu structure of the mobile phone's graphical user interface. In the menu created by the SIM card program, the subscriber can, for example, request a current news overview. When the subscriber enters the menu, all user input via the keypad is forwarded by the mobile phone to the SIM card. The program on the SIM card in this example would

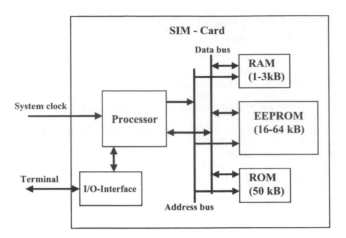

Figure 1.48 Block diagram of SIM card components

react to the news request by generating an SMS, which it then instructs the mobile phone to send to the network. The network replies with one or more SMS messages which contain a news overview. The SIM card can then extract the information from the SMS messages and present the content to the subscriber.

A much more complex application of the SIM application toolkit is in use by O2 Germany for a service called 'Genion'. If a user has subscribed to 'Genion', he can make cheaper calls to fixed-line phones if the subscriber is currently located in his so-called 'homezone'. To define the homezone, the SIM card contains information about its size and geographical location. In order to inform the user if he is currently located in his homezone, the SIM card receives information about the geographical position of the current serving cell. This information is broadcast to the mobile phone via the short message service broadcast channel (SMSCB) of the cell. When the program on the SIM card receives this information, it compares the geographical location contained on the SIM card with the coordinates received from the network. If the user is inside his homezone, the SIM card then instructs the mobile phone to present a text string ('home' or 'city') in the display for the user.

From a logical point of view, data is stored on a GSM SIM card in directories and files in a similar way as on a PC's hard drive. The file and folder structure is specified in 3GPP TS 31.102 [22]. In the specification, the root directory is called the main file (MF) which is somewhat confusing at first. Subsequent directories are called dedicated files (DF) and normal files are called elementary files (EF). As there is only a very limited amount of memory on the SIM card, files are not identified via file and directory names. Instead, hexadecimal numbers with a length of four digits are used which require only two bytes of memory. The standard nevertheless assigns names to these numbers which are, however, not stored on the SIM card. The root directory for example is identified via ID 0x3F00, the GSM directory is identified by ID 0x7F20, and the file containing the IMSI for example is identified via ID 0x6F07. In order to read the IMSI from the SIM card, the mobile station thus has to open the following path and file: 0x3F00 0x7F20 0x6F07.

To simplify access to the data contained on the SIM card for the mobile phone, a file can have one of the following three file formats:

- Transparent: the file is seen as a sequence of bytes. The file for the IMSI for example is of this format. How the mobile station has to interpret the content of the files is again specified in 3GPP TS 31.002 [22].
- Linear fixed: this file type contains records of a fixed length and is used for example for the file that contains the telephone book records. Each phone record uses one record of the linear fixed file.
- Cyclic: this file type is similar to the linear fixed file type but contains an additional pointer which points to the last modified record. Once the pointer reaches the last record of the file, it wraps over again to the first record of the file. This format is used for example for the file in which the phone numbers are stored which have previously been called.

A number of different access right attributes are used to protect the files on the SIM card. By using these attributes, the card manufacturer can control if a file is read or write only when accessed by the mobile phone. A layered security concept also permits network operators to change files which are read only for the mobile phone over the air by sending special provisioning SMS messages.

The mobile phone can only access the SIM card if the user has typed in the PIN when the phone is started. The mobile phone then uses the PIN to unlock the SIM card. SIM cards of some network operators, however, allow deactivating the password protection and thus the user does not have to type in a PIN code when the mobile phone is switched on. Despite unlocking the SIM card with the PIN, the mobile phone is still restricted to only being able to read or write certain files. Thus, it is not possible for example to read or write the file which contains the secret key Ki even after unlocking the SIM card with the PIN.

Details on how the mobile station and the SIM card communicate with each other has been specified in ETSI TS 102 221 [23]. For this interface, layer 2 command and response messages have been defined which are called application protocol data units (APDU). When a mobile station wants to exchange data with the SIM card, a command APDU is sent to the SIM card. The SIM card analyzes the command APDU, performs the requested operation, and returns the result in a response APDU. The SIM card only has a passive role in this communication as it can only send response APDUs back to the mobile phone.

If a file is to be read from the SIM card, the command APDU contains among other information the file ID and the number of bytes to read from the file. If the file is of type cyclic or linear fixed, the command also contains the record number. If access to the file is allowed, the SIM card then returns the requested information in one or more response APDUs.

If the mobile phone wants to write some data into a file on the SIM card, the command APDUs contain the file ID and the data to be written into the file. In the response APDU, the SIM card then returns a response as to whether the data was successfully written to the file.

Figure 1.49 shows the format of a command APDU. The first field contains the class of instruction, which is always 0xA0 for GSM. The instruction (INS) field contains the ID of the command that has to be executed by the SIM card.

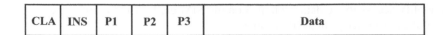

CLA	INS	P1	P2	P3	Data

Figure 1.49 Structure of a command APDU

Table 1.8 shows some commands and their IDs. The fields P1 and P2 are used for additional parameters for the command. P3 contains the length of the following data field which contains the data that the mobile phone would like to write to the SIM card.

The format of a response APDU is shown in Figure 1.50. Apart from the data field, the response also contains two fields called SW1 and SW2. These are used by the SIM card to inform the mobile station if the command was executed correctly.

An example: to open a file for reading or writing, the mobile station sends a SELECT command to the SIM card. The SELECT APDU is structured as shown in Figure 1.51.

As a response, the SIM card replies with a response APDU which contains a number of fields. Some of them are shown in Table 1.9.

For a complete list of information returned for the example, see [23]. In a next step, the READ BINARY or WRITE BINARY APDU can be used to read or modify the file.

In order to physically communicate with the SIM card, there are six contact areas on the top side of the SIM card. Only four of those contacts are required:

- C1: power supply;
- C2: reset;
- C3: clock;
- C7: input/output.

Table 1.8 Examples for APDU commands

Command	ID	P1	P2	Length
Select (open file)	A4	00	00	02
Read Binary (read file)	B0	Offset High	Offset Low	Length
Update Binary (write file)	D6	Offset High	Offset Low	Length
Verify CHV (check PIN)	20	00	ID	08
Change CHV (change PIN)	24	00	ID	10
Run GSM algorithm (RAND, SRES, Kc,...)	88	00	00	10

Data	SW1	SW2

Figure 1.50 Response APDU

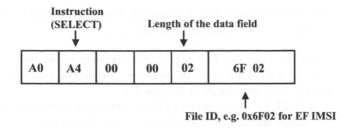

Figure 1.51   Structure of the SELECT command APDU

Table 1.9   Some fields of the response APDU for a SELECT command

Byte	Description	Length
3–4	File size	2
5–6	File ID	2
7	Type of file (transparent, linear fixed, cyclic)	1
9–11	Access rights	3
12	File status	1

As only a single line is used for input and output of command and status APDUs, the data is transferred in half-duplex mode only. The clock speed for the transmission has been defined as C3/327. At a clock speed of 5 MHz on C3, the transmission speed is thus 13,440 bit/s.

1.11 The Intelligent Network Subsystem and CAMEL

All components that have been described in this chapter are mandatory elements for the operation of a mobile network. Mobile operators, however, usually offer additional services beyond simple post-paid voice services for which additional logic and databases are necessary in the network. Here are a number of examples:

- Location based services (LBS) are offered by most network operators in Germany in different variations. One LBS example is to offer cheaper phone calls to fixed-lines phones in the area in which the mobile subscriber is currently located. In order to be able to apply the correct tariff for the call, the LBS service in the network checks if the current location of the subscriber and the dialed number are in the same geographical area. If so, additional information is attached to the billing record so the billing system can later calculate the correct price for the call.
- Prepaid services have become very popular in many countries since their introduction in the mid-1990s. Instead of receiving a bill once a month, a prepaid subscriber has an account with the network operator which is funded in advance with a certain amount of money determined by the subscriber. The amount on the account can then be used for phone calls and other services. During every call, the account is continually charged. If the account runs out of credit, the connection is interrupted. Furthermore, prepaid systems are also connected to the SMSC, the multimedia messaging server (MMS-Server, see

Chapter 2), and the GPRS network (see Chapter 2). Therefore, prepaid subscribers can also be charged in real time for the use of these services.

These and many other services can be realized with the help of the intelligent network (IN) subsystem. The logic and the necessary databases are located on a service control point (SCP), which has already been introduced in Section 1.4.

In the early years of GSM, the development of these services had been highly proprietary due to the lack of a common standard. The big disadvantage of such solutions was that they were customized to work only with very specific components of a single manufacturer. This meant that these services did not work abroad as foreign network operators used components of other network vendors. This was especially a problem for the prepaid service as prepaid subscribers were excluded from international roaming when the first services were launched.

In order to ensure the interoperability of intelligent network components between different vendors and in networks of different mobile operators, industry and operators standardized an IN network protocol in 3GPP TS 22.078 [24] which is called customized applications for mobile enhanced logic, or CAMEL for short. While CAMEL also offers functionality for SMS and GPRS charging, the following paragraph only describes the basic functionality necessary for circuit-switched connections.

CAMEL is not an application or a service, but forms the basis to create services (customized applications) on an SCP, which is compatible with network elements of other vendors and between networks. Thus, CAMEL can be compared with the HTTP protocol for example. HTTP is used for transferring web pages between a web server and a browser. HTTP ensures that any web server can communicate with any browser. If the content of the data transfer is a web page or a picture is of no concern to HTTP because this is managed on a higher layer directly by the web server and the web client. Transporting the analogy back to the GSM world, the CAMEL specification defines the protocol for the communication between the different network elements such as the MSC and the SCP, as well as a state model for call control.

The state model is called the basic call state model (BCSM) in CAMEL. A circuit-switched call for example is divided into a number of different states. For the originator (O-BCSM) the following states, which are also shown in Figure 1.52, have been defined:

- call establishment;
- analysis of the called party number;
- routing of the connection;
- notification of the called party (alerting);
- call is ongoing (active);
- disconnection of the call;
- no answer of the called party;
- called party busy.

For a called subscriber, CAMEL also defines a state model which is called the terminating BCSM (T-BCSM). T-BCSM can be used for prepaid subscribers who are currently roaming in a foreign network in order to control the call to the foreign network and to apply real-time charging.

For every state change in the state model, CAMEL defines a detection point (DP). If a DP is activated for a subscriber, the SCP is informed of the particular state change. Information

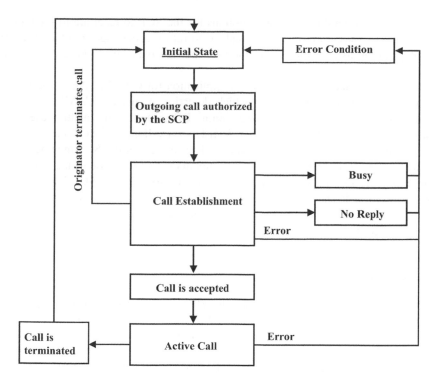

Figure 1.52   Simplified state model for an originator (O-BCSM) according to 3GPP TS 23.078 [25]

contained in this message is for example the IMSI of the subscriber, the current position (MCC, MNC, LAC, and cell-ID), and the number that was called. Whether a detection point is activated is part of the subscriber's HLR entry. This allows creating specific services on a per subscriber basis. When the SCP is notified that the state model has triggered a detection point, the SCP is able to influence how the call should proceed. The SCP can take the call down, change the number that was called, or return information to the MSC, which is put into the billing record of the call for later analysis on the billing system.

For the prepaid service for example the CAMEL protocol can be used between the MSC and the SCP as follows.

If a subscriber wants to establish a call, the MSC detects during the setup of the call, that the 'authorize origination' detection point is activated in the subscriber's HLR entry. Therefore, the MSC sends a message to the SCP and waits for a reply. As the message contains the IMSI of the subscriber as well as the CAMEL service number, the SCP recognizes that the request is for a prepaid subscriber. By using the destination number, the current time and other information, the SCP calculates the price per minute for the connection. If the subscriber's balance is sufficient, the SCP then allows the call to proceed and informs the MSC for how many minutes the authorization is valid. The MSC then continues and connects the call. At the end of the call, the MSC sends another message to the SCP to inform it of the total duration of the call. The SCP then modifies the subscriber's balance. If the time which the SCP initially granted for the call expires, the MSC has to contact the SCP again. The SCP then has the possibility to send an additional authorization to the MSC which is again limited to a certain duration. Other options for the SCP to react are to send a reply in which the MSC is asked to terminate the call or to return a message in which the MSC is asked to play a tone as an indication to the user that the balance on the account is almost depleted.

Location based services (LBS) are another application for CAMEL. Again the HLR entry of a subscriber contains information at which detection points the CAMEL service is to be invoked. For LBS, the 'authorize origination' DP is activated. In this case, the SCP determines, by analyzing the IMSI and the CAMEL service ID, that the call has been initiated by a user that has subscribed to an LBS service. The service on the SCP then deduces from the current location of the subscriber and the national destination code of the dialed number which tariff to apply for the connection. The SCP then informs the MSC of the correct tariff by returning a 'furnish charging information' (FCI) message. At the end of the call, the MSC includes the FCI information in the billing record and thus enables the billing system to apply the correct tariff for the call.

1.12 Questions

1. Which algorithm is used to digitize a voice signal for transmission in a digital circuit-switched network and at which data rate is the voice signal transmitted?
2. Name the most important components of the GSM network subsystem (NSS) and their tasks.
3. Name the most important components of the GSM radio network (BSS) and their tasks.
4. How is a BTS able to communicate with several subscribers at the same time?
5. Which steps are necessary in order to digitize a speech signal in a mobile phone before it can be sent over the GSM air interface?
6. What is a handover and which network components are involved?
7. How is the current location of a subscriber determined for a mobile terminated call and how is the call forwarded through the network?
8. How is a subscriber authenticated in the GSM network? Why is an authentication necessary?
9. How is an SMS message exchanged between two subscribers?
10. Which tasks are performed by the RISC processor and which tasks are performed by the DSP in a mobile phone?
11. How is data stored on the SIM card?
12. What is CAMEL and for which services can it be used?

Answers to these questions can be found on the companion website for this book at http://www.wirelessmoves.com.

References

[1] European Technical Standards Institute (ETSI), website, http://www.etsi.org.
[2] The 3rd Generation Partnership Project, website, http://www.3gpp.org.
[3] 3GPP, 'Mobile Application Part (MAP) Specification', TS 29.002.
[4] 3GPP, 'AT Command Set for 3G User Equipment', TS 27.007.
[5] 3GPP, 'Call Forwarding (CF) Supplementary Services – Stage 1', TS 22.082.

[6] 3GPP, 'Call Barring (CB) Supplementary Services – Stage 1', TS 22.088.

[7] 3GPP, 'Call Waiting (CW) and Call Hold (HOLD) Supplementary Services – Stage 1', TS 22.083.

[8] 3GPP, 'Multi Party (MPTY) Supplementary Services – Stage 1', TS 22.084.

[9] 3GPP, 'Man–Machine Interface (MMI) of the User Equipment (UE)', TS 22.030.

[10] 3GPP, 'Mobile Radio Interface Layer 3 Specification; Core Network Protocols – Stage 3', TS 24.008.

[11] 3GPP, 'Technical Realisation of Short Message Service (SMS)', TS 23.040.

[12] 3GPP, 'Voice Group Call Service (VGCS) – Stage 2', TS 43.068.

[13] 3GPP, 'Voice Broadcast Service (VGS) – Stage 2', TS 43.069.

[14] 3GPP, 'Enhanced Multi-Level Precedence and Preemption Service (eMLPP) – Stage 2', TS 23.067.

[15] Union Internationale des Chemins de Fer, GSM-R website, http://gsm-r.uic.asso.fr.

[16] 3GPP, 'Multiplexing and Multiple Access on the Radio Path', TS 45.002.

[17] 3GPP, 'AMR Speech CODEC: General Description', TS 26.071.

[18] 3GPP, 'Full Speech Transcoding', TS 46.010.

[19] 3GPP, 'Basic Call Handling: Technical Realization', TS 23.018.

[20] Sun Microsystems, The Java 2 Micro Edition, http://java.sun.com/j2me/.

[21] 3GPP, 'USIM Application Toolkit', TS 31.111.

[22] 3GPP, 'Characteristics of the USIM Application', TS 31.102.

[23] ETSI, 'Smart Cards; UICC-Terminal Interface; Physical and Logical Characteristics', TS 102 221.

[24] 3GPP, 'Customised Applications for Mobile Network Enhanced Logic (CAMEL): Service Description – Stage 1', TS 22.078.

[25] 3GPP, 'Customised Applications for Mobile Network Enhanced Logic (CAMEL): Service Description – Stage 2', TS 23.078.

2

General Packet Radio Service (GPRS)

In the mid-1980s voice calls were the most important service for fixed and wireless networks. This is the reason why GSM was initially designed and optimized for voice transmission. Since the mid-1990s, however, the importance of the Internet has been constantly increasing. GPRS, the General Packet Radio Service, enhances the GSM standard to transport data in an efficient manner and thus allows wireless devices to access the Internet. The first part of this chapter shows the advantages and disadvantages of GPRS compared to data transmission in classic GSM and fixed-line networks. The second part of the chapter focuses on how GPRS has been standardized and implemented. At the end of the chapter, some applications for GPRS are discussed and an analysis is presented on how the network behaves for a web-browsing session.

2.1 Circuit-Switched Data Transmission over GSM

As we have seen in Chapter 1, the GSM network was initially designed as a circuit-switched network. All resources for a voice or data session are set up at the beginning of the call and are reserved for the user until the end of the call as shown in Figure 2.1. The dedicated resources assure a constant bandwidth and end-to-end delay time. This has a number of advantages for the subscriber:

- Data that is sent does not need to contain any signaling information such as information about the destination. Every bit simply passes through the established channel to the receiver. Once the connection is established no overhead, e.g. addressing information, is necessary to send and receive the information.
- As the circuit-switched channel has a constant bandwidth the sender does not have to worry about a permanent or temporary bottleneck in the communication path. This is especially important for a voice call. As the data rate is constant, any bottleneck in the communication path would lead to a disruption of the voice call.

Communication Systems for the Mobile Information Society   Martin Sauter
© 2006 John Wiley & Sons, Ltd

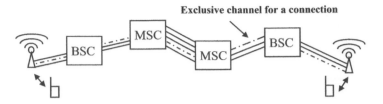

Figure 2.1 Exclusive connections of a circuit-switched system

- Furthermore, circuit-switched connections have a constant delay time. This is the time between sending a bit and receiving it at the other end. The greater the distance between the sender and receiver the longer the delay time. This makes a circuit-switched connection ideal for voice applications as they are extremely sensitive to a variable delay time. If a constant delay time cannot be guaranteed, a buffer at the receiving end is necessary. This adds additional unwanted delay especially for applications like voice calls.

While circuit-switched data transmission is ideally suited for voice transmissions, there are a number of grave disadvantages for data transmission with variable bandwidth usage. Web browsing is a typical application with variable or 'bursty' bandwidth usage. For sending a request to a web server and receiving the web page, as much bandwidth as possible is desired to receive the web page as quickly as possible. As the bandwidth of a circuit-switched channel is constant there is no possibility of increasing the data transmission speed while the page is being downloaded. After the page has been received no data is exchanged while the subscriber reads page. The bandwidth requirement during this time is zero. The resources are simply unused during this time and are thus wasted.

2.2 Packet-Switched Data Transmission over GPRS

For bursty data applications it would be far better to request resources to send and receive data and release them again after the transmission, as shown in Figure 2.2. This can be done by collecting the data in packets before it is sent over the network. This method of sending data is called 'packet switching'. As there is no longer a logical end-to-end connection, every packet has to contain a header. The header for example contains information about the sender (source address) and the receiver (destination address) of the packet. This information is used in the network to route the packets through the different network elements. In the Internet for example the source and destination addresses are the IP addresses of the sender and receiver.

To be able to send packet-switched data over existing GSM networks, the General Packet Radio Service (GPRS) was conceived as a packet-switched addition to the circuit-switched GSM network. It should be noted that IP packets can be sent over a circuit-switched GSM data connection as well. However, until they reach the Internet service provider they are transmitted in a circuit-switched channel and thus cannot take advantage of the benefits described below. GPRS on the other hand is an end-to-end packet switched network and IP packets are sent packet switched from end to end.

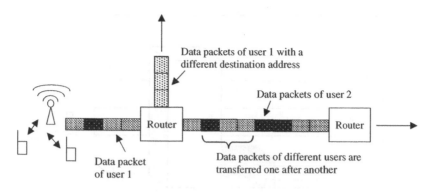

Figure 2.2   Packet-switched data transmission

The packet-switched nature of GPRS also offers a number of other advantages for bursty applications over GSM circuit-switched data transmission:

- By flexibly allocating bandwidth on the air interface, GPRS exceeds the slow data rates of GSM circuit-switched connections of 9.6 or 14.4 kbit/s. Data rates of up to 170 kbit/s are theoretically possible. Today (2006) multislot class 10 mobiles (see below) reach speeds of up about 50 kbit/s and are thus in the range of a fixed-line analog modem.
- With the enhanced data rates for GSM evolution (EDGE) update of the GSM system, further speed improvements have been made. The enhancements of EDGE for GPRS are called EGPRS in the standards. With an EGPRS class 10 mobile it is possible to reach transmission speeds of up to 230 kbit/s in operational networks. While GPRS is offered in most GSM networks today not all operators have chosen to upgrade to EGPRS. Some operators have decided to go directly to UMTS and leave the GPRS system as it is. A comparison of the speed of the different technologies is shown in Figure 2.3.
- GPRS is usually charged by volume and not by time as shown in Figure 2.4. For subscribers this has the advantage that they pay for downloading a web page but not for the time reading it, as would be the case with a circuit-switched connection. For the operator of a wireless network it has the advantage that the scarce resources on the air interface are not wasted by 'idle' data calls because they can be used for other subscribers.
- GPRS dramatically reduces the call set-up time. Similar to a fixed-line analog modem, a GSM circuit-switched data call takes about 20 seconds to establish a connection with the Internet service provider. GPRS accomplishes the same in less than 5 seconds.
- As the subscriber does not pay for the time when no data is transferred, the call does not have to be disconnected to save costs. This is called 'always-on' and enables applications like email programs to poll for incoming emails in certain intervals or allows messaging clients like Yahoo or MSN messenger to wait for incoming messages.
- When the subscriber is moving, by train for example, it happens quite frequently that the mobile has bad network coverage or even loses the network completely for some time. When this happens, circuit-switched connections are disconnected and have to be manually re-established once network coverage is available again. GPRS connections on the other hand are not dropped as the logical GPRS connection is independent of the physical connection to the network. After regaining coverage the interrupted data transfer simply resumes.

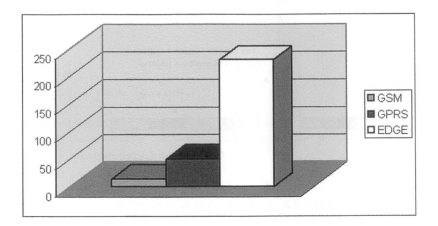

Figure 2.3   GSM, GPRS, and EGPRS data transmission speed comparison

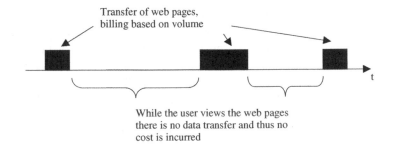

Figure 2.4   Billing based on volume

2.2.1 GPRS and the IP Protocol

GPRS was initially designed to support different types of packet-switching technologies. With the great success of the Internet, which exclusively uses the Internet Protocol (IP) for packet switching, it is the only supported protocol today. Therefore, whenever this chapter uses the terms 'user data transfer', 'user data transmission', or 'packet switching', it always refers to 'transferring IP packets'.

2.2.2 GPRS vs. Fixed-Line Data Transmission

Despite the potential cost savings for the subscriber if he is charged for the transferred data volume and not for connection time, transferring data via GPRS and EGPRS is still more expensive than transferring data from a PC connected to the Internet via a fixed-line connection. It can be observed that the higher data rates of EGPRS and especially UMTS help to close the gap in combination with falling prices. Many websites today also offer their information in a format that is more suitable for screens of smaller devices like PDAs and mobile phones. As those web pages are tailored for smaller screens the pages are usually quite compact. This means that the amount of data that has to be transferred per page is much lower compared to a standard web page with lots of graphical banners and advertisements. Pictures are usually downsized as well and a higher compression factor is used to further reduce the amount of data that has to be transferred. This somewhat compensates for the higher transmission costs. As those pages are often plain HTML pages that are just optimized for smaller devices it is also possible to view them using a normal web browser on a notebook and thus also benefit from their smaller size and reduced transmission cost.

As a conclusion it can be said that (E)GRPS will not be able to replace fixed-line technologies that provide similar speeds like modems, ISDN connections, or ADSL. For classical Internet applications like web browsing or email, however, (E)GPRS is an ideal technology while on the move. GPRS has also laid the foundation for completely new applications such as mobile messaging clients, which benefits from the 'always-on' functionality of a packet-switched wireless network.

2.3 The GPRS Air Interface

2.3.1 GPRS vs. GSM Timeslot Usage on the Air Interface

Circuit-Switched TCH vs. Packet-Switched PDTCH

As shown in Chapter 1, GSM uses timeslots on the air interface to transfer data between the subscribers and the network. During a circuit-switched call a subscriber is assigned exactly one traffic channel (TCH) which is mapped to a single timeslot. This timeslot remains allocated for the duration of the call and cannot be used for other subscribers even if there is no data transferred for some time.

In GPRS, the smallest unit that can be assigned is a block which consists of four bursts of a packet data traffic channel (PDTCH). A PDTCH is similar to a TCH as it also uses one physical timeslot. If the subscriber has more data to transfer, the network can assign more blocks on the same PDTCH right away. The network can also assign the following block(s) to other subscribers or for logical GPRS signaling channels. Figure 2.5 shows how the blocks of a PDTCH are assigned to different subscribers.

Instead of using a 26 or 51 multiframe structure as in GSM (see Section 1.7.3), GPRS uses a 52 multiframe structure for its timeslots. Frames 24 and 51 are not used for sending data as they are used to allow the mobile to perform signal strength measurements on neighboring cells. Frames 12 and 38 are used for timing advance calculations as will be described in more detail later on. All other frames in the 52 multiframe are collected into blocks of four frames (one burst per frame), which is the smallest unit to send or receive data.

Timeslot Aggregation

To increase the transmission speed, a subscriber is no longer bound to a single traffic channel as in circuit-switched GSM. If more than one timeslot is available when a subscriber wants to send or receive data, the network can allocate several timeslots (multislot) to a single subscriber.

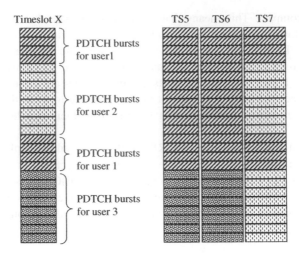

Figure 2.5 Simplified visualization of PDTCH assignment and timeslot aggregation

Multislot Classes

Depending on the multislot class of the terminal, two, three, four or more timeslots can
be aggregated for a subscriber at the same time. Thus, the transmission speed for every
subscriber is increased providing that not all of them want to transmit data at the same
time. Table 2.1 shows some multislot classes. Today, most mobiles on the market support
either multislot class 8 or 10. As can be seen in the table, multislot class 10 supports four
timeslots in the downlink direction and two in the uplink. This means the speed in the

Table 2.1 Some GPRS multislot classes

Multislot class	Possible timeslots		
	Rx	Tx	Sum
1	1	1	2
2	2	1	3
3	2	2	3
4	3	1	4
5	2	2	4
6	3	2	4
7	3	3	4
→ **8**	**4**	**1**	**5**
9	3	2	5
→ **10**	**4**	**2**	**5**
11	4	3	5
12	4	4	5
13	3	3	NA-2
14	4	4	NA-2
15	5	5	NA-2

uplink is substantially slower than in the downlink. For applications like web browsing it is no big disadvantage to have more bandwidth in the downlink than in the uplink direction. Requests for web pages that are sent in the uplink direction are usually quite small while web pages and the embedded pictures require a fast speed in the downlink direction. Thus, web browsing benefits from the higher data rates in downlink and does not suffer very much from the limited uplink speed. For applications like sending emails with file attachments or MMS messages with large pictures or video content, two timeslots in the uplink direction are a clear limitation and increase the transmission time considerably. Only a few networks and terminals today are able to make use of more than two timeslots in the uplink direction. On the terminal side this is mostly due to the fact that using four timeslots requires a lot more transmission power than what the GSM hardware was initially designed for. For GPRS PCMCIA cards this is not a big problem as they get their power from the notebook. Thus, some of those cards are GPRS class 12 capable and can make use of up to four timeslots in the uplink direction if supported by the network as well. Furthermore, the antenna is not close to the user, which also allows the use of a higher power class than for handheld devices.

Also important to note in Table 2.1 is that for most classes the maximum number of timeslots used simultaneously is lower than the combined number of uplink and downlink timeslots. For GPRS class 10 for example, which is widely used today, the sum is five timeslots. This means that if four timeslots are allocated by the network in the downlink, only one can be allocated in the uplink. If the network detects that the mobile stations want to send a larger amount of data to the network it can reconfigure the connection to use two timeslots in the uplink and three in the downlink thus again resulting in the use of five simultaneous timeslots. During a web-browsing session for example it can be observed that the network assigns two uplink timeslots to the subscriber when the web page request is initially sent. As soon as data arrives to be sent to the subscriber, the network quickly reconfigures the connection to use four timeslots in the downlink direction and only a single timeslot if required in the uplink direction.

In order for the network to know how many timeslots the terminal supports it has to inform the network of its capabilities. This so-called mobile station classmark also contains other information such as ciphering capabilities. The classmark information is sent every time the terminal accesses the network. It is then used by the network together with other information like available timeslots to decide how many of them can be assigned to the user. The network also stores the classmark sent in the uplink direction and is thus able to assign resources in the downlink direction immediately without asking the mobile for its capabilities first.

2.3.2 Mixed GSM/GPRS Timeslot Usage in a Base Station

As GPRS is an addition to the GSM network, the eight timeslots available per carrier frequency on the air interface can be shared between GSM and GPRS. Therefore, the maximum GPRS data rate decreases the more GSM voice/data connections are needed. The network operator can choose how to use the timeslots as shown in Figure 2.6. Timeslots can be assigned statically which means some timeslots are reserved for GSM and some for GPRS. The operator also has the possibility to dynamically assign timeslots to GSM or GPRS. If there is a high amount of GSM voice traffic more time slots can be used for GSM. If voice traffic decreases more time slots can be given to GPRS. It is also possible to

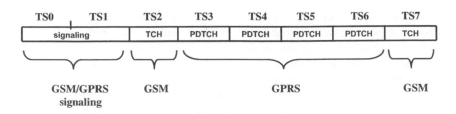

Figure 2.6   Shared use of the timeslots of a cell for GSM and GPRS

assign a minimum number of timeslots for GPRS and dynamically add and remove timeslots depending on voice traffic.

2.3.3 Coding Schemes

Another way to increase the data transfer speed besides timeslot aggregation is to use different coding schemes. If the user is at close range to a base station the data transmitted over the air is less likely to be corrupted during transmission than if the user is further away and the reception is weak. As has been shown in Chapter 1, the base station adds error detection and correction to the data before it is sent over the air. This is called coding and the method used to code the user data is called the coding scheme. In GRPS, four different coding schemes (CS-1 to 4) can be used to add redundancy to the user data depending on the quality of the channel [1]. Table 2.2 shows the properties of the different coding schemes.

While CS-1 and CS-2 are commonly used, CS-3 and CS-4 are not implemented in today's GPRS networks. This is because data that is carried over one timeslot on the air interface is carried in one-quarter of an E-1 timeslot between BTS and BSC which can only carry 16 kbit/s. When the overhead created by the packet header, which is not shown in Table 2.2, is included, CS-3 and CS-4 exceed the amount of data that can be carried over one-quarter of an E-1 timeslot. In order to use these coding schemes it is no longer possible to use a fixed mapping. Unfortunately this requires a costly software and possibly also hardware redesign of the BTS, BSC, and PCU (packet control unit). This is why many operators will not introduce these coding schemes as it would require costly replacement of their BSCs.

Figure 2.7 shows how CS-2 and CS-3 encode the data before it is transmitted over the air interface. CS-4 does not add any redundancy to the data. Therefore, CS-4 can only be used when the signal quality between the network and the mobile station is very good.

Table 2.2   GPRS coding schemes

Coding scheme	Number of user data bits per block (4 bursts with 114 bits each)	Transmission speed per timeslot (kbit/s)
CS-1	160	8
CS-2	240	12
CS-3	288	14.4
CS-4	400	20

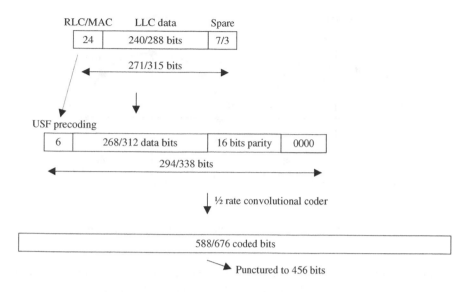

Figure 2.7　CS-2 and CS-3 channel coder

GPRS uses the same 1/2-rate convolutional decoder as already used for GSM voice traffic. The result of the convolutional coding in CS-2 and CS-3 are more coded bits than can be transmitted over a radio block. To compensate for this some of the bits are simply not transmitted. This is called 'puncturing'. As the receiver knows which bits were punctured it can insert 0 bits at the correct positions and then use the convolutional decoder to recreate the original data stream. This of course reduces the effectiveness of the channel coder as not all the bits that were punctured were 0 bits at the sender side.

2.3.4 Enhanced Data Rates for GSM Evolution (EDGE) – EGPRS

In order to further increase the data transmission speeds a new modulation and coding scheme, which uses 8PSK, has been introduced into the standards. The new coding scheme forms the basis of the 'enhanced data rates for GSM evolution' package, which is also called EDGE. The packet-switched part of EDGE is also referred to in the standard as enhanced GPRS or EGPRS. In the GPRS context, EGPRS and EDGE are often used interchangeably. By using 8PSK, EDGE puts three bits into a single transmission step. This way, data transmission speeds can be up to three times faster compared to GSM and GPRS which both use GMSK modulation which only transmits a single bit per transmission step. Figure 2.8 shows the differences between GMSK and 8PSK modulation. While with GMSK the two possibilities 0 and 1 are coded as two positions in the I/Q space, 8PSK codes the three bits in eight different positions in the I/Q space.

Together with the highest of the nine new coding schemes introduced with EGPRS it is possible to transfer up to 60 kbit/s per timeslot. Similarly to CS-3 and CS-4, new hardware components are necessary in the radio network to cope with the higher data rates, which will no longer fit in a quarter of an E-1 timeslot as described before. Furthermore, new terminals are necessary to use these new modulation schemes. As network and terminal inform each

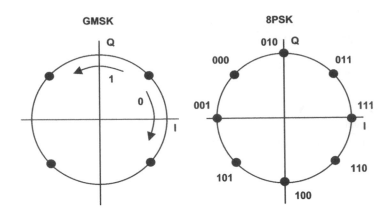

Figure 2.8   GMSK (GPRS) and 8PSK (EGPRS) modulation

other of their capabilities it is possible to use the standard GMSK modulation with older terminals and the new 8PSK modulation with new terminals at the same time in the same cell. From the network side, the terminal is informed of the EGPRS capability of a cell by the EGPRS capability bit in the GPRS cell options of the system information 13 message which is broadcast on the broadcast common control channel (BCCH). From the mobile side, the network is informed of the terminal's EDGE capability during the establishment of a new connection. Therefore, EGPRS is fully backward compatible to GPRS and allows the mixed use of GPRS and EDGE terminals in the same cell. EDGE terminals are also able to use the standard GMSK modulation for GPRS and can thus also be used in networks that do not offer EDGE functionality.

Another advantage of the new modulation and the nine different coding schemes compared to the four different coding schemes of GPRS is a precise use of the best modulation and coding for the current radio conditions. This is done in the terminal by constantly calculating the current bit error probability (BEP) and reporting the values to the network. The network in turn can then adapt its current downlink modulation and coding to the appropriate value. For the uplink direction the network can measure the error rate of data that was recently received and instruct the mobile to change its MCS accordingly. As both network and terminal can report the BEP very quickly it is possible to also quickly adapt to changing signal conditions especially when the terminal is moving in a car or train. This reduces the error rate and ensures the highest transmission speed in every radio condition. In practice it can be observed that this control mechanism allows the use of MCS-8 and MCS-9 if reception conditions are good and a quick fallback to other MCS if the situation deteriorates. Therefore, transmission speeds of over 200 kbit/s can be reached with a class 10 EDGE terminal under real conditions. Table 2.3 gives an overview of the possible modulation and coding schemes and the data rates that can be achieved per timeslot.

Despite the ability to react quickly to changing transmission conditions it is of course still possible that a block contains too many errors and thus the data cannot be reconstructed correctly. To some extent this is even desired as retransmitting a few faulty blocks is preferred over switching to a slower coding scheme. In order to preserve the continuity of the data flow on higher layers, EGPRS introduces a number of enhancements in this area as well. In order to correct transmission errors a method called 'incremental redundancy' has been introduced. As

Table 2.3 EGPRS modulation and coding schemes (MCS)

	Modulation	Speed per timeslot (kbit/s)	Coding rate (user bits to error correction bits)	Coding rate with one retransmission
MCS-1	GMSK	8.8	0.53	0.26
MCS-2	GMSK	11.2	0.66	0.33
MCS-3	GMSK	14.8	0.85	0.42
MCS-4	GMSK	17.6	1.00	0.50
MCS-5	8PSK	22.4	0.37	0.19
MCS-6	8PSK	29.6	0.49	0.24
MCS-7	8PSK	44.8	0.76	0.38
MCS-8	8PSK	54.4	0.92	0.46
MCS-9	8PSK	59.2	1.00	0.50

is already the case with the GPRS coding schemes, some error detection and correction bits produced by the convolutional decoder are punctured and therefore not put into the final block that is sent over the air interface. With the incremental redundancy scheme it is possible to send the previously punctured bits in a second or even third attempt. On the receiver side the original block is stored and the additional redundancy information received in the first and second retry is added to the information. Usually only a single retry is necessary to be able to reconstruct the original data based on the additional information received. Figure 2.9 shows how

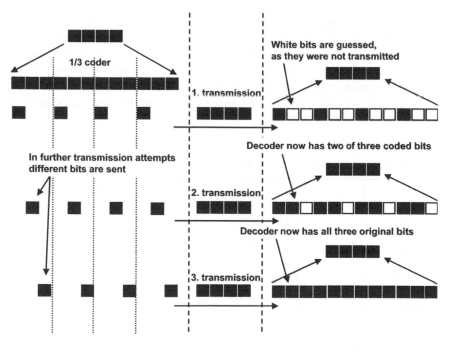

Figure 2.9 MCS-9 convolutional coding and incremental redundancy

MCS-9 uses a 1/3 convolutional decoder to generate three output bits for a single input bit. For the final transmission, however, only one of those three bits are sent. In case the block was not received correctly, the sender will use the second bits that were generated by the convolutional decoder for each input bit to form the retry block. In the unlikely event that it is still not possible for the receiver to correctly decode the data, the sender will send another block containing the third bit. This further increases the probability that the receiver can decode the data correctly by combining the information that is contained in the original block with the redundancy information in the two additional retransmissions.

Another way of retransmitting faulty blocks is to split them up into two blocks for a retransmission that uses a different MCS. This method is called re-segmentation. As can be seen in Table 2.4 the standard defines three code families. If for example a block coded with MCS-9 has to be retransmitted the system can decide to send the content of this block embedded in two blocks which are then coded by using MCS-6. As MCS-6 is more robust than MCS-9, it is much more likely that the content can be decoded correctly. In a real network though, it can be observed that the incremental redundancy scheme is preferred over re-segmentation.

The interleaving algorithm, which re-orders the bits before they are sent over the air interface in order to disperse consecutive bit errors, has been changed for EGRPS as well. GSM voice packets and GPRS data blocks are always interleaved over four bursts as described in Section 1.7.3. As EGPRS notably increases the number of bits that can be sent in a burst it has been decided to decrease the block size for MCS-7, -8 and -9 to fit in two bursts instead of four. This reduces the number of bits that need to be retransmitted after a block error has occurred and thus helps the system to recover more quickly. The block length reduction is especially useful if frequency hopping is used in the system. When frequency hopping is used, every burst is sent on a different frequency in order to avoid using a constantly jammed channel. While the approach is good for voice services that can hide badly damaged blocks from the user up to a certain extent, it poses a retransmission risk for packet data if one of the frequencies used in the hopping sequence performs very badly. Thus, limiting the size of MCS-7, -8 and -9 blocks to two bursts helps to better cope with such a situation.

Table 2.4   Re-segmentation of EGPRS blocks using a different MCS

MCS	Family	Speed (kbit/s)	Re-segmentation
MCS-9	A	59.2 (2 × 29.2)	2 × MCS-6
MCS-8	A	54.4 (2 × 29.2 + padding)	2 × MCS-6 (+ padding)
MCS-6	A	29.2 (2 × 14.8)	2 × MCS-3
MCS-3	A	14.8	—
MCS-7	B	44.8 (2 × 22.4)	2 × MCS-5
MCS-5	B	22.4 (2 × 11.2)	2 × MCS-2
MCS-2	B	11.2	—
MCS-4	C	17.6	2 × MCS-1
MCS-1	C	8.8	—

2.3.5 Mobile Station Classes

The GPRS standard defines three different classes of mobile stations: class C mobiles can only be attached to GPRS or GSM at a time. As this is quite inconvenient for a user, class C mobiles are only suited for embedded applications that only need either GSM or GPRS to transfer data.

Today, all mobiles available on the market are class B. They can be attached to both GPRS and GSM at the same time. However, there is one important limitation: GSM and GPRS cannot be used at the same time. This means that while a voice call is ongoing it is not possible to transfer data via GPRS. Likewise, during data transmission no voice call is possible. For outgoing calls this is not a problem. If a GPRS data transmission is ongoing it will be interrupted when the user starts a telephone call and is automatically resumed once the call is finished. There is no need to reconnect to GPRS as only the data transfer is interrupted. The logical GPRS connection remains in place during the voice call.

During data transmission the mobile station is not able to listen to the GSM paging channel. This means that the mobile is not able to detect incoming voice calls or SMS messages. For bursty Internet applications this is not a problem in most cases. Once the current data transfer is finished the PDTCHs are released and the mobile is again able to listen to the paging channel. As a paging message is usually repeated after a couple of seconds the probability of overhearing a paging message and thus missing a call depends on the ratio between active data transmission time and idle time. Optionally, the circuit-switched and packet-switched core networks can exchange information about incoming calls or SMS messages (NOM I, see below). This way, no paging message is lost during an ongoing data transfer.

The GPRS standard has also foreseen class A mobiles that can be active in both GSM and GPRS at the same time. This means that a GPRS data transfer and a GSM voice call can be active at the same time. Today, there are no such mobiles on the market as the practical implementation would require two sets of independent transceivers in the mobile station. As this has been deemed impractical an enhancement was put into the GPRS standard that is called the 'dual transfer mode' or DTM for short. DTM synchronizes the circuit- and packet-switched parts of the GSM/GPRS network and thus allows using GSM and GPRS functionality in the handset simultaneously with a single transceiver.

2.3.6 Network Mode of Operation

Similar to GSM, the data transferred over the GPRS network can be both user data and signaling data. Signaling data is exchanged for example during the following procedures:

- The network pages the mobile station to inform it of incoming packets.
- The mobile station accesses the network to request resources (PDTCHs) to send packets.
- Modification of resources assigned to a subscriber.
- Acknowledgment of correct reception of user data packets.

This can be done in a number of ways:

In GPRS, network operation mode I (NOM I) signaling for packet- and circuit-switched data is done either via the GSM paging channel (PCH) or the GPRS packet paging channel

(PPCH) if it is available. To make sure incoming voice calls are not missed by class B mobile stations during an active data transfer, an interface between the circuit-switched part (MSC) and the packet-switched part (serving GPRS support node – SGSN) of the network is used. This interface is called the Gs interface. Paging for incoming circuit-switched calls will be forwarded to the packet-switched part and then sent to the mobile as shown in Figure 2.10. If a packet data transfer is in progress when a paging needs to be sent, the mobile will be informed via the packet associated control channel (PACCH) to which the circuit-switched GSM part of the network does not have access. Otherwise, the paging is done via the PCH or the PPCH. The Gs interface can also be used for combined GSM/GPRS attach procedures and location updates. NOM I is the only mode in which the Gs interface is available and thus the only mode in which the mobile is capable of receiving the paging during an ongoing data transfer. As the Gs interface is optional, it is not widely used in networks today. However, most network vendors have implemented this interface and a number of operators have begun using it to enhance their network's behavior for subscribers during GPRS sessions.

The GPRS NOM II is the simplest of the three network modes and therefore commonly used today. There is no signaling connection between the circuit-switched and packet-switched part of the network and therefore the PPCH is not present. For more on this see the next section. This has the disadvantage that the mobile will not see incoming circuit-switched calls during packet-switched data transmission as described before.

And finally there is NOM III. In this mode the Gs interface is not available and thus the circuit-switched paging has to be done over the PCH. In this mode, the GPRS common control channel with its subchannels PPCH, PRACH, and PAGCH is available and the packet-switched side performs its signaling via its own channels. This mode might be

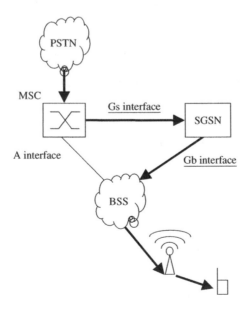

Figure 2.10   Paging for an incoming voice call via the Gs interface

preferable to NOM II in some situations as it reduces the traffic load on the PCH, which is used heavily in operational networks for paging messages, random access requests and assignment requests for circuit-switched services.

To inform users which of these GPRS network modes is used, GPRS uses the GSM broadcast common control channel (BCCH).

2.3.7 GPRS Logical Channels on the Air Interface

GPRS defines a number of new logical channels on the air interface. They are used for sending user and signaling data in the up- and downlink direction. The following logical channels that are shown in Figure 2.12 are mandatory for the operation of GPRS:

- The packet data traffic channel (PDTCH): This is a bi-directional channel which means it exists in the up- and downlink direction. It is used to send user data across the air interface. The PDTCH is carried over timeslots that are dedicated for GPRS in a 52-multiframe structure that was introduced in Section 2.3.1.
- The packet associated control channel (PACCH): This channel is also bi-directional and is used to send control messages. These are necessary in order to acknowledge packets that are transported over the PDTCH. When a mobile receives data packets from the network via a downlink PDTCH it has to acknowledge them via the uplink PACCH. Like the PDTCH the PACCH is also carried over the GPRS dedicated timeslots in blocks of the 52-multiframe structure that was introduced in Section 2.3.1. In order for the terminal and the network to distinguish between PDTCH and PACCH that are carried over the same physical resource, the header of each block contains a logical channel information field as shown in Figure 2.11.
- The packet timing advance control channel (PTCCH). This channel is used for timing advance estimation and control of active terminals. In order to calculate the timing advance, the network can instruct an active terminal to send a short burst in a regular interval on the PTCCH. The network then calculates the timing advance and sends the result back in the downlink direction of the PTCCH.

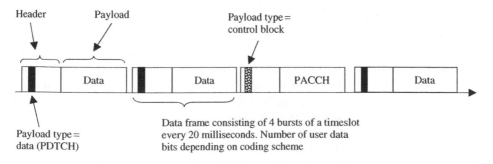

Figure 2.11   PDTCH and PACCH are sent on the same timeslot

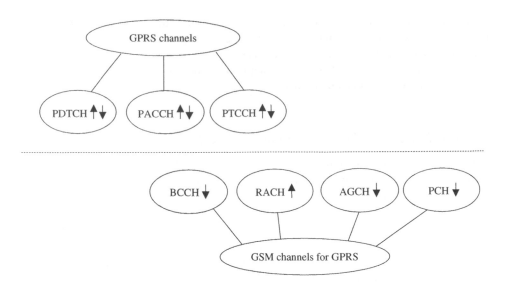

Figure 2.12 Logical channels of GPRS NOM II

In NOM II, GPRS uses the channels described below for network access that are already used in GSM for circuit-switched tasks. An overview of how a mobile uses these channels to request resources on the air interface is given in Figure 2.13.

- The random access channel (RACH): If the mobile wants to send data blocks to the network it has to contact the network and ask for uplink resources. This is done the same way as already described in Chapter 1 for voice calls. The only difference is the content of the channel request message. Instead of asking for a circuit-switched resource the message asks for packet resources on the air interface.
- The access grant channel (AGCH): The network will answer to a channel request on the RACH with an immediate packet assignment message that contains information about the PDTCH packet resources the mobile is allowed to use in the uplink. The PCU can also assign resources in the downlink direction for the mobile if there is data to be sent to the mobile. If the mobile is in ready state, the network can send an immediate packet assignment message right away without paging the mobile first.
- The paging channel (PCH): In case the mobile is in standby state, only the location area of a subscriber is known. As the cell itself is not known, resources cannot be assigned right away and the subscriber has to be paged first. GPRS uses the GSM PCH to do this.
- The broadcast common control channel (BCCH): A new system information message (SYS_INFO 13) has been defined on the BCCH to inform mobiles about GPRS parameters of the cell. This is necessary to let mobiles know for example if GPRS is available in a cell, which NOM is used, if EGPRS is available, etc.

In busy networks, the common control channels like RACH, PCH and AGCH are already used heavily for circuit-switched applications. Because GPRS also makes heavy use of these channels, the situation is further aggravated. In order to remove the GPRS load on these

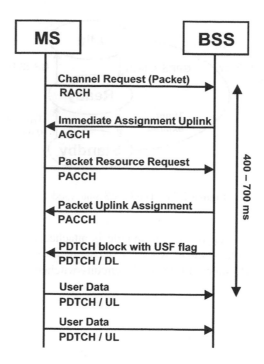

Figure 2.13 Request of uplink resources, NOM II

channels and on the BSC, which controls these channels, an optional packet common control channel (PCCCH) was defined. The PCCCH uses its own timeslot and is transparently forwarded by the BSC to the PCU. The PCCCH is an umbrella term for the following logical channels: The packet random access channel (PRACH), the packet paging channel (PPCH) and the packet access grant channel (PAGCH). The PCCCH is mandatory in NOM III and optional in NOM I. The disadvantage of the PCCCH is the fact that it takes a complete timeslot of a cell and thus reduces the capacity for user data.

2.4 The GPRS State Model

When the mobile is attached to the GSM network it can be either in IDLE mode as long as there is no connection, or in DEDICATED mode during a voice call or exchange of signaling information. For GPRS, a new state model was introduced to address the needs of a packet-switched connection which is shown in Figure 2.14.

The Idle State

In this state the mobile is not attached to the GPRS network at all. This means that the serving GPRS support node (SGSN) is not aware of the user's location, no packet data protocol (PDP) context is established and the network cannot forward any packets for the user. It was very unfortunate by the standards body to name this state 'idle' because in

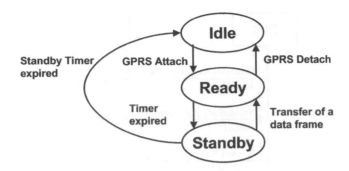

Figure 2.14   The GPRS state model

the GSM circuit-switched 'idle mode' the mobile is attached to the circuit-switched side
of the network and reachable by the network. Great care therefore has to be taken not to mix
up the packet-switched idle state with the GSM circuit-switched idle mode.

The Ready State

If the user wants to attach to the GPRS network the mobile enters the ready state as soon
the first packet is sent. While in ready state the mobile station has to report every cell
reselection to the network so the SGSN can update the user's position in its database.
This process is called 'cell update'. It enables the network to send any incoming data for
a user directly to the mobile station instead of having to page the mobile first to locate
the user's serving cell. The mobile will remain in the ready state while signaling or user
data is transferred and for a certain time afterwards. The timer that controls how long the
mobile will remain in this state after the last block of data was transferred is called T3314.
The value of this timer is broadcast on the BCCH or PBCCH as part of the GPRS system
information. A typical value for this timer that is used in many networks is 44 seconds. The
timer is reset to its initial value in both the mobile station and the SGSN whenever data is
transmitted. When the timer reaches 0 the logical connection between the mobile and the
network automatically falls back into the standby state which is further described below.

It is important to note that the ready state of a mobile is not synonymous with ability of the
mobile station to transfer data to and from the Internet. In order to transfer user data, a
so-called PDP context is necessary which is further described in Section 2.8.2. The ready
state simply represents the fact that both signaling and possibly also user data can be sent to
the terminal without prior paging by the network.

The ready state resembles in some ways the GSM dedicated mode. However, it should
be noted that in the GPRS ready state the network is not responsible for the user's mobility
as would be the case in the GSM dedicated mode. The decision to select a new cell for an
ongoing data transfer is not made by the network (see GSM handover) but by the mobile
station. When the signal quality deteriorates during an ongoing data transfer and the mobile
sees a better cell, it will interrupt the ongoing data transfer and change to the new cell. After
reading the system information on the BCCH it re-establishes the connection and informs the
network of the cell change. The complete procedure takes about 2 seconds. Afterwards, the
communication resumes. Data of the aborted connection might have to be resent if it was not

acknowledged by the network or the mobile before the cell change. While the disruption of the communication path is just a small annoyance for interactive services like web browsing, this behavior is a great disadvantage for real-time services such as voice over IP.

In order to minimize the impact of cell changes, an optional method requiring support of both terminal and network has been added to the GPRS standard, which is called network assisted cell change (NACC). If implemented, the terminal has the possibility to send a packet cell change notification message to the network when it wants to change into a different cell. The network responds with a packet neighbor cell data message alongside the ongoing user data transfer that contains all necessary parts of the system information of the new cell to perform a quick reselection. Afterwards the network stops the user data transfer in the downlink direction and instructs the terminal to switch to the new cell. The mobile then moves to the new cell and re-establishes the connection to the network without having to read the system information messages from the broadcast channel first. By skipping this step the data traffic interruption is reduced to a few hundred milliseconds. The network can then resume data transfer in the downlink direction from the point the transmission was interrupted. While there is usually some loss of data during the standard cell change procedure in the downlink this is not the case with NACC. Thus, this additional benefit also contributes to speeding up the cell change. In order to complete the procedure the terminal asks the network for the remaining system information via a provide system information message while the user data transfer is already ongoing again.

While the implementation of NACC in the terminals is quite simple there are a number of challenges on the network side. While old and new cells are in the same location area and thus controlled from the same radio network node the procedure is straightforward. If the new and old cells are in different location areas, however, they might be controlled by different network elements. Therefore, an additional synchronization between the elements in the network is necessary in order to redirect the downlink data flow to the new cell before the terminal performs the cell reselection. Unfortunately, this synchronization was not included when NACC was first introduced into the GPRS standards in Release 4. This will probably limit NACC to cell changes inside location areas for some time to come. As NACC was standardized rather late, there are currently no operational networks and terminals that support the functionality. Several vendors, however, have announced the availability of this feature so operators will most likely be able to activate it in the near future.

The Standby State

In case no data is transferred for some time, the ready timer expires and the terminal changes into the standby state. In this state the terminal only informs the network of a cell change if the new cell belongs to a different routing area then the previous one. If data arrives in the network for the terminal after it has entered the standby state the data needs to be buffered and the network has to page the subscriber in the complete routing area to get the current location. Only then can the data be forwarded as shown in Figure 2.15. A routing area is a part of a location area and thus also consists of a number of cells. While it would have been possible to use location areas for GPRS as well, it was decided that splitting location areas into smaller routing areas enables operators to better fine tune their networks by being able to control GSM and GPRS signaling messages independently.

If the terminal detects after a cell change that the routing area is different from the previous cell it starts to perform a routing area update which is similar to a GSM location area update.

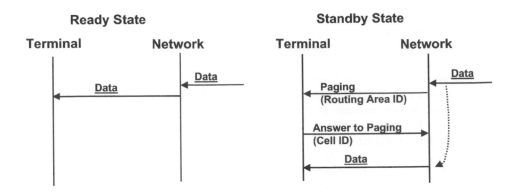

Figure 2.15 Difference between ready and standby state

In case the location area has changed as well the terminal needs to perform both a location update and a routing area update.

The advantage of the standby state for the network is the reduced signaling overhead as not every cell change has to be reported. Thus, scarce resources on the RACH, the AGCH and the PDTCH can be saved. For the terminal the advantage of the standby state lies in the fact that it can stop the continuous monitoring of the AGCH and only infrequently monitor the PCH as described in more detail below. Most operators have set the PCH monitoring interval to around 1.5 seconds (e.g. 6–8 multiframes), which helps to significantly reduce power consumption.

In the uplink direction there is no difference between ready and standby state. If a terminal wants to send data while being in standby state, it implicitly switches back to ready state once the first frame is sent to the network.

2.5 GPRS Network Elements

As has been shown in the previous paragraphs, GPRS works in a very different way compared to the circuit-switched GSM network. This is why three new network components were introduced into the mobile network and software updates had to be made to some of the existing components. Figure 2.16 gives an overview of the components of a GPRS network, which are described in more detail below.

2.5.1 The Packet Control Unit (PCU)

The BSC has been designed to switch 16 kbit/s circuit-switched channels between the MSC and the subscribers. It is also responsible for the handover decisions for those calls. As GPRS subscribers no longer have a dedicated connection to the network, the BSC and its switching matrix are not suited to handle packet-switched GPRS traffic. Therefore, this task has been assigned to a new network component, the packet control unit (PCU). The PCU is the packet-switched counterpart of the BSC and fulfills the following tasks:

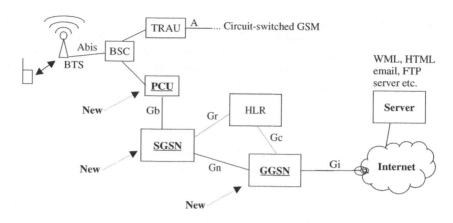

Figure 2.16 GPRS network nodes

- Assignment of timeslots to subscribers in the uplink direction when requested by the mobile via the RACH or the PRACH.
- Assignment of timeslots to subscribers in the downlink direction for data arriving from the core network.
- Flow control of data in the up- and downlink and prioritization of traffic.
- Error checking and retransmission of lost or faulty frames.
- Subscriber paging.
- Supervising entity for subscriber timing advance during data transmission.

In order for the PCU to control the GPRS traffic the BSC turns over control for some of the timeslots to the PCU. This is done by redirecting timeslots in the BSC switching matrix away from the MSC and TRAU towards the PCU. The BSC then simply forwards all data contained in these timeslots to and from the PCU without any processing.

As GPRS uses GSM signaling channels like the RACH, PCH and AGCH to establish the initial communication, a control connection has to exist between the PCU and the BSC. If the mobile requests GPRS resources from the network, the BSC receives a channel request message for packet access. The BSC forwards such packet access request messages straight to the PCU without further processing. It is then the PCU's task to assign uplink blocks on a PDTCH and return an immediate packet assignment command which contains a packet uplink assignment for the subscriber. The BSC just forwards this return message from the PCU to the BTS without further processing. Once GPRS uplink resources have been assigned to a user by the PCU, further signaling will be handled by the PCU directly over the GPRS timeslots and no longer via the GSM signaling channels.

Monitoring GSM and GPRS Signaling Messages

In GSM it is quite easy to use a network tracer to monitor all signaling messages being exchanged between the BSC, BTS and the mobile stations communicating over a BTS. All messages use the same logical LAPD channel which is transmitted over dedicated LAPD timeslots on the Abis interface. The traffic channels used for the voice data are transmitted on different timeslots. In order to monitor the signaling messages it is only necessary to monitor the LAPD timeslots. Monitoring GPRS signaling messages is a far more complex task, as they can be sent over any GPRS timeslot and in between user data blocks. Therefore, it is also necessary to trace user data and empty packets on the Abis interface which requires more processing power and memory on the network tracer.

PCU Positioning

The GSM standards allow for a number of different positions of the PCU in the network. The most common implementation is to have the PCU behind the BSC as shown in Figure 2.16. Based on the design of the PCU some network suppliers deliver it as one or more cards that can be plugged into the BSC. Others have chosen to base the PCU on a more powerful computing architecture that is able to process the GPRS traffic of more then just one BSC. In such an architecture the PCU is implemented in a physically independent cabinet from the BSC. Several BSCs are then connected to a single PCU.

The interface between the PCU and BSC has not been standardized. This means that the PCU and BSC have to be from the same supplier. If a network operator has BSCs from multiple suppliers he is constrained to also buy the PCUs from the same number of network suppliers.

2.5.2 The Serving GPRS Support Node (SGSN)

The SGSN can be seen as the packet-switched counterpart to the MSC in the circuit-switched core network. It is responsible for the following tasks:

User Plane Management

The user plane combines all protocols and procedures for the transmission of user data frames between the subscriber and external networks like the Internet or a company intranet. All frames that arrive for a subscriber at the SGSN are forwarded to the PCU, which is responsible for the current cell of the subscriber. In the reverse direction the PCU delivers data frames of a subscriber to the SGSN, which will in turn forward them to the next network node, which is called the gateway GPRS support node (GGSN). The GGSN is further described in the next section.

The Internet protocol (IP) is used as the transport protocol in the GPRS core network between the SGSN and GGSN. This has the big advantage that on lower layers a great number of different transmission technologies can be used (Figure 2.17). For short distances between the network elements, 100 Mbit/s Ethernet twisted pair links can be used while over long distances ATM over optical STM (e.g. STM-1 with 155 Mbit/s) is more likely to be used. By using IP it is ensured that the capacity of the core network can be flexibly enhanced in the future.

To connect the SGSN with the PCU, the frame relay protocol was selected. The decision not to use IP on this interface is somewhat difficult to understand from today's perspective. At the time frame relay was selected because the data frames between SGSN and PCU are usually transported using E-1 links, which are quite common in the GSM BSS. Frame relay

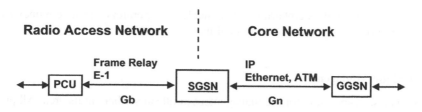

Figure 2.17 Interfaces and protocols of the SGSN on layers 2 and 3

with its similarities to ATM is well suited for transmitting packet data over 2 Mbit/s E-1 channels and had already been used for many years in wide area network communication systems. The disadvantage of using frame relay, however, is the fact that besides the resulting complicated network architecture, the SGSN has to extract the user data frames from the frame relay protocol and forward them via IP to the GGSN and vice versa.

As ATM and IP have become more common since the conception of GPRS, the UMTS radio network described in the next chapter no longer uses frame relay on this interface but ATM and IP instead which significantly simplifies the network architecture.

While ciphering for circuit-switched traffic is terminated in the BTS, ciphering for packet-switched traffic is terminated in the SGSN (Figure 2.18). This has a number of advantages. In GPRS, the mobile station and not the network has control over cell changes during data transfers. If ciphering were done on the BTS, the network would first have to supply the ciphering information to the new BTS before the data transfer could resume. As this step is not necessary when the ciphering is terminated in the SGSN the procedure is accelerated. Furthermore, the user data remains encrypted on all radio network links. From a security point of view this is a great improvement. The link between BTS and BSC is often carried over microwave links, which are not very difficult to spy on. The drawback of this solution is that the processing power necessary for ciphering is not distributed over many BTS but concentrated on the SGSN. Some SGSN suppliers therefore offer not only ciphering in software but also hardware-assisted ciphering. As ciphering is optional it can be observed

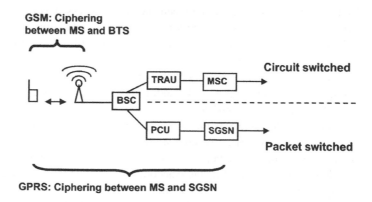

Figure 2.18 Ciphering in GSM and GPRS

that in some networks GPRS ciphering is not enabled as operators can save money by using the processing power they save for additional user data traffic.

Signaling Plane Management

The SGSN is also responsible for the management of all subscribers in its area. All protocols and procedures for user management are handled on the signaling plane.

To be able to exchange data with the Internet it is necessary to establish a data session with the GPRS network. This procedure is called packet data protocol (PDP) context activation and is part of the session management (SM) tasks of the SGSN. From the user point of view this procedure is invoked to get an IP address from the network.

Subscribers can change their location in a mobile network frequently. When this happens the SGSN needs to change its routing of packets to the radio network accordingly. This task is done by the GPRS mobility management (GMM) sublayer.

When a subscriber leaves the area of the current SGSN, GMM also contains procedures to change the routing for a subscriber in the core network to the new SGSN. This procedure is called inter SGSN routing area update (IRAU).

In order to charge the subscriber for usage of the GPRS network, the SGSN and the GGSN, which is described in more detail in the next paragraph, collect billing information in so-called call detail records (CDRs). These are forwarded to the billing server which collects all CDRs and generates an invoice for each subscriber once a month. The CDRs of the SGSN are especially important for subscribers that roam in a foreign network. As will be described in Section 2.8.2, the SGSN is the only network node in the foreign network that can generate a CDR for a GPRS session of a roaming subscriber for the foreign operator. For roaming subscribers the CDRs of the SGSN are then used by the foreign operator to charge the home operator for the data traffic the subscriber has generated. For GPRS data traffic generated in the home network, the GGSN usually generates the CDR and the billing information of the SGSN is not used.

2.5.3 The Gateway GPRS Support Node (GGSN)

While the SGSN routes user data packets between the radio access network and the core network, the GGSN connects the GPRS network to the external data network. The external data network will in most cases be the Internet. For business applications, the GGSN can also be the gateway to a company intranet [2].

The GGSN is also involved in setting up a PDP context. In fact it is the GGSN that is responsible for assigning a dynamic or static IP address to the user. The user keeps this IP address while the PDP context is established.

As shown in Figure 2.19, the GGSN is the anchor point for a PDP context and hides the mobility of the user towards the Internet. When a subscriber moves to a new location, a new SGSN might become responsible and data packets are sent to the new SGSN (IRAU). In this scenario the GGSN has to update its routing table accordingly. To the Internet this is invisible as the GGSN always remains the same. It can thus be seen as the anchor point of the connection and ensures that despite user mobility, the assigned IP address does not have to be changed.

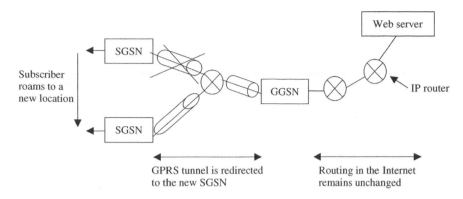

Figure 2.19 Subscriber changes location within the GPRS network

2.6 GPRS Radio Resource Management

As has been shown in Figure 2.5, a GPRS timeslot can be assigned to several users at the same time. It is also possible to assign several timeslots to a single subscriber in order to increase his data transmission speed. In any case, the smallest transmission unit that can be assigned to a user is one block, which consists of four bursts on one timeslot on the air interface for GPRS and two bursts for EGPRS MCS 7-9. A block is also called a GPRS RLC/MAC (radio link control/medium access control) frame.

Temporary Block Flows (TBF) in the Uplink Direction

Every RLC/MAC frame on the PDTCH or PACCH consists of an RLC/MAC header and a user data field. When a user wants to send data on the uplink, the terminal has to request resources from the network by sending a packet channel request message via the RACH or the PRACH as shown in Figure 2.13.

The PCU then answers with a packet uplink assignment message on the AGCH. The message contains information in which timeslots the terminal is allowed to send data. As a timeslot in GPRS may not only be used exclusively by a single subscriber, a mechanism is necessary to indicate to a terminal when it is allowed to send on the timeslot. Therefore, the uplink assignment message contains a parameter called the uplink state flag (USF). A different USF value is assigned to every subscriber that is allowed to send on the timeslot. The USF is linked to the so-called temporary flow identity (TFI) of a temporary block flow (TBF). A TBF identifies data to or from a user for the time of the data transfer. Once the data transfer is finished the TFI is reused for another subscriber. In order to know when it can use the uplink timeslots, the terminal has to listen to the timeslots it has been assigned in the downlink direction. Every block that is sent in the downlink to a subscriber contains a USF in its header as shown in Figure 2.20. It indicates who is allowed to send in the next uplink block. By including the USF in each downlink block the PCU can dynamically schedule who is allowed to send in the uplink. Therefore, this procedure is also called 'dynamic allocation'. The GPRS standard also defines two other methods of allocation: 'fixed allocation' and 'extended dynamic allocation'. As they are not used very widely today they are not further discussed.

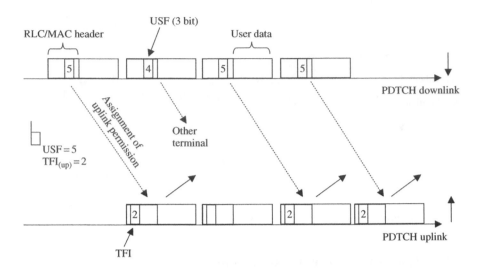

Figure 2.20 Use of the uplink state flag

Note that the USF information in the header and data portion of a downlink block is usually not intended for the same user. This is due to the fact that the assignments of up- and downlink resources are independent. This makes sense when considering web surfing for example where it is usually not necessary to already assign downlink resources at the time the universal resource locator (URL) of the web page is sent to the network.

For mobiles that have an uplink TBF established the network needs to send control information from time to time. This is necessary for example to acknowledge the receipt of uplink radio blocks. The logical PACCH that can be sent in a radio block instead of a PDTCH is used to send control information. The mobile recognizes its own downlink PACCH blocks because the header of the block contains its TFI value.

The PCU will continue to assign uplink blocks until the mobile station indicates that it no longer requires blocks in the uplink direction. This is done with the so-called 'countdown procedure'. Every block header in the uplink direction contains a four-bit countdown value. The value is decreased by the mobile for every block sent at the end of the data transfer. The PCU will no longer assign uplink blocks for the mobile once this value has reached 0.

While coordinating the use of the uplink is quite efficient, this way it creates a high latency time if data is only sent sporadically. This is especially problematic during a web-browsing session for two reasons: As shown at the end of this chapter, high latency has a big impact on the time it takes to establish TCP connections which are necessary before a web page can be requested. Furthermore, several TCP connections are usually opened to download the different elements like text, pictures, etc., of a web page so high latency slows down the process in several instances. To reduce this effect the GPRS standard was enhanced by a method called the 'extended uplink TBF'. In case both network and mobile device support the functionality, the uplink TBF is not automatically closed at the end of the countdown procedure but is kept open by the network until the expiry of an idle timer, which is usually set in the order of several seconds. While the uplink TBF is open, the network continues to assign blocks in the uplink direction to the mobile device. This enables the mobile to send data in the uplink quickly without requesting a new uplink TBF. The first mobile devices and networks that support extended uplink TBF appeared on the market in 2005 and a substantial improvement of web page download and delay times can be observed as shown at the end of the chapter.

Temporary Block Flows in the Downlink Direction

If the PCU receives data for a subscriber from the SGSN it will send a packet downlink assignment message to the mobile station similar to the one shown in Figure 2.22 in the AGCH or the PAGCH. The message contains a TFI of a TBF and the timeslots the mobile has to monitor. The mobile will then start to monitor the timeslots immediately. In every block it receives it will check if the TFI included in the header equals the TFI assigned to it in the packet downlink assignment message as shown in Figure 2.21. If they are equal it will process the data contained in the data portion of the block. If they are not equal the mobile discards the received block. Once the PCU has sent all data for the subscriber currently in its queue it will set the 'final block indicator' bit in the last block it sends to the mobile. Afterwards the mobile stops listening on the assigned timeslots and the TFI can be reused for another subscriber. In order to improve performance, the network can also choose to keep the downlink TBF established for several seconds so no TBF establishment is necessary if further data for the user arrives.

In order to acknowledge blocks received from the network, the mobile station has to send control information via the logical PACCH. For sending control information to the network it is not necessary to assign an uplink TBF. The network informs the mobile in the header of downlink blocks which uplink blocks it can use to send control information.

Timing Advance Control

The further a mobile is away from a BTS the sooner it has to start sending its data bursts to the network in order for them to arrive at the BTS at the correct time. As the position of the user

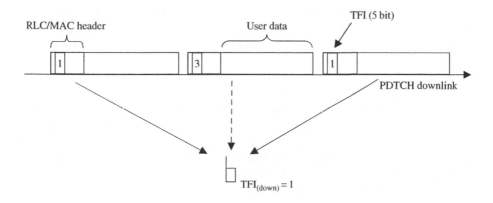

Figure 2.21   Use of the temporary flow identifier (TFI) in the downlink direction

```
[ ... ]
                          RLC/MAC PACKET TIMESLOT RECONFIGURE
000111--                  Message Type : 7 = packet timeslot reconfigure
------00                  Page Mode : 0 = normal paging
                          Global TFI:
--01111-                      Uplink Temporary Flow Identifier : 15
00------                  Channel Coding Command : Use CS-1 in Uplink
                          Global Packet Timing Advance:
----0001                      Uplink TA Index : 1
101-----                      Uplink TA Timeslot Number : 5
----0001                      Downlink TA Index : 1
101-----                      Downlink TA Timeslot Number : 5
---0----                  Downlink RLC Mode : RLC acknowledged mode
----0---                  CTRL ACK : 0 = downlink TBF already established
xxxxxxxx                  Downlink Temporary Flow ID: 11
xxxxxxxx                  Uplink Temporary Flow ID: 15
                          Downlink Timeslot Allocation:
-0------                      Timeslot Number 0 : 0
--0-----                      Timeslot Number 1 : 0
---0----                      Timeslot Number 2 : 0
----0---                      Timeslot Number 3 : 0
-----1--                      Timeslot Number 4 : 1 = assigned
------1-                      Timeslot Number 5 : 1 = assigned
-------1                      Timeslot Number 6 : 1 = assigned
0-------                      Timeslot Number 7 : 0
                          Frequency Parameters:
--000---                      Training Sequence Code : 0
xxxxxxxx                      ARFCN : 067
[ ... ]
```

Figure 2.22   Packet timeslot reconfiguration message according to 3GPP TS 44.060, 11.2.31 [3]

can change during the data exchange it is necessary for the network to constantly monitor how far the user is away from the serving base station. If the user moves closer to the BTS the network has to inform the mobile to delay sending its data compared to the current timing. If the user moves farther away it has to start sending its bursts earlier. This process is called timing advance control.

As we have seen in the previous paragraph the assignment of uplink and downlink resources is independent from each other. When downloading a large web page for example it might happen that a downlink TBF is assigned while no uplink TBF is established because the mobile has no data to send.

Even though no uplink TBF is established, it is necessary from time to time to send layer 2 acknowledgment messages to the network for the data that has been received in the downlink. To send these messages quickly, no uplink TBF has to be established. In this case the PCU informs the mobile in the downlink TBF from time to time which block to use to send the acknowledgment. As this only happens infrequently the network cannot take the previous acknowledgment bursts for the timing advance calculation for the following bursts. Because of this, a number of new methods have been standardized to measure and update the timing advance value while the mobile is engaged in exchanging GPRS data.

The Continuous Timing Advance Update Procedure

In a GPRS 52-multiframe, frames 12 and 38 are dedicated to the logical PTCCH uplink and downlink. The PTCCH is further divided into 16 subchannels. When the PCU assigns a TBF to a mobile station the assignment message also contains an information element that instructs the mobile to send access bursts on one of the 16 subchannels in the uplink with timing advance 0. These bursts can be sent without a timing advance because they are much shorter than a normal burst. For more information about the access burst see Chapter 1. The BTS monitors frames 12 and 38 for access bursts and calculates the timing advance value for every subchannel. The result of the calculation is sent in the PTCCH in the following downlink block. As the PTCCH is divided in 16 subchannels, the mobile sends an access burst on the PTCCH and receives an updated value every 1.92 seconds.

2.7 GPRS Interfaces

As can be seen in Figure 2.16, the GPRS standards define a number of interfaces between components. Apart from the PCU, which has to be from the same manufacturer as the BSC, all other components can be selected freely. Thus, it is possible for example to connect a Nokia PCU to a Nortel SGSN, which is in turn connected to a Cisco GGSN.

The Abis Interface

The Abis interface connects the BTS with the BSC. The protocol stack as shown in Figure 2.23 is used on all timeslots of the radio network which are configured as (E)GPRS PDTCHs. Usually, all data on these timeslots is sent transparently over the non-standardized interface between the BSC and PCU. However, as the link is also used to coordinate the BSC and PCU with each other it is still not possible to connect the BSC and PCUs of two different vendors. On the lower layers of the protocol stack the RLC/MAC protocol is used for the radio resource management. On the next protocol layer the logical link control (LLC) protocol is responsible for the framing of the user data packets and signaling messages of the mobility management and session management subsystems of the SGSN. Optionally, the

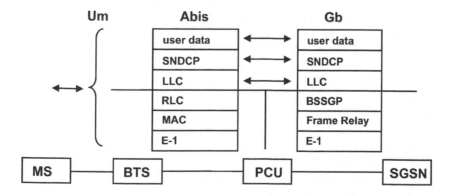

Figure 2.23 GPRS protocol stacks in the radio network

LLC protocol can also ensure a reliable connection between the mobile terminal and the SGSN by using an acknowledgment mechanism for correctly received blocks (acknowledged mode). On the next higher layer the subnetwork dependent convergence protocol (SNDCP) is responsible for framing IP user data to send it over the radio network. Optionally, SNDCP can also compress the user data stream. The LLC layer and all layers above are transparent for the PCU, BSC and BTS as they are terminated in the SGSN and the mobile terminal respectively.

The Gb Interface

The Gb interface connects the SGSN with the PCU as shown in Figure 2.23. On layer 1, mostly 2 Mbit/s E-1 connections are used. An SGSN is usually responsible for several PCUs in an operational network and they are usually connected by several 2 Mbit/s connections to the SGSN. On layers 2 and 3 of the protocol stack the frame relay protocol is used, which is a standard packet-switched protocol used for many years in the telecoms world. Frame relay is also a predecessor of the ATM protocol which has gained a lot of popularity for packet-based long-distance transmission in the telecoms world and which is heavily used in the UMTS network as will be shown in Chapter 3. Thus, its properties were very well known at the time of standardization especially for packet-switched data transfer over 2 Mbit/s E-1 connections. The disadvantage is that the user data has to be encapsulated into frame relay packets which make the overall protocol stack more complex as this protocol is only used on the Gb interface.

The Gn Interface

This is the interface between the SGSNs and GGSNs of a GPRS core network and is described in detail in 3GPP TS 29.060 [4]. Usually, a GPRS network comprises more than one SGSN because a network usually has more cells and subscribers then can be handled by a single SGSN. Another reason for having several GGSNs in the network is to assign them different tasks. While one GGSN for example could handle the traffic of post-paid subscribers, a different one could be specialized on handling the traffic of pre-paid subscribers. Yet another GGSN could be used to interconnect the GPRS network with companies that want to offer direct intranet access to their employees without sending the data over the Internet. Of course all of these tasks can also be done by a single GGSN if it has enough processing power to handle the number of subscribers for all these different tasks.

On layer 3, the Gn interface uses IP as the routing protocol (Figure 2.24). If the SGSN and GGSN are deployed close to each other, 100 Mbit/s Ethernet over twisted pair cables can be used for the interconnection. If larger distances need to be overcome, ATM over various transport technologies (e.g. STM-1 with 155 Mbit/s) is used to carry the IP frames. To increase capacity or due to redundancy purposes, several physical ATM connections are usually needed between two network nodes.

User data packets are not sent directly on the IP layer of the Gn interface but are encapsulated into GPRS tunneling protocol (GTP) packets. This creates some additional overhead which is needed for two reasons: Each router in the Internet between the GGSN and the destination makes its routing decision for a packet based on the destination IP address and its routing table. In the fixed-line Internet this approach is very efficient as

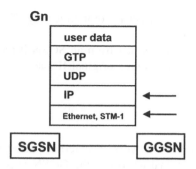

Figure 2.24   The Gn interface protocol stack

the location of the destination address never changes and thus the routing tables can be static. In the GPRS network, however, subscribers can change their location at any time as shown in Figure 2.19 and thus the routing of the packets must be flexible. As there are potentially many IP routers between the GGSN and SGSN these would have to change their routing tables whenever a subscriber changes its location. In order to avoid this, the GPRS network does not use the source and destination IP address of the user's IP packet. Instead, the IP addresses of the current SGSN and GGSN are used for the routing process. As a consequence, the user data packets need to be encapsulated into GTP packets to be able to tunnel them transparently through the GPRS network. If the location of a subscriber changes the only action that needs to be taken in the core network is to inform the GGSN of the IP address of the new SGSN that has become responsible for the subscriber. The big advantage of this approach is the fact that only the GGSN has to change its routing entry for the subscriber. All IP routers between the GGSN and SGSN can therefore use their static routing tables and no special adaptation of those routers is necessary for GPRS. Figure 2.25 shows the most important parameters on the different protocol layers on the Gn interface. The IP addresses on layer 3 are those of the SGSN and GGSN while the IP addresses of the user data packet which is encapsulated into a GTP packet belong to the subscriber and the server

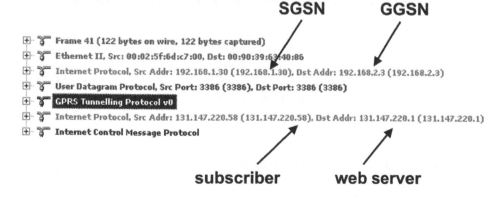

Figure 2.25   GTP packet on the Gn interface

in the Internet with which the subscriber is communicating. This means that such a packet contains two layers on which IP is used.

When the GGSN receives a GTP packet from an SGSN it removes all headers including the GTP header. Afterwards the remaining original IP packet is routed via the Gi interface to the Internet.

The Gi Interface

This interface connects the GPRS network to external packet networks, e.g. the Internet. From the perspective of the external networks, the GGSN is just an ordinary IP router. As on the Gn interface, a number of different transmission technologies from 'ordinary' twisted pair 100 Mbit/s Ethernet to ATM over STM-1 optical interface can be used. To increase bandwidth or to add redundancy, several physical interfaces can be used simultaneously.

The Gr Interface

This interface connects the SGSN with the HLR, which contains information about all subscribers on the network (Figure 2.26). It was enhanced with a software upgrade to also act as a central database for GPRS subscriber data. The following list shows some examples:

- GPRS service admission on a per user (IMSI) basis.
- Which GPRS services the user is allowed to use (access point names, APNs).
- GPRS international roaming permissions and restrictions.

As has been shown in Chapter 1, the HLR is a SS7 service control point (SCP). Therefore, the Gr interface is based on E1 trunks, SS7 on layer 3 and MAP on the application layer. The MAP protocol was also extended to be able to exchange GPRS specific information.s The following list shows some of the messages that are exchanged between SGSN and HLR:

- Send authentication information: This message is sent from the SGSN to the HLR when a subscriber attaches to the network for which the SGSN does not yet have authentication information.
- Update location: The SGSN informs the HLR that the subscriber has roamed into its area.
- Cancel location: When the HLR receives an update location message from an SGSN, it sends this message to the SGSN to which the subscriber has previously been attached.
- Insert subscriber data: As a result of the update location message sent by the SGSN, the HLR will forward the subscriber data to the SGSN.

The Gc Interface

This interface connects the GGSN with the HLR. It is optional and thus not widely used in networks today. There is one scenario, though, for which this interface is quite interesting: A mobile device, such as a wireless measurement device, offers services to clients on the Internet. To be able retrieve data from the device it is connected via GPRS to the Internet and

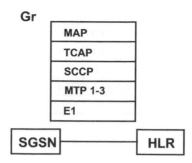

Figure 2.26 The Gr interface

gets assigned the same IP address whenever it activates a PDP context. This is called a static IP address. This is necessary because the device would be very difficult to reach if the IP address changed all the time. For this purpose, the GPRS network offers the possibility to assign a fixed IP address to a subscriber. When somebody wants to reach the device from the Internet it will send a packet to the IP address assigned to the device. The packet will be routed through the Internet to the GGSN. In case the GGSN detects that the device to which this fixed IP address belongs has established no connection to communicate with the Internet so far, it can query the HLR via the Gc interface for its location. If the device is attached to the network, the HLR returns the address of the SGSN to which the device is currently attached to. The GGSN can then go along and inform the SGSN that there are incoming packets for a subscriber that is attached but for which no PDP context is established so far. The SGSN then contacts the device and informs it that there are packets waiting for it. Afterwards, the device has the possibility to establish a PDP context to enable the transfer of packets to and from the Internet. This process is called 'network initiated PDP context activation'. A more detailed description of the GPRS attach and PDP context activation procedures can be found in Section 2.8.2.

The Gp Interface

This interface is described in 3GPP TS 29.060 [4] and connects GPRS networks of different countries or different operators with each other for GTP traffic (Figure 2.27). It enables a subscriber to roam outside the coverage area of the home operator and still use GPRS to connect to the Internet. The user's data will be tunneled via the Gp interface just like on the Gn interface from the SGSN in the foreign network to the GGSN in the subscriber's home network and from there to the Internet or a company intranet. At first it seems some-what complicated not to use a GGSN in the visited GPRS network as the gateway to the Internet. From the end-user perspective though, this redirection has a big advantage as no settings in the device have to be changed. This is a great advantage of GPRS over any other fixed or mobile Internet connectivity solution available today while roaming.

Note that the Gp interface is for GTP traffic only. For signaling with the HLR the two networks also need an SS7 interconnection so the visited SGSN can communicate with the HLR in the home network.

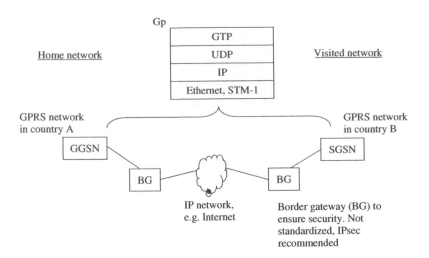

Figure 2.27 The Gp interface

The Gs Interface

3GPP TS 29.018 [5] describes this interface which is also optional. It connects the SGSN and the MSC/VLR. The functionality and benefits of this interface in conjunction with GPRS NOM I is discussed in Section 2.3.6.

2.8 GPRS Mobility Management and Session Management (GMM/SM)

Apart from forwarding data packets between GPRS subscribers and the Internet, the GPRS network is also responsible for the mobility management of the subscribers and the session management to control the individual connections between subscribers and the Internet. For this purpose signaling messages and signaling flows have been defined that are part of the GMM/SM protocol.

2.8.1 Mobility Management Tasks

Before a connection to the Internet can be established, the user has to first connect to the network. This is similar to attaching to the circuit-switched part of the network. When a subscriber wants to attach, the network usually starts an authentication procedure, which is similar to the GSM authentication procedure. If successful, the SGSN sends a location update message to the HLR to update the location information of that subscriber in the network's database. The HLR acknowledges this operation by sending an 'insert subscriber data' message back to the SGSN. As the name of the message suggests, it not only acknowledges the location update but also returns the subscription information of the user to the SGSN so no further communication with the HLR is necessary as long as the subscriber does not change location. Afterwards, the SGSN will send an attach accept message to the

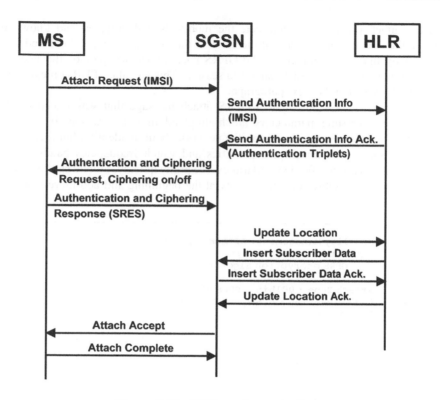

Figure 2.28 GPRS attach message flow

subscriber. The attach procedure is complete when the subscribers returns an attach complete message to the SGSN. Figure 2.28 shows the message flow for this procedure.

If the subscriber was previously attached to a different SGSN the procedure is somewhat more complex. In this case, the new SGSN will ask the old SGSN for identification information of the subscriber. Once the subscriber has authenticated successfully the SGSN will send the location update message as above to the HLR. As the HLR knows that the subscriber was previously attached to a different SGSN it sends a cancel location message to the old SGSN. Afterwards, it returns the insert subscriber data message to the new SGSN.

It is also possible to do a combined GSM/GPRS attach procedure in case the Gs interface is available. To inform the mobile of this possibility the network broadcasts the GPRS network operation mode on the BCCH. Should the mobile thus request a combined attach from the SGSN, it is the new SGSN's task to inform the new MSC of the location of the subscriber. The new MSC will then send an update location to the HLR for the circuit-switched part of the network. The HLR will then cancel the location in the old MSC and send an insert subscriber data back to the new MSC. Once all operations have been performed, the new MSC sends back a location update accept to the SGSN which will then finally return the attach accept message to the mobile station. While this message flow is quite complicated from the point of view of the core network it allows the mobile to attach in both circuit- and packet-switched network parts with only a single procedure. This speeds up the process for the mobile and reduces the signaling load in the radio network.

Once the attach procedure is complete the mobile is authenticated and known to the network. In the circuit-switched part of the network the user can now go ahead and establish a voice call by dialing a number. In the GPRS packet-switched part of the network the mobile can now go ahead and establish a data session. This so-called PDP context activation procedure is described in the next paragraph.

Figure 2.29 shows an example of a GPRS attach message that was traced on the Gb interface. Some interesting parameters are highlighted in bold. As can be seen in the message, the mobile does not only inform the network about its identity, but it also includes its capabilities such as its multislot capabilities and which frequencies bands it supports (900, 1800, 1900 MHz), etc. While standards evolve pretty quickly, mobile station developers often only implement a subset of functionality at the beginning and add more features over

```
[...]          Mobility Management: ATTACH REQUEST
                  MS Network Capability:
1-------             GPRS encryption algorithm GEA/1: 1 = available
[...]
-----001       Attach Type : 001bin = GPRS attach
-100----       GPRS Ciphering Key Sequence Number : 100bin
                  DRX Parameter
01000000          Split PG cycle code : 64 = 64
-----011          Non-DRX timer: max. 4 sec non-DRX mode after transfer state
----0---          SPLIT on CCCH: not supported
               Mobile Identity
-----100          Type of identity: TMSI
----0---          Parity: 0 = even
xxxxxxxx          TMSI: D4CC3EC4h
               Old Routing Area Identification
xxxxxxxx          Mobile Country Code: 232
xxxxxxxx          Mobile Network Code: 03
xxxxxxxx          Location area code: 6F32h
00000001          Routing area code: 0Fh
               MS Radio Access Capability
0001----          Access technology type: 1 = GSM E (900MHz Band)
               Access capabilities
---100--          RF power capability: 4h
                  A5 bits
-------1             A5/1: 1 = Encryption algorithm available
1-------             A5/2: 1 = Encryption algorithm available
-0------             A5/3: 0 = Encryption algorithm not available
[...]
------1-          ES IND : 1h = early Classmark Sending is implemented
[...]
               Multislot capability
xxxxxxxx          GPRS multi slot class: 10 (4 downlink + 2 uplink)
--0-----          GPRS extended dynamic allocation: not implemented
----1101             Switch-measure-switch value: 0
1000----             Switch-measure value: 8
xxxxxxxx       Access technology type: 3 = GSM 1800
xxxxxxxx       Access capabilities
001-----          RF power capability: 1
----1---          ES IND: 1 = early Classmark Sending is implemented
[...]
```

Figure 2.29 GPRS attach message on the Gb interface

time in new software versions or even only in new models. This flexibility and thus fast time to market is only possible if networks and mobile stations are able to exchange information about their capabilities.

A good example of such an approach is the multislot capability. Early GPRS mobiles were able to aggregate only two downlink timeslots and use only a single one in the uplink. Current mobile terminals support up to four timeslots in the downlink and two in the uplink (multislot class 10).

Once the mobile is attached, the network has to keep track of the location of the mobile. As has been shown in Chapter 1, this is done by dividing the GSM network into location areas. When a mobile in idle mode changes to a cell in a different location area it has to do a so-called location update (LU). This is necessary in order for the network to be able to find the subscriber for incoming calls or SMS messages. In GPRS, the same principle exists. In order to be more flexible the location areas are subdivided into GPRS routing areas. If a mobile in ready or standby state crosses a routing area border it reports to the SGSN. This procedure is called routing area update (RAU).

If the new routing area is administered by a new SGSN the process is called inter SGSN routing area update (IRAU). While from the mobile point of view there is no difference between a RAU and IRAU there is quite a difference from the network point of view. This is due to the fact that the new SGSN does not yet know the subscriber. Therefore, the first task of the new SGSN is to get the subscriber's authentication and subscription data. As the routing area update contains information about the previous routing area, the SGSN can then contact the previous SGSN and ask for this information. At the same time this procedure also prompts the previous SGSN to forward all incoming data packets to the new SGSN in order not to lose any user data while the procedure is ongoing. Next, the GGSN is informed about the new location of the subscriber so further incoming data is from now on sent directly to the new SGSN. Finally, the HLR is also informed about the new location of the subscriber and his information is deleted in the old SGSN. Further information about this procedure can be found in 3GPP TS 23.060, 6.9.1.2.2 [6].

2.8.2 GPRS Session Management

In order to communicate with the Internet a PDP context has to be requested after the attach procedure. For the end user this in effect means getting an IP address from the network. As this procedure is in some ways similar to establishing a voice call it is sometimes also referred to as 'establishing a packet call'.

While there are some similarities between a circuit-switched call and a packet call there is one big difference which is important to remember: For a circuit-switched voice or data call the network reserves resources on all interfaces. A timeslot is reserved for this connection on the air interface, in the radio network and also in the core network. These timeslots cannot be used by anyone else while the call is established even if no data is transferred by the user. When a GPRS packet call is established there are no resources dedicated to the PDP context. Resources on the various interfaces are only used during the time data is transmitted. Once the transmission is finished (e.g. after the web page has been downloaded) the resources are used for other subscribers. Therefore, the PDP context represents only a logical connection with the Internet. It remains active even if no data is transferred for a prolonged amount

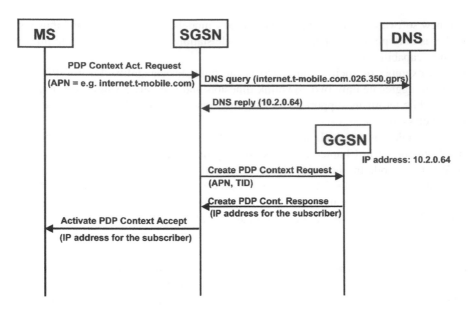

Figure 2.30 The PDP context activation procedure

of time. For this reason a packet call can remain established indefinitely without blocking resources. This is sometimes also referred to as 'always on'.

Figure 2.30 shows the PDP context activation procedure. At the beginning, the subscriber sends a PDP context activation request message to the SGSN. The most important parameter of the message is the access point name (APN). The APN is the reference which GGSN uses as a gateway to an external network. The network operator could have one APN to connect to the Internet transparently, one to offer WAP services, several other APNs to connect to corporate intranets, etc. The SGSN compares the requested APN to the list of allowed APNs for the subscriber that has been received from the HLR during the attach procedure. The APN is a fully qualified domain name like 'internet.t-mobile.com' or simply 'internet' or 'wap'. The names of the APN can be freely chosen by the GPRS network operator.

In a second step, the SGSN uses the APN to find the IP address of the GGSN that will be used as a gateway. To do this the SGSN performs a domain name service (DNS) lookup with the APN as the domain name to be queried. The DNS lookup is identical to a DNS lookup a web browser has to perform in order to get the IP address of a web server. Therefore, a standard DNS server can be used for this purpose in the GPRS network. To get an internationally unique qualified domain name, the SGSN adds the mobile country code (MCC) and mobile network code (MNC) to the APN, which is deduced from the subscriber's IMSI. As a top level domain, '.gprs' is added to form the complete domain name. An example of domain name for the DNS query is 'internet.t-mobile.com.026.350.gprs'. Adding the MCC and MNC to the APN by the SGSN enables the subscriber to roam to any country that has a GPRS roaming agreement with the subscriber's home network and use the service without having to modify any parameters. The foreign SGSN will always receive the IP address of the home GGSN from the DNS server and all packets will be routed to and from

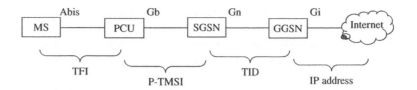

Figure 2.31 Identification of user data packets on different GPRS interfaces

the home GGSN and from there to the external network. Of course it is also possible to use a GGSN in the visited network. In order to do that, however, the user would have to change the settings in his device which is very undesirable. Therefore, most operators prefer to always route the traffic back to the home GGSN and thus offer a seamless service to the user.

After the DNS server has supplied the GGSN's IP address the SGSN can then forward the request to the correct GGSN. The APN and the user's IMSI are included in the message as mandatory parameters. In order to tunnel the user data packets through the GPRS network later on, the SGSN assigns a so-called tunnel identifier (TID) for this virtual connection that is also part of the message. The TID consist of the user's IMSI and a two-digit network subsystem access point identifier (NSAPI). This allows a user to have more then a single active PDP context at a time. This is quite useful to access the Internet via a notebook and at the same time send or receive an MMS message via the mobile terminal.

If the GGSN grants access to the external network (e.g. the Internet) it will assign an IP address out of an address pool for the subscriber. For special purposes it is also possible to assign a fixed IP address for a subscriber. Next, the GGSN responds to the SGSN with a PDP context activation response message that contains the IP address of the subscriber. Furthermore, the GGSN will store the TID and the subscriber's IP address in its PDP context database. This information is needed later on in order to forward packets between the subscriber and the Internet and of course for billing purposes.

Once the SGSN receives the PDP context activation response message from the GGSN it also stores the context information in its database and forwards the result to the subscriber. The subscriber then uses the IP address to communicate with the external network.

Different IDs are used for packets of a certain user on each network interface due to the different nature of the protocols and due to the different packet sizes. On the GPRS air interface with its small data frames of only 456 bits or 57 bytes, which even include the overhead for error detection and correction, the three-bit TFI is used to route the frame to the correct terminal. In the radio network the P-TMSI/TLLI is used to identify packets of a user. Finally in the core network, the GPRS TID is used as identification. Figure 2.31 shows the different interfaces and IDs used on them at a glance.

2.9 Session Management from a User Point of View

GPRS can also be used to connect PDAs or notebooks to the Internet. This can be done by connecting the PDA or notebook to the mobile phone via a serial or USB cable, via infrared or via Bluetooth. In all cases, the mobile phone acts as a wireless 'modem' for the external device.

Before we take a look at how the mobile can be used as a wireless 'modem' to establish a GPRS connection to the Internet or company intranet, let's first look at the process that is widely used today to establish a circuit-switched connection to the Internet. The process is the same for a fixed-line modem or mobile phone acting as a wireless circuit-switched modem.

Modems communicate with external devices with the standardized AT command set which is an ASCII command and response language. For example: To dial a telephone number the notebook sends a dial command which includes the telephone number (e.g. 'ATD 0399011782') to the modem. The modem will then try to establish a circuit-switched data connection with the other end. If successful, the modem will return a connect message (e.g. 'CONNECT 38400'). Afterwards, it will enter transparent mode and forward all data that is sent by the notebook to the other end of the connection instead of interpreting it as a command.

Once the connection has been established it is up to the connected device to interpret the data that is sent and received. In the case of an Internet connection the point-to-point protocol (PPP) is widely used to send and receive IP packets over modem connections. An example of a service that uses PPP is the 'dial up network' of the Microsoft Windows operating system. After call establishment, the PPP is responsible for establishing an IP connection with the PPP server on the other side. Usually, a username and password are exchanged before the server at the other side accepts the PPP connection and returns an IP address. The PPP stack on the PDA and notebook will then connect to the IP stack and encapsulate all IP packets in PPP frames for sending them over the serial interface via the modem to the other side (Figure 2.32).

For GPRS, using this approach in the same way is not possible as there are a number of differences to a circuit-switched dial-up connection:

- GPRS is a packet-switched connection to the Internet.
- There is no telephone number to dial.
- There is no PPP server at the other side of the connection.

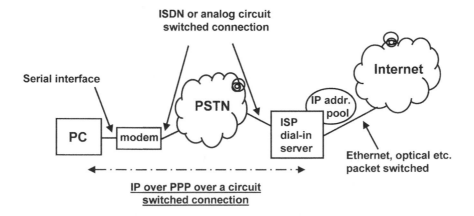

Figure 2.32 IP over PPP for Internet connections

In order to add GPRS functionality to notebooks and PDAs their operating systems could have been extended to support GPRS as part of the network stack. This would have meant a substantial amount of software development on many different devices and operating systems. As this was seen as not practical the following approach has been chosen: The already existing PPP stack of the dial-up network on the user's device as described before is used for GPRS with slight modifications in the configuration. This means that no software changes had to be made to existing devices. As the PPP stack of the user's device requires a PPP server on the other side it was decided to implement the PPP server inside the mobile phone to terminate the PPP connection. The PPP server in the mobile phone then converts the PPP commands and PPP data frames into GPRS commands and GPRS data frames. This approach is shown in Figure 2.33.

In order to establish a packet-switched GPRS connection from a user's device the following parameters have to be configured in the dial-up network settings:

The APN is a GPRS specific parameter and needs to be sent from the user's device to the mobile phone during the connection establishment. This is done via a new AT command that has been standardized so all mobile terminal vendors that offer GPRS via a serial connection can implement the command in the same way. The AT command for example to use 'internet.t-mobile.com' as the APN for the connection is: `AT+CGDCONT=1,"IP","internet.t-mobile.com"`. In order to send the command during the connection establishment, it has to be entered in the dial-up networking configuration in the advanced modem settings dialog box as shown in Figure 2.34.

The next step in the process is to instruct the mobile to connect to the Internet via GPRS and not via a circuit-switched connection. This is done by using `*99***1#` as a telephone number in the dial-up networking instead of an ordinary telephone number. When the mobile phone receives this unusual telephone number with the ATD command, it will establish a GPRS connection instead of a circuit-switched data connection. It will do this by starting the internal PPP server and sending a PDP context activation request message to the SGSN with the APN that was previously given to it via the `AT+CGDCONT` command. Once it receives a PDP context activation acknowledge message it will forward the IP address to

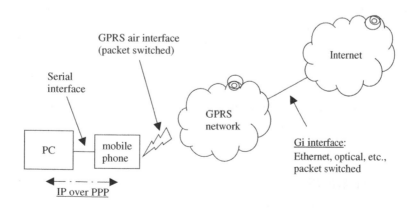

Figure 2.33 PPP termination in the mobile phone for GPRS

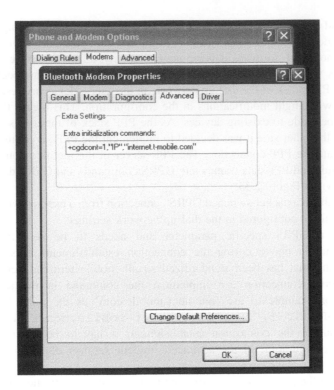

Figure 2.34 The advanced settings dialog box for entering the APN

the external device via the PPP connection. Afterwards, the external device can send and receive IP packets over the connection.

A 'real' PPP server can supply the user's device with all the settings necessary for the IP connection. Apart from the IP address that was assigned to the user it is also necessary to configure the IP address of the DNS server. This is necessary in order to convert domain names into IP. Unfortunately, the IP address of the DNS server is only an optional information element in the PDP context activation procedure. This means that if the operator does not return the DNS server IP address, the user needs to ask the network operator not only for the APN but also which IP address to use for the DNS server settings in the dial-up networking 'IP settings' dialog box.

While for network experts, these additional steps are quite simple to execute, the average user will probably be quite challenged by this. Therefore, many mobile phone suppliers or network operators offer utilities to configure the GPRS connection automatically. These utilities, however, will only create a new modem and dial-up connection entry and use the AT+CDGCONT and ATD *99***1# as described above.

2.10 WAP over GPRS

While GPRS is a bearer for IP packets it has some properties that distinguish it from fixed-line connections. These include longer latency, varying latency if used in moving environments and even loss of service for some time if the user moves outside the coverage area of the network. Furthermore, many devices that use GPRS for communication have limited abilities such as small screens and relatively low processing power when compared to notebooks or desktop PCs. Therefore, a number of applications for which the fixed-line Internet is widely used have been adapted for mobile environments. Web browsing is certainly the most popular Internet application. It benefits from a fast connection and depends on the reliability of the bearer especially if web pages are big and thus take some time to be transferred during which no transfer interruptions should occur. Modern web browsers are also designed to make use of big displays and processing power of notebooks or desktop workstations. EGPRS offers sufficient bandwidth today to allow the use of web browsers running on such equipment. If the user, however, wants to access information with his mobile phone a different approach is necessary to adapt to the limited capabilities of the handset. For this reason, the wireless application protocol (WAP) standard was created by the Wapforum which was later consolidated into the Open Mobile Alliance (OMA) forum. Basically, the standard adopts the concepts of hypertext transfer protocol (HTTP) and hypertext markup language (HTML) and adapts them for the use in a mobile environment. iMode is a rivaling standard to WAP, initially designed by NTT DoCoMo in Japan. Since its creation it has also spread to other countries, but WAP remains the most widely used protocol for mobile devices outside of Japan.

Several different versions of the WAP standard exist today and are used in operational networks.

WAP 1.1 was designed for web browsing in very constrained environments. Special attention was given to the following limitations:

- Very limited bandwidth of the connection, which has an impact on the speed a page can be downloaded.
- Very limited processing power of the mobile device, which has an impact on how quickly pages can be rendered on the screen.
- Reliability of the connection. Pages should be loaded as quickly as possible to reduce the effects of transmission interruptions and lost network coverage on the user experience.

HTML and its successor XHTML are used today to describe how web pages are to be rendered in the browser. While the text and layout of a page are directly embedded in the document, pictures and other elements are usually referenced and have to be requested separately. As these languages are quite complex and offer many possibilities that cannot be used in mobile devices due to the small displays and limited processing capabilities, WAP defines its own page description language which is called the wireless markup language (WML). Therefore, using a WAP browser on a mobile device is sometimes also called WAP browsing. Figure 2.35 shows a simple WML description to show a text on the display.

While at first the WML source looks quite similar to HTML there are some differences apart from the limited functionality. The main difference is the use of so-called 'cards' inside a single page. Inside the text of each card a link to other cards can be included so a user can navigate between the cards. The advantage of this approach is so download several cards that are related to each other in a single transaction rather than having to access the network every time the user clicks on a link. This is helpful to break down

```
<?xml version= "1.0"?>
<!DOCTYPE wml PUBLIC "-//WAPFORUM//DTD WML 1.1//EN"
"http://www.wapforum.org/DTD/wml.dtd">
<wml>
  <card id="main" title="First Card">
    <p mode="wrap">This is a WML page including only this
    sentence.</p>
  </card>
</wml>
```

Figure 2.35 Simple WML page

long texts into several cards and bind them together with a referring link at the bottom of each card. Devices with small displays benefit from this approach as the user doesn't have to scroll down a lot of text but can click on a link. Separating a long text into several cards also speeds up the embedded WAP browser as only a single card of the downloaded document needs to be rendered when downloading a page. Apart from formatted text and hyperlinks, WML also supports references to images. WAP 1.1 only supports black and white images in the wireless application bitmap protocol (WBMP) format as mobile devices at the time were limited to black and white screens. Like HTML, WML also supports 'forms' elements to allow users to input text that can be sent to a server for processing.

To transfer WML pages, a new protocol called the wireless session protocol (WSP) was created instead of using the well-known hypertext transfer protocol (HTTP). In order to be able to request WML pages from ordinary web servers, a gateway is necessary that acts as a translator between the WSP and the HTTP world. The concept of the WAP gateway, which is sometimes also referred to as WAP proxy, is shown in Figure 2.36. Between the mobile device and the WAP gateway, WSP is used for requesting the web page, while HTTP is used to request the WML page from a web server in the Internet.

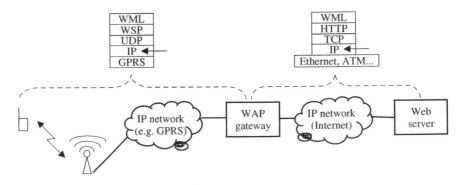

Figure 2.36 Different protocol stacks on the two sides of the WAP gateway

In order to speed up the request for a page and the subsequent data transfer, there are a number of differences between WSP and HTTP:

- While HTTP uses the session-oriented TCP (transfer control protocol) on layer 4 of the protocol stack, WSP uses the less known UDP (user datagram protocol) and the wireless transaction protocol (WTP) to compensate for some of the functionality of TCP that cannot be found in UDP. In order to simplify the terminal configuration, UDP port 9201 was standardized for the use of UDP/WTP. This way, WAP 1.1 avoids the three-way TCP handshake for the session establishment, which saves a lot of time in wireless environments with long round-trip delay times. In order to ensure no data is lost during the download of a WAP page, WTP includes a packet number in each packet. Thus, the receiver can reassemble arriving UDP/WTP packets in the correct order and can ask for a retransmission of missing packets.
- Every time an ordinary web browser requests a web page it includes a lot of information in the request such as a list of supported file formats. This takes a lot of space and increases the overall transaction time for every request. WSP therefore uses a different approach. At the beginning of a WAP session, the WAP client registers with the WAP gateway and informs the gateway once about its capabilities. When requesting WAP pages afterwards, only the URL needs to be included in the request. This reduces the size of the page request from over 1000 bytes, which are typically needed for a HTTP request, to about 100 bytes for a WSP request.

Since the adoption of the WAP 1.1 standard, capabilities of both network and terminals have improved significantly. Mobile phone processing power has increased which allows among other things fast data downloads by aggregating several GPRS timeslots. Mobile phone manufacturers have also moved to color displays and their size and resolution has increased compared to earlier WAP phones as well. As a consequence, the WAP standard had to be adapted and WAP 2.0, which was approved in 2002, takes advantage of the enhanced capabilities of terminals and network.

Instead of using a proprietary language for describing WAP pages, WAP 2.0 makes use of a subset of XHTML called XHTML Mobile Profile. As XHTML is backward compatible to HTML this allows viewing ordinary web pages as well. It still makes sense to adapt standard web pages for the smaller screens of mobile phones but development of these pages has been greatly simplified by this change as standard development tools can now be used. Some WAP browsers such the Opera mobile phone web browser even support the complete XHTML syntax and intelligently reflow the layout to adapt standard web pages to the smaller screen of a mobile phone.

WAP 2.0 browsers now also support additional graphics formats such as the widely used graphics interchange format (GIF). Again, standard graphics formats are used compared to the previous proprietary approach. This reduces the overhead of the page creation as pictures no longer have to be converted into a special format and allows using color images instead of black and white images only.

The protocol stack between the mobile station and the WAP gateway has also changed with WAP 2.0. As today's networks offer higher bandwidths it was decided to use TCP and HTTP instead of the proprietary WSP and WTP. The concept of the WAP gateway was preserved so operators can continue to charge for WAP usage as before such as per page or by applying different tariffs depending on whether the user accesses content provided by the operator or from external servers in the Internet. Apart from the billing and control functionality, which is transparent for the connection, the WAP 2.0 gateway acts as a simple HTTP proxy. TCP port 8080 has been standardized for this use. From the mobile point of view, the WAP gateway is no longer strictly necessary and some mobile phones allow skipping the gateway settings in the WAP set up completely. If no gateway is configured, the WAP browser directly communicates with the server in the Internet. As some operators charge much more for WAP traffic via the gateway compared to the direct Internet access via a different APN it make sense in these cases to remove the gateway settings in the mobile device and to change the APN. If the gateway entry cannot be removed another option to save money is to set the gateway IP address to that of a public HTTP proxy that supports TCP port 8080. Some phones even allow setting the gateway port so any free HTTP proxy can be used.

Changing the proprietary gateway approach in WAP 1.1 to standard HTTP proxy functionality has one disadvantage: Instead of registering once with the gateway and including all capability information only at the beginning of a session, the HTTP proxy concept requires sending the capability information in every request. As HTTP requests with capability information easily use more than 1000 bytes compared to the small WAP 1.1 requests, which use around 100 bytes, the data volume consumed during a WAP session increases substantially especially when viewing only small WAP pages. From a technical point of view, there is only a small impact to the page load time due to the enhanced transmission speeds of GPRS and EGPRS. Depending on the amount an operator charges per transferred kilobyte via GPRS, however, the increased request size could have a more notable impact.

The quality of experience of a WAP session from the user's point of view mainly consists of short page load times and a high click success rate. The click success rate is defined in this context as the percentage of pages that start to get displayed within a certain amount of time after the user has selected a link to another page. A typical maximum value for this reaction time is seven seconds. Beyond that time the user will get the impression that there is a problem and either abort or repeat the request. The transmission speed on the air interface is one factor that influences the user's quality of experience of a WAP session with a mobile terminal. Additionally, there are a number of other factors that have to be optimized in the network and handset to increase the user's quality of experience:

- Screen size: As WAP browsers are included even in very small phones today, user experience is limited by small displays which require the user to scroll through pages a lot more often then on bigger displays. Thus, a good approach for WAP page design is to find a compromise that suits both big and small mobile phone displays.
- A fast processor can render WAP pages a lot faster then a slower one. Therefore, a processor architecture that offers high processing power and a good power saving mode once the page has been rendered greatly increases the user experience while preserving battery time. Processor speed also has an influence of how fast pages can be scrolled up and down.
- Good integration of the WAP browser into the mobile phone operating system and wireless stack. As has been discussed in Chapter 1, different companies produce different parts of the overall software of a mobile device. The web browser for example is one of the most

visible parts of the mobile phone software that is often outsourced by phone manufacturers to third-party companies. In order to quickly react to user input the browser needs to be integrated very tightly with the phone's operating system and especially the GPRS stack.

- Sufficient capacity in the radio network. Especially during busy hours, resources are scarce as voice calls usually take precedence over GPRS traffic if the radio network is not dimensioned correctly. This reduces the available bandwidth that has to be shared by all GPRS users of a cell and thus increases the page load times.
- Good network coverage is essential for a good WAP browsing experience especially if the mobile device is used in cars, trains, buses or the subway. While roads are usually covered quite well, the same can often not be said of train tracks especially inside tunnels. Also, subways are either only partly covered like only the stations themselves or not at all.
- In order to minimize the impact of a cell change, the network and terminal should implement the network assisted cell change functionality to decrease the time the data transfer is interrupted. As cell changes happen quite often in moving environments this optimization has a big impact on the click success rate.
- Sufficient capacity on the WAP gateway to handle incoming requests quickly. Especially during busy hours, under-dimensioned gateways are unable to cope with the traffic which also results in requests being delayed or dropped.
- Sufficient capacity on the operator's content platform to be able to quickly process the high number of simultaneous requests during busy hours.
- Sufficient capacity on the Gi and Gp interfaces of the GPRS network as these points of interconnection with other networks can quickly become a bottleneck if not dimensioned properly.
- The size of the web or WAP page also has an influence on the quality of experience especially in moving environments with bad network coverage. The smaller the page to be downloaded, the higher the chance that the complete page is downloaded before the coverage degradation interrupts the ongoing data transfer.

2.11 The Multimedia Messaging Service (MMS) over GPRS

Another mobile data application that has become very popular is the multimedia messaging service or MMS. MMS is advertised by mobile phone operators as the multimedia successor of the text-based short messaging service (SMS) that can transport not only text but also pictures, music and videos. The architectures of SMS and MMS, however, are fundamentally different. The SMS service is based on the SS-7 signaling channels of the network and is thus fully integrated into the GSM system and the GSM standards. MMS on the other hand is based on the Internet protocol (IP) and has many similarities with the Internet's email system. As can be seen in Figure 2.37 the MMS system uses the GPRS network only as a transparent IP network. It therefore does not rely on GPRS and can be used with any other wireless IP network technology such as UMTS or CDMA.

If a mobile terminal wants to send an MMS message, it establishes an IP connection to the MMS server via the GPRS network. The PDP context activation procedure that is required to get an IP address in the first place has already been described before. Instead of using the same APN as for a transparent connection to the Internet, the MMS service

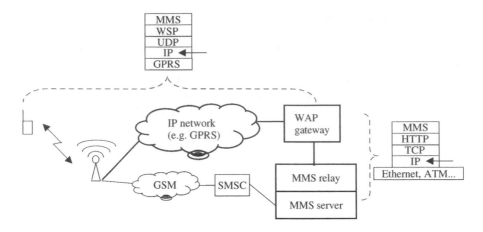

Figure 2.37 MMS architecture overview as defined in 3GPP TS 23.140 [7]

usually requires its own APN. This enables the operator to charge separately for the MMS traffic. As is shown in Figure 2.37 the system makes use of the WAP gateway, which seems somewhat strange at first. The WAP gateway was initially developed to optimize the transfer of WAP pages through a bandwidth limited wireless network. By putting the WAP gateway between the terminal and the MMS server, the MMS service also benefits from this functionality. This is also the reason for the two different protocol stacks seen in the figure. With the introduction of WAP 2.0, however, the protocol stacks on the two sides of the WAP gateway have been harmonized and new mobile terminals use HTTP and TCP for the MMS service.

If an MMS is exchanged between two subscribers of the same network the mobile phone number (MSISDN) is used to identify the recipient. Once the MMS has been sent to the MMS server it needs to notify the recipient that a new MMS message is waiting. This is done by sending an SMS message to the recipient's terminal. If the recipient has a non-MMS capable terminal the SMS contains a text message that informs the user of the MMS which he can then access via a web page. If the recipient has registered his handset with the MMS server as MMS capable by having previously sent an MMS message, the SMS is formatted differently. Instead of a text message intended for the subscriber, the SMS is formatted in a standardized way so the terminal recognizes on its own that there is a new MMS message waiting on the server. Depending on the terminal settings the MMS message is then either retrieved automatically or the terminal will inform the user that a new MMS message is waiting in the network. The user notification prior to the MMS download can be useful in cases when receiving an MMS is not free, for example with some operators when roaming abroad. If the terminal is allowed to download the MMS automatically, there is no user interaction and the user is only notified once the MMS message has been fully retrieved.

If the recipient is a subscriber of a different network, the MMS server cannot deliver the MMS directly, but has to forward the message to the MMS server of the home network for the recipient. This is only possible if the MMS servers are interconnected.

The MMS standard also allows direct MMS delivery to email addresses. As further described below, this is easily possible as the MMS format is quite similar to an email. In this case, no SMS notification of the receiver is necessary as the MMS is sent in a slightly modified way directly to the email account of the recipient.

In order to be able to send and receive MMS messages a number of settings are necessary in the mobile terminal. As MMS is an IP-based service, the first step in sending an MMS is activating a PDP context. Therefore, the MMS configuration on the mobile terminal has to contain an APN and optionally a login name and password. As a WAP gateway is used between the terminal and the MMS server it is also necessary to specify the IP address of the gateway. Finally, the terminal also has to be configured with the MMS server address which is specified as a URL. Table 2.5 shows the MMS settings for one of the mobile operators in the UK.

Similar to an email, MMS messages not only contain text but also attachments like pictures, sound files and video sequences. Unlike an email, however, the user can decide during the creation of an MMS in which order, at which position and for how long pictures and texts are shown on the display and when and for how long sounds and videos are played. On the network side, the MMS server has the possibility of adapting the multimedia elements to the capabilities of the recipient's terminal [8]. This might be necessary for example if the receiving terminal is unable to process an MMS message beyond a certain size.

While creating an MMS, the MMS software in the terminal converts the user's design of the message into a text-based layout and event description language which is called synchronized multimedia integration language (SMIL, pronounced 'Smile'). SMIL was standardized by the World Wide Web Consortium (http://www.w3c.org) and has many similarities with HTML, which is used to describe web pages. Figure 2.38 shows a sample SMIL description of an MMS. The general framework first of all describes the layout, number of pages and the presentation flow. The content of each page such as texts, pictures, sounds, video etc. is not part of the SMIL description. These are referenced via 'src=' tags and are included in the message after the SMIL description.

Similar to an email the different parts of the MMS including the SMIL description, texts, pictures, etc., are not sent separately from each other but together in a single transaction. In order to be able to differentiate the different parts of the message, the MMS standard makes use of the multipart Internet mail extension (MIME) protocol that is also used in the email standard. As can be seen in Figure 2.39 the MIME header contains a general description of the information that is to be transferred. Afterwards the SMIL description of the MMS and the content of the message such as texts, pictures, etc. are attached to form the complete body of the message. In order to separate the different parts of the message, a boundary marker is inserted between the different elements.

The boundary markers also contain the reference tags, which have been set in the SMIL description, as well as the description of the content type of the next part of the

Table 2.5 Sample MMS settings

APN	orangemms
Username	Orange
Password	Multimedia
IP address of WAP gateway	192.168.224.10
MMS server URL	http://mms.orange.co.uk

```
<smil>
 <head>
  <layout>
   <root-layout height = "80" width = "101"/>
    <region id = "Image" fit = "meet" height = "40"
              left ="0" top ="0" width ="101"/>
    <region id = "Text" fit ="meet" height="40"
              left = "0" top ="40" width ="101"/>
  </layout>
 </head>

 <body>
  <par dur = "10000ms">
   <img region = "Image" src = "cid : AA"/>
   <text region = "Text" src = "cid : AC"/>
  </par>

  <par dur = "10000ms">
   <text region = "Text" src ="cid : AD"/>
  </par>

 </body>
</smil>
```

page layout

first page with a picture and some text

second page, text only

Figure 2.38 SMIL description of the layout of an MMS message

```
Content-Type: multipart/related;
             start=<mmsdescription1>;
             boundary="boundary123456789";
```

MIME Header

```
--boundary123456789
Content-ID: <mmsdescription1>
Content-Type: application/smil;
             charset="US-ASCII"
< smil >
 [see previous picture]
< /smile >
```

1. part:
SMIL description

```
--boundary123456789
Content-ID: <mmsstuff1>
Content-Location: cid:AA
Content-Type: image/jpeg

[JPEG picture in binary format]
```

2. part:
Picture for page 1

```
--boundary123456789
Content-ID: <mmsstuff2>
Content-Location: cid:AC
Content-Type: text/plain

Hey, I am on vacation, the beach is great!
```

3. part:
Text for page 1

```
--boundary123456789
Content-ID: <mmsstuff2>
Content-Location: cid:AD
Content-Type: text/plain

I am having lots of fun!
```

4. part:
Text for page 2

Figure 2.39 MIME boundaries of the different parts of an MMS message

message. The following formats have so far been specified for MMS messages in 3GPP
TS 26.140 [9]:

- Pictures: JPEG, GIF, WBMP. The maximum guaranteed image size is 160×120 pixels.
  This corresponds to the display resolution of small mobile phones. If pictures are sent
  with a higher resolution they might have to be downsized in the receiving terminal or
  by the MMS server [8]. Nevertheless, it makes sense to include pictures with a higher
  resolution in the MMS as the user might later on transfer the picture to a PC via Bluetooth
  for example to benefit from the higher resolution of the PC's display.
- Text: ASCII 8-bit, UTF-8 or UTF-16.
- Audio: AMR (adaptive multi rate).
- Video: 3GPP MPEG-4 format. Many PC audio/video players offer plugins for this format
  so received videos can be transferred to PCs for playback as well.

In order to be able to send an MMS, an overall header is necessary that includes general
information like the address of the receiver and the subject line. Again, MMS leverages
already existing standards and uses a standard email header. The only adaptation that was
made in order to shrink the header size was to not use plain text parameter names. Instead,
each parameter name (e.g. `To:` or `Subject:`) was given a binary representation. The
parameter value, however, is not compressed. In order to be able to include some additional
information that is not necessary for sending emails, a number of MMS specific extension
tags were defined. They are all named `X-MMS...`and some of them are shown in Figure 2.40.
As MMS messages are usually exchanged between mobile subscribers, an MSISDN is used
in the `To:` field of the header. For clear identification, the `/Type=PLMN` suffix is added to

```
From: <insert address>
Date: Thu, 10 Juni 2004 10:49:55 +0100

To: +4916014867651/TYPE=PLMN
CC: <John Doe> jdoe@cm-networks.de [optional]

Subject: Still kicking! [optional]              MMS Header
MIME-Version: 1.0 [optional]


X-MMS-Version: 1.0
X-MMS-Message-Type: m-send-req
X-MMS-Transaction-ID: 867634563
X-MMS-Read-Reply: Yes [optional]

Content-Type: multipart/related;
            start=<mmsdescription1>;
            boundary= "boundary123456789";

--boundary123456789
Content-ID: <mmsdescription1>                    See previous picture
Content-Type: application/smil; charset= "US-ASCII"

<smil>
 [see previous picture]
</smil>
[...]
```

Figure 2.40 Uncompressed view of an MMS header

the MSISDN. PLMN stands for public land mobile network, the technical term for ground-based mobile telecommunication network. As can also be seen in the figure, the `From:` field of the header does not contain the sender's identification. This was done on purpose in order to prevent the sender from using a random originator address and thus conceal his true identity. The MMS server therefore has to query the GPRS network for the identification of the user.

While there are many similarities between MMS and email, they use different protocols for the actual transfer of the message. While email uses the simple mail transfer protocol (SMTP) it was decided to use the HTTP POST method for MMS. This protocol was initially designed to send the user supplied content of input boxes on web pages to the web server for processing. Using HTTP POST for the MMS transfer has the advantage that no handshake procedure is necessary on the application layer when the MMS server is contacted, which speeds up the MMS transaction.

As has been shown above, the MMS sever uses an SMS message to notify the receiver of a waiting message. The message contains a URL that identifies the stored MMS. In order to save time, it was decided not to use the POP3 or IMAP protocols which are used for retrieving email but to retrieve the MMS via the HTTP GET protocol. HTTP GET was initially designed to allow web browsers to request web pages from a server.

Due to the many similarities of MMS and email messages, it is quite simple to forward an MMS to an email address as well. As SMIL is not understood by the majority of email readers today, the MMS server removes the SMIL description of the message and sends the user's multimedia content (text, pictures, etc.) as file attachments. More sophisticated MMS server implementations even use HTML formatting for the presentation of the converted MMS and allow the recipient to return an email to the sender which is then converted back into an MMS message and delivered to the mobile terminal.

As MMS is a pure IP application and only uses open standards, the bearer over which it is transferred is no longer relevant. This means that theoretically MMS messages could also be used for messaging purposes between PCs in the Internet. Practically this does not make a lot of sense as the enhancements over the email standard that were put into the MMS specification do not entail many advantages for fixed-line use. In Europe MMS is used in both GSM and UMTS networks which are based on the same core architecture. The use of IP and open standards also allows using the MMS system with mobile networks that use a different network architecture, like CDMA. Unlike SMS messages, which require a special gateway to allow message exchange between networks that use different standards, MMS messages can be exchanged between different network standards like UMTS and CDMA with only minor adaptations for some multimedia elements.

2.12 Web Browsing via GPRS

2.12.1 Impact of Delay on the Web Browsing Experience

While a high bandwidth connection is certainly one of the most important factors for a good web browsing experience the round-trip delay time of the connection must also not be underestimated [10][11]. The round-trip delay (RTD) time is defined in this context as the time it takes to receive a response to a transmitted frame. This RTD can be measured for example with the `ping` command. The following example shows how delay impacts the web browsing experience. When requesting a new page the following delays are experienced before the page can be downloaded and displayed:

- The URL has to be converted into the IP address of the web server that hosts the requested page. This is done via a DNS query that causes a delay of one RTD of the connection.
- Once the IP address of the server has been determined, the web browser needs to establish a TCP connection. This is done via a three-way handshake. During this handshake the client sends a synchronization packet to the server which is answered by a synchronization-ack packet. This is in turn acknowledged by the client by sending an acknowledgment packet. As three packets are sent before the connection is established the whole operation causes a delay of 1.5 times the RTD of the connection. As the first packet containing user data is sent right after the acknowledgment packet, however, the time is reduced to approximately a single RTD time.
- Only after the TCP connection has been established, the first packet can be sent to the web server which usually contains the actual request of the web browser. The server then analyzes the request and sends back a packet that contains the beginning of the requested web page. As the request (e.g. 300–500 bytes) and the first response packet (1400 bytes) are quite large, the network requires somewhat more time to transfer those packets then the simple RTD time.

There are three different RTD times experienced in GPRS and EGPRS networks. If there is no TBF established at the time the web browser starts the request, then the mobile first needs to set up a connection in the up- and downlink direction. This typically results in a RTD time of the DNS query of 750 to 800 milliseconds in EGPRS networks. As the network expects further data to be transferred, the downlink TBF is usually kept open by the network, which reduces the RTD time to about 550 milliseconds for any subsequent communication. If the network supports the extended uplink TBF functionality the delay time is considerably less. As this functionality is not widely available today, however, the following calculation does not take it into account. The third delay time can be experienced when big frames are sent and received. As sending big frames takes some time by itself, the RTD time is increased for the example to about 1000 milliseconds for a 300 byte request and 1400 byte response frame in an EGPRS network with three to four timeslots assigned to the downlink TBF. Figure 2.41 shows a message trace for the example above with the exception of the DNS query which is not included. It also should be noted that the trace was generated while roaming so the RTD times are generally 200 milliseconds higher then the values that are experienced in the home network. This is due to the extra latency introduced by not sending the packets to the Internet from the visited network right away but routing them from and to the home network and from there to the Internet.

Due to the processes presented above, the time between entering a URL and being shown the first part of the web page can be calculated as follows:

$$\text{Total Delay (EGPRS)} = \text{Delay DNS query} + \text{Delay TCP Establish}$$
$$+ \text{Delay Request/Response}$$
$$= 750\,\text{ms} + 550\,\text{ms} + 1200\,\text{ms}$$
$$= 2500\,\text{ms}$$

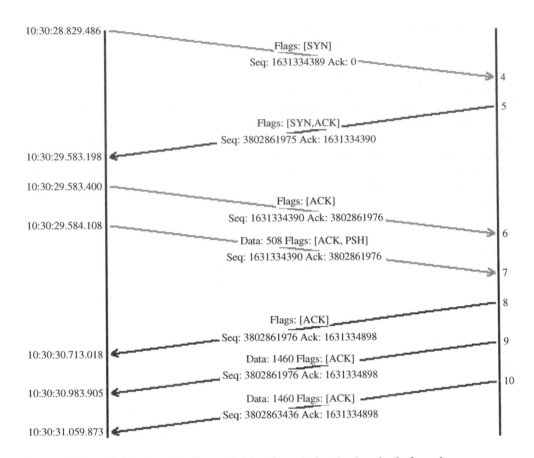

Figure 2.41   IP packet flow and delay times during the download of a web page

After this time, the page is not fully loaded but the browser usually starts to present the text of the main page. If the user then clicks on a link to a different page of the same website no further DNS query is necessary as the IP address of the server is still stored in the DNS cache. Thus, the delay time for loading subsequent pages is reduced by about 550 milliseconds.

Compared to the page download delay of only 370 milliseconds of a fixed-line DSL connection, the EGPRS delay appears to be rather big. This difference is caused by the fact that no up- and downlink connections have to be established prior to the transfer of data when using a DSL connection. The higher bandwidth of the DSL connection also allows downloading the remainder of the page quicker then via EGPRS. Nevertheless, a delay of about two to two and a half seconds before the first parts of a page are displayed is still acceptable. Together with the higher bandwidth offered by EGPRS compared to the standard GPRS network it is possibly to comfortably use the wireless connection for web browsing.

An excellent source for further information about (E)GPRS performance aspects can be found in [10].

2.12.2 Web Browser Optimization for Mobile Web Browsing

As has been shown, GPRS and EGPRS connections behave differently when used for web surfing compared to fixed-line connections. As most users access the Internet via fixed-line connections, web browsers are usually optimized for short latency and very high bandwidth connections. However, some browsers allow changing their network access settings which can help to improve the download times. The Firefox browser for example allows changing the values for pipelining of web requests. Pipelining is a method which was introduced with HTTP 1.1 to request several elements of a page at once in a single TCP connection instead of only sending the next request once the previous element was downloaded. In this way it is possible to reduce the effect of the higher latency of the wireless connection.

Pipelining is activated in the Firefox browser as follows:

- In the URL field, `about:config` has to be entered to get the list of all parameters that can be changed.
- Pipelining is activated by setting the `network.http.pipelining` parameter to TRUE.
- A good value for the number of accumulated requests that should be set in `network.http.pipelining.maxrequests` is eight.

When comparing the page download times of a large web page with and without pipelining in an EGPRS network a big difference can be observed. While the delay to display the first parts of the page is almost unchanged the time to download the complete page including all pictures was reduced from 60 seconds to 40 seconds.

2.13 Questions

1. What are the differences between circuit-switched and packet-switched data transmission?
2. What are the advantages of the data transmission over GPRS compared to GSM?
3. Why are different modulation and coding schemes used?
4. What is the difference between the GPRS ready state and the GPRS standby state?
5. Does the GPRS network perform a handover if a cell change is required while data is transferred?
6. Which are the new network elements that have been introduced with GPRS and what are their responsibilities?
7. What is a temporary block flow?
8. What actions are performed during an inter-SGSN routing area update (IRAU)?
9. Why is the IP protocol used twice in the protocol stack of the Gn interface?
10. Why is it not necessary to change any settings on the terminal for GPRS when roaming abroad?
11. What is the different between a GPRS attach and a PDP context activation?
12. Why is an access point name (APN) necessary for the PDP context activation procedure?
13. How are MMS messages sent and received via GPRS?
14. Name the different parts of an MMS message.

Answers to these questions can be found on the companion website for this book at http://www.wirelessmoves.com.

References

[1] 3GPP, 'Radio Access Network: Channel Coding', TS 45.003.
[2] Yuan-Kai Chen and Yi-Bing Lin, 'IP Connectivity for Gateway GPRS Support Node', *IEEE Wireless Communications Magazine*, pp. 37–46, February 2005.
[3] 3GPP, 'General Packet Radio Service (GPRS); Mobile Station (MS) – Base Station System (BSS) Interface; Radio Link Control/Medium Access Control (RLC/MAC) protocol', TS 44.060.
[4] 3GPP, 'General Packet Radio Service (GPRS); GPRS Tunnelling Protocol (GTP) across the Gn and Gp Interface', TS 29.060.
[5] 3GPP, 'General Packet Radio Service (GPRS); Serving GPRS Support Node (SGSN) – Visitors Location Register (VLR); Gs Interface Layer 3 Specification', TS 29.018.
[6] 3GPP, 'General Packet Radio Service (GPRS); Service Description; Stage 2', TS 23.060.
[7] 3GPP, 'Multimedia Messaging Service (MMS); Functional Description; Stage 2', TS 23.140.
[8] Stéphane Coulombe and Guido Grassel, 'Multimedia Adaptation for the Multimedia Messaging Service', *IEEE Communications Magazine*, pp. 120–6, July 2004.
[9] 3GPP, 'Multimedia Messaging Service (MMS); Media Formats and Codecs', TS 26.140.
[10] Peter Benko, Gabor Malicsko, and Andras Veres, 'A Large-scale, Passive Analysis of End-to-End TCP Performance over GPRS', IEEE Infocom Conference 2004.
[11] Rajiv Chakravorty, Joel Cartwright, and Ian Pratt, 'Practical Experience with TCP over GPRS', IEEE Globecom 2002.

3

Universal Mobile Telecommunications System (UMTS)

The Universal Mobile Telecommunications System (UMTS) is a third-generation wireless telecommunication system and follows in the footsteps of GSM and GPRS. Since GSM was standardized in the 1980s, huge progress has been made in many areas of telecommunication. This allowed system designers at the end of the 1990s to design a new system that goes far beyond the capabilities of GSM and GPRS. UMTS combines the properties of the circuit-switched voice network with the properties of the packet-switched data network and offers a multitude of new possibilities compared to the earlier systems. UMTS was not defined from scratch and reuses a lot of GSM and GPRS. Therefore, this chapter will first give an overview of the advantages and the enhancements of UMTS compared to its predecessors, which have been described in the previous chapters. After an end-to-end system overview, the focus of the chapter will then be on the functionality of the UMTS radio access network. New concepts like the radio resource control mechanisms as well as changes in mobility, call control and session management will also be described in detail.

3.1 Overview, History, and Future

Similar developments as seen in fixed-line networks are also appearing in mobile networks with a delay of about five years. In fixed networks the number of people using the network not only for voice telephony but also to connect to the Internet is increasing as steadily as the transmission speeds. While the first modems that were used in the mid-1990s featured speeds of about 14.4 kbit/s, the latest generation of modems achieve over 50 kbit/s in the downlink direction. Another incredible step forward was made around the year 2002 when technologies like cable and ADSL modems reached the mass market and dramatically increased transmission speed for end users. With these technologies, transmission speeds of several megabits per second are easily achieved.

Communication Systems for the Mobile Information Society   Martin Sauter
© 2006 John Wiley & Sons, Ltd

In the mobile world, GPRS (see Chapter 2) with its packet-oriented transmission scheme was the first step towards the mobile Internet. With data rates of about 50 kbit/s in the downlink direction in operational networks, similar speeds as those of fixed-line modems are achieved. New air interface modulation schemes like EDGE (enhanced data rates for GSM evolution) have increased the speed to about 150–200 kbit/s per user in operational networks. However, even with EDGE some limitations of the radio network such as the timeslot orientation of a 200 kHz narrowband transmission channel, GSM medium access schemes, and longer transmission delays compared to fixed-line data transmission cannot be overcome. Therefore, further transmission speed increases are difficult to achieve with the GSM air interface.

Since the first GSM networks went into service at the beginning of the 1990s, the increase in computing power and memory capacity has not stopped. According to Moore's law, the number of transistors in integrated circuits grows exponentially (Figure 3.1). Therefore, today's processors used in mobile networks offer 80–100 times the processing power compared to the processors of the early GSM days. This in turn enables the use of much more computing intensive air interface transmission methods that use the scarce bandwidth on the air interface considerably better than the comparatively simple GSM air interface.

For UMTS these advances were consistently used. While voice communication was the most important application for a wireless communication system when GSM was designed, data services are foreseen to play a major role in the next generation of wireless networks. Therefore, the convergence of voice and high-speed data services into a single system has been a driving force in the UMTS standardization from the beginning.

As will be shown in the following chapter, UMTS is as much an evolution as it is a revolution. Many components of the GSM core network were reused for UMTS in a first step with a simple software upgrade. A migration towards new technologies will only happen in the years to come. The radio access network, however, has been redesigned from scratch on the basis of a new air interface technology called WCDMA (wideband code division multiple access).

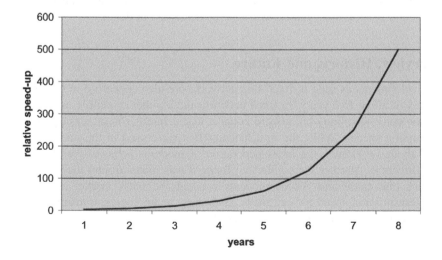

Figure 3.1 Processor speed increase in the time between standardization of GSM and UMTS

3.1.1 UMTS Release 99: A New Radio Access Network

The different steps of the UMTS specifications have been organized by the 3GPP (3rd Generation Partnership Project) standardization body in a number of consecutive versions. These versions are called 'Releases'. During the process the numbering scheme was changed which led to a lot of confusion. At the beginning a Release was named after the year of ratification while later on, a single number was used. This is why the first UMTS Release is called Release 99 while the following evolutions are called Release 4, Release 5, Release 6 and so on. At the time this book was written, 3GPP was in the process of working on Release 7.

Release 99 contains all the specifications for the first step of UMTS. The main improvement of UMTS compared to GSM in this first step is the completely redesigned radio access network, which the UMTS standards call the UMTS terrestrial radio access network (UTRAN). Instead of using the time- and frequency-multiplexing method of the GSM air interface, a new method called WCDMA was introduced. In WCDMA, users are no longer separated from each other by timeslots and frequencies but are assigned a unique code. Furthermore, the bandwidth of a single carrier was substantially increased compared to GSM, which enables a much faster data transfer than previously possible. This allows a Release 99 UTRAN to send data with a speed of up to 384 kbit/s per user in the downlink (network to user) direction and up to 64–128 kbit/s in the uplink direction (Figure 3.2). The standard also foresees uplink speeds of up to 384 kbit/s. However, so far, in current operational networks such uplink speeds have not been observed.

For the overall design of the UTRAN, the concept of base stations and controllers was adopted from GSM. However, while they are called BTS and BSC in the GSM network, the corresponding UTRAN network elements are called Node-B and radio network controller (RNC). Furthermore, the mobile station (MS) has also received a new name and is now called user equipment (UE).

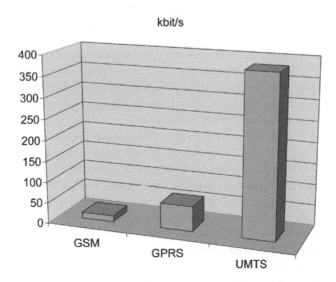

Figure 3.2 Speed comparison between GSM, GPRS and UMTS (Release 99)

In Europe and Asia, 12 blocks of 5 MHz each have been assigned to UMTS in the frequency range between 1920 MHz and 1980 MHz. This frequency range is just above the range used by DECT (cordless phones) in Europe. For the downlink direction, i.e. from the network to the user, another 12 blocks of 5 MHz each have been reserved in the frequency range between 2110 MHz and 2170 MHz.

In North America, no dedicated frequency blocks have been assigned for third-generation networks at the time this book is published. Instead, UMTS networks share the frequency band between 1850 MHz and 1910 MHz in the uplink direction, and between 1930 MHz and 1990 MHz in the downlink direction with second-generation networks such as GSM and CDMAOne networks.

Despite being in use for many years and thus not really up to date, the current technology for the GSM circuit-switched core network continues to be the basis for UMTS. It was decided not to specify major changes in this area but rather concentrate on the access network. The changes in the circuit core network to support UMTS Release 99 are therefore mainly software enhancements in order to support the new Iu(cs) interface between the MSC and the UTRAN. While it is quite similar to the GSM A-interface on the upper layers, the lower layers have been completely redesigned and are now based on ATM. Furthermore, the HLR and authentication center software have been enhanced in order to support the new UMTS features.

The GPRS packet core network, which connects users to the Internet or a company intranet, was adopted from GSM with only minor changes. No major changes were necessary for the packet core because GPRS was a relatively new technology at the time of the Release 99 specification, and was already ideally suited to a high-speed packet-oriented access network. Changes mostly impact the interface between the SGSN and the radio access network, which is now called the Iu(ps) interface. The biggest difference to its GSM/GPRS counterpart, the Gb interface, is the use of ATM instead of frame relay on lower layers of the protocol stack. Furthermore, the SGSN software has been modified in order to tunnel GTP user data packets transparently to and from the RNC instead of analyzing the contents of the packets and reorganizing them onto a new protocol stack as was previously done in GSM/GPRS.

As no major changes were necessary in the core network for Release 99, it is possible to connect the new UTRAN to an existing GSM and GPRS core network (Figure 3.3). The MSCs and SGSNs only require a software update and new interface cards in order to support the Iu(cs) and Iu(ps) interfaces. This is an advantage especially for those operators that already have an existing network infrastructure.

A common GSM and UMTS network furthermore simplifies the seamless roaming of users between GSM and UMTS. This is especially important during the first few years after the initial rollout of UMTS, as the new networks only cover big cities at first and expand into smaller cities and the rest of the country afterwards. While GSM network rollouts in the mid-1990s were done in a very similar way, it was of little consequence at that time not to offer nationwide coverage immediately due to the fact that the number of users of the previous generation wireless networks were quite small. Today, however, nobody would change to a new network that offers inferior coverage compared to the previous system at the time it is taken into service.

Due to the necessity of seamless UMTS to GSM roaming and vice versa, dual mode mobile devices that are capable of using both the GSM and UMTS networks are necessary to

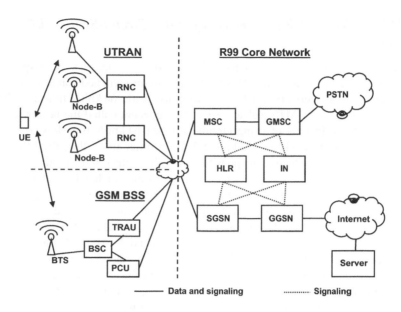

Figure 3.3 Common GSM/UMTS network, Release 99

meet customer expectations. These devices also allow a seamless handover of ongoing voice calls from UMTS to GSM if a user leaves the UMTS coverage area. Similar mechanisms were implemented for data sessions. However, due to the lower speed of the GSM/GPRS network, the process is not seamless.

UMTS Release 99 networks can of course be used for voice telephony, but the main goal of UMTS beyond this service was the introduction of fast packet data services. Since the first UMTS networks became operational in late 2002, network operators are able to offer high-speed Internet access on the one hand for their business customers to enable, for example, a true mobile office any time, anywhere across national boundaries. Fast access to email with large file attachments and fast access to databases and resources on their company intranets make the mobile office a much more practical option on UMTS compared to GPRS with its comparably slow transmission rates. On the other hand, UMTS Release 99 also enables network operators to offer compelling new high-speed services to private users. The main applications for private users are, for example, news and information with graphical content, MMS messages with large images, videos and music, mobile television and entertainment applications like mobile Java games that can be downloaded and played on the handset. Another compelling application for UMTS is music downloading. With file sizes of about 1.5 Mb per song and 200–500 kb per game, UMTS is fast enough to download a song in less than 40 seconds and a game in less than 10 seconds. As music and game downloads via the fixed-line Internet are also subject to charges, this kind of content offers a good opportunity for network operators to increase their revenue, as users are much more willing to download content for which they would also have to pay on the fixed-line Internet.

3.1.2 UMTS Release 4: Enhancements for the Circuit-Switched Core Network

A major enhancement for circuit-switched voice and data services has been specified with UMTS Release 4. Up to and including Release 99, all circuit-switched connections have been routed through the core network via E-1 connections inside 64 kbit/s timeslots. The most important enhancement of UMTS Release 4 is a new concept called the bearer independent core network (BICN). Instead of using circuit-switched 64 kbit/s timeslots, traffic is now carried inside ATM or IP packets (see Figure 3.4). In order to do this, the MSC has been split into an MSC server which is responsible for call control and mobility management (see Chapter 1) and a media gateway which is responsible for handling the actual bearer (user traffic). The media gateway is also responsible for the transcoding of the user data for different transmission methods. This way it is possible for example to receive voice calls via the GSM A-interface via E-1 64 kbit/s timeslots at the MSC media gateway which will then convert the digital voice data stream onto a packet-switched ATM or IP connection towards another media gateway in the network. The remote media gateway will then again convert the incoming user data packets if necessary, to send them for example to a remote party via the UMTS radio access network (Iu(cs) interface) or back to a circuit-switched E-1 timeslot if a connection is established into the fixed-line telephone network.

The introduction of this new architecture is driven by network operators that want to combine the circuit- and packet-switched core networks into a single converged network for all traffic. This is desirable as mobile network operators no longer only need a strong circuit-switched backbone but also have to invest in packet-switched backbones for the GPRS and UMTS user data traffic. As packet-switched data continues to increase so does the need for investment into the packet-switched core network. By using the packet-switched

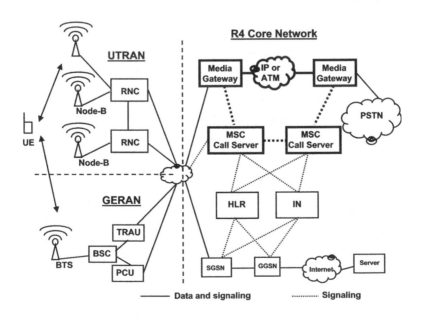

Figure 3.4   UMTS Release 4 (BICN)

core network for the voice traffic as well, operators expect noticeable cost reductions. At the time of publication, many infrastructure vendors have shown that their R4 BICN MSCs have become stable and mature and first deployments in real networks are likely to be seen in 2006.

3.1.3 UMTS Release 5: Introduction of the IP Multimedia Subsystem (IMS)

UMTS Release 5 takes the core network one step further and defines an architecture for an end-to-end all-IP network. The circuit-switched MSC and the Iu(cs) interface are no longer required in a pure Release 5 network. As can be seen in Figure 3.5, the MSC is replaced by the IP multimedia subsystem (IMS) with which the user equipment communicates via the SGSN and GGSN. The core of the IMS comprises a number of nodes that form the call session control function (CSCF). The CSCF is basically a SIP (session initiation protocol) architecture which was initially developed for the fixed-line world and is one of the core protocols for most voice over IP telephony services available on the market today. The CSCF takes the concept one step further and enhances the SIP standard with a number of functionalities necessary for mobile networks. This way, Release 5 enables voice calls to be transported via IP not only in the core network but from end to end, i.e. from mobile phone to mobile phone. While the CSCF is responsible for the call setup and call control, the user data packets which for example include voice or video conversations are directly exchanged

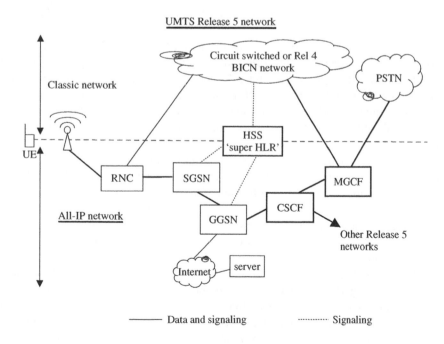

Figure 3.5 UMTS Release 5 architecture

between the end-user devices. A media gateway control function (MGCF) is only necessary if one of the users still uses a circuit-switched phone.

With the UMTS radio access network it is possible for the first time to implement an IP-based mobile voice and video telephony architecture. This is not only due to the fact that UMTS offers enough bandwidth on the air interface for such applications but also due to a new way of handling cell changes for packet connections. With GPRS in the GSM access network, the roaming from one cell to another (mobility management) for packet-switched connections is controlled by the mobile station. This results in an interruption of packet traffic of 2–3 seconds at every cell change. For voice or video calls this is not acceptable. With UMTS, the mobility management for packet-switched connections can now also be controlled by the network. This ensures uninterrupted packet traffic even while the user is roaming from one cell to another. The overhead of an IP connection for voice telephony, however, remains a problem for the wireless world. As the delay must be as short as possible, only a few bytes of voice data are put into a single IP packet. This means that the overhead for the header part of the IP packet is about 50%. Circuit-switched voice connections on the other hand do not need any header information and are transported very efficiently over the UMTS network today. While the IP overhead in a fixed-line network is still important it does not prevent the use of IP for voice calls over ADSL or cable due to the high bandwidth available on these services. However, on the UMTS air interface with its limited bandwidth, a connection which requires twice as much resource is not desirable as it *de facto* cuts the number of simultaneous calls per cell in half. It should be noted at this point that video telephony which is currently offered in UMTS Release 99 networks is not based on IP but on a 64 kbit/s circuit-switched channel established between two users via the MSC. This has only become possible with UMTS as the GSM access network was limited to 9.6 or 14.4 kbit/s channels on the air interface.

Despite the evolution of voice telephony towards IP it has to be ensured that every user can talk to every other user regardless of which kind of telephony architecture they use. As can be seen in Figures 3.4 and 3.5 this is achieved by using media gateways which convert between IMS voice over IP (VoIP), BICN and the classic circuit-switched approach. As optimizing and improving mobile networks for IMS VoIP calls is an evolutionary process, the different architectures will coexist in operational networks for many years to come.

As the IMS has been designed to serve as a universal communication platform, the architecture offers a far greater variety of services then just voice and video calls, which are undoubtedly the most important applications for the IMS in the long term. Due to the complexity involved to compete against wireless circuit-switched voice and video calls, a number of other services will drive the first introduction of the IMS in public networks in 2006. Push to talk (PTT), which is already very popular in the United States, is one of those applications. By using the IMS as a platform for a standardized PTT application it is possible to include people in talk groups who have subscriptions with different operators. Other interesting IMS services include mobile presence and messaging capabilities like the Yahoo or Microsoft Messenger offer today for PCs, standardized access to video content or mobile TV as well as enabling multimedia multi-user applications such as giving presentations to several remote persons or multi-player games across networks and country boundaries.

3.1.4 UMTS Release 5: High Speed Downlink Packet Access (HSDPA)

Equally important in UMTS Release 5 is the introduction of a new data transmission scheme called high speed downlink packet access (HSDPA) which increases data transmission speeds from the network to the user. While 384 kbit/s is the maximum speed offered by a Release 99 UTRAN, HSDPA enables speeds of 1.4 to 3.6 Mbit/s per user, depending on the capability of the user equipment and up to 14.4 Mbit/s with evolved terminals. Even under less ideal radio conditions and under heavy load of a cell, speeds of 800 kbit/s can still be reached per user. By further increasing the overall data rate available per cell, HSDPA allows for new bandwidth-hungry services, and so as the total bandwidth requirements of the network dramatically increases, the number of cell sites can remain the same. As the main cost for HSDPA is to increase the capacity of the backhaul connection of the cells to the network, the transmission cost per bit will further decrease due to the fact that the same number of base stations are able to support a much higher overall bandwidth. The introduction of HSDPA in 2006 therefore enables UMTS network operators to compete directly with DSL, cable and WIMAX Internet access for home and office use. For example, some operators in Austria, Switzerland, Italy and Germany are already positioning UMTS Release 99 as an alternative to DSL or cable Internet access and surely welcome HSDPA as it will improve their competitive position, allow higher data speeds per user and increase the total number of high-speed connections the network can support simultaneously.

3.1.5 UMTS Release 6: High Speed Uplink Packet Access (HSUPA)

The IMS and HSDPA continue to be evolved in UMTS Release 6. However, this revision of the specification is best known for the introduction of yet another enhancement of the radio access network. While HSDPA substantially increases the overall bandwidth available per cell and per user in the downlink direction, uplink speeds have not increased since Release 99. Thus the uplink is still limited to 64–128 kbit/s and to 384 kbit/s in some networks under ideal conditions. The emergence of the IMS, however, triggers the widespread use of a number of direct user-to-user applications such as multimedia conferencing. These applications send as much data as they receive and therefore the uplink will become the bottleneck of the system over time. Therefore, UMTS Release 6 introduces an uplink transmission speed enhancement called high speed uplink packet access (HSUPA). In theory HSUPA allows data rates of several Mbit/s for a single user under ideal conditions. Taking realistic signal conditions, the number of users per cell and terminal capabilities into consideration, HSUPA will still be able to deliver speeds of around 800 kbit/s. Furthermore, HSUPA also increases the maximum number of users that can simultaneously send data via the same cell and thus further reduces the overall cost of the network. Other non-IMS applications like sending email messages with very large file attachments and MMS messages with large video content also benefit from HSUPA.

3.1.6 UMTS Release 7 and Beyond: Even Higher Data Rates

While HSDPA already increases data rates far beyond initial UMTS speeds the race for more bandwidth and user data speeds continues. Ever more sophisticated transmission techniques like OFDM (orthogonal frequency division multiplexing) and MIMO (multiple input and multiple output) are discussed in the 3GPP working groups for UMTS Release 7. The aim is to again increase the data rate by a factor of 10 compared to HSDPA to enable UMTS networks to be able to compete against other wireless and fixed-line technologies of the future.

3.2 Important New Concepts of UMTS

As shown in the previous paragraphs, UMTS on the one hand introduces a number of new functionalities compared to GSM and GPRS. On the other hand many properties, procedures and methods of GSM and GPRS, which are described in Chapters 1 and 2, have been kept. Therefore, this chapter focuses mainly on the new functionalities and changes UMTS has introduced compared to its predecessors. In order not to lose the end-to-end overview, references are made to Chapters 1 and 2 for methods and procedures that UMTS continues to use.

3.2.1 The Radio Access Bearer (RAB)

An important new concept that has been introduced with UMTS is the radio access bearer (RAB), which is a description of the transmission channel between the network and a user. The RAB is divided into the radio bearer on the air interface and the Iu bearer in the radio network (UTRAN). Before data can be exchanged between a user and the network it is necessary to establish an RAB between them. This channel is then used for both user and signaling data. A RAB is always established by request of the MSC or SGSN. In contrast to the establishment of a channel in GSM, the MSC and SGSN do not specify exactly how the channel has to look. Instead the RAB establishment requests only contain a description of the required channel properties. How these properties are then mapped to a physical connection is up to the UTRAN. The following properties are specified for an RAB:

- service class (conversational, streaming, interactive or background);
- maximum speed;
- guaranteed speed;
- delay;
- error probability.

The UTRAN is then responsible for establishing an RAB that fits the description. The properties not only have an impact on the bandwidth of the established RAB but also on parameters like coding scheme, selection of a logical and physical transmission channel as well as on the behavior of the network in the event of erroneous or missing frames on different layers of the protocol stack. The UTRAN is free to set these parameters as it sees fit; the standards merely contain examples. As an example, for a voice call (service class conversational) it does not make much sense to repeat lost frames. For other services like web browsing, such behavior is beneficial as delay times are shorter if lost packets are only retransmitted in the radio network instead of end to end.

3.2.2 The Access Stratum and Non-Access Stratum

UMTS strives to separate functionalities of the core network from the access network as much as possible in order to be able to independently evolve the two parts of the network

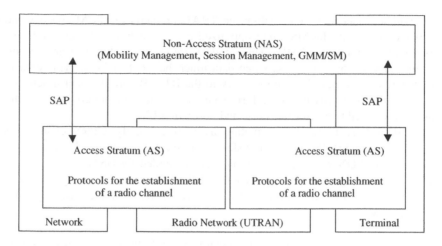

Figure 3.6 Separation of protocols between the core and radio network into access stratum (AS) and non-access stratum (NAS)

in the future. Therefore, UMTS strictly differentiates between functionalities of the access stratum (AS) and the non-access stratum (NAS) as shown in Figure 3.6.

The access stratum contains all functionalities that are associated with the radio network ('the access') and the control of active connections between a user and the radio network. The handover control, for example, for which the RNC is responsible in the UTRAN is part of the access stratum.

The non-access stratum contains all functionalities and protocols which are used directly between the mobile device (user equipment or UE) and the core network. These have no direct influence on the properties of the established RAB and its maintenance. Furthermore, NAS protocols are transparent to the access network. Functionalities like call control, mobility and session management as well as supplementary services (e.g. SMS), which are controlled via the MSC and SGSN, are considered NAS functionalities.

While the NAS protocols have no direct influence on an existing RAB, it is nevertheless necessary for NAS protocols like call control or session management to request the establishment, modification, or termination of a bearer. To enable this, three different service access points (SAPs) have been defined between NAS and AS:

- notification SAP (Nt, e.g. for paging);
- dedicated control SAP (DC, e.g. for RAB setup);
- general control SAP (GC, e.g. for modification of broadcast messages, optional).

3.2.3 Common Transport Protocols for CS and PS

In GSM networks, data is transferred between the different nodes of the radio network with three different protocols. The most important task of these protocols is to split incoming data into smaller frames, which can be transferred over the air interface. While these protocols are described in more detail in Chapters 1 (GSM) and 2 (GPRS) here is a short overview.

- Circuit-switched data (e.g. voice calls): the TRAU converts the PCM-coded voice data which it receives from the MSC into optimized codecs such as enhanced full rate, half rate, or adaptive multi rate (AMR). These codecs are more suitable for transmission over the air interface as they compress voice data much better then PCM. This data is then sent transparently through the radio network to the BTS. Before the data is sent over the air interface, the BTS only has to perform some additional channel coding (e.g. increase of redundancy by adding error detection and correction bits).
- Signaling data (circuit-switched signaling as well as partly GPRS channel request messaging and paging): this data is transferred via the LAPD protocol, which is already known from the ISDN world and which has been extended for GSM.
- Packet-switched user and signaling data for GPRS: while user and signaling data are separated in GSM, GPRS combines the two data streams into a single lower layer protocol called RLC/MAC.

In UMTS, these different kinds of data streams are combined into a single lower layer protocol called the radio link control/medium access control (RLC/MAC) protocol. Giving this protocol the same name as a protocol in the GPRS network was quite intentional. Both protocols work quite similarly in areas, e.g. breaking up large data packets from higher layers into smaller chunks for transmission over the air interface. Due to the completely different transmission methods of the UMTS air interface compared to GSM/GPRS, there are, however, also big differences as will be shown in the next section.

3.3 Code Division Multiple Access (CDMA)

To be able to better comprehend the differences between the UMTS radio access network and its predecessors, the next paragraph gives a short overview about the basic principles of the GSM/GPRS network and its limitations.

In GSM, data for different users is simultaneously transferred by multiplexing them on different frequencies and timeslots (frequency and time division multiple access, FTDMA). A user is assigned one of eight timeslots on a specific frequency. To increase the number of users that can simultaneously communicate with a base station the number of simultaneously used frequencies can be increased. However, it must be ensured that two neighboring base stations do not use the same frequencies as they would otherwise interfere with each other. As the achievable speed with only a single timeslot is limited, GPRS introduced the concept of timeslot bundling on the same carrier frequency. While this concept enables the network to transfer data to a user much faster then before, there are still a number of shortcomings that have been solved by UMTS.

With GPRS, it is only possible to bundle timeslots on a single carrier frequency. Therefore, it is theoretically possible to bundle up to eight timeslots. In an operational network, however, it is rare that a mobile station can bundle more than four timeslots, as some of them are also necessary for voice calls of other users. Furthermore, on the terminal side today, most phones can only handle four timeslots at a time in the downlink direction. This is because bundling more timeslots would require more complex hardware in the mobile station.

A GSM base station was initially designed for voice traffic, which only requires a modest amount of transmission capacity. This is why GSM base stations are usually connected to the BSC via a single 2 Mbit/s E-1 connection. Depending on the number of carrier frequencies and sectors of the base station, only a fraction of the capacity of the E-1 connection is used. The remaining 64 kbit/s timeslots are therefore used for other base stations. Furthermore, the processing capacity of GSM base stations was only designed to support the modest requirements for voice processing compared to the computing intensive high-speed data transmission capabilities required today.

With GPRS, a user is only assigned resources (i.e. timeslots) in the uplink and downlink directions for the time they are required. In order for uplink resources to be assigned, the mobile station has to send a request to the network. A consequence of this is unwanted delays ranging from 500 to 700 milliseconds when data needs to be sent.

Likewise, resources are only assigned in the downlink direction if data has to be sent from the core network to a user. Therefore, it is necessary to assign resources before they can be used by a specific user, which takes another 200 ms.

These delays, which are compared in Figure 3.7 to the delays experienced with ADSL and UMTS, are tolerable if a large chunk of data has to be transferred. For short and bursty data transmissions as in a web-browsing session, however, the delay is noticeable.

UMTS solves these shortcomings as follows. In order to increase the data transmission speed per user, UMTS increases the bandwidth per carrier frequency from 200 kHz to 5 MHz. This approach has advantages over just adding more carriers to a data transmission which are dispersed over the frequency band, as mobiles can be manufactured much more cheaply when only a single frequency is used for the data transfer.

The most important improvement of UMTS is the use of a new medium access scheme on the air interface. Instead of using a frequency and time division multiple access scheme as GSM, UMTS uses code multiplexing to allow a single base station to communicate with many users at the same time. This method is called code division multiple access (CDMA).

Contrary to the frequency and time multiplexing of GSM, all users communicate on the same carrier frequency and at the same time. Before transmission, the data of a user is multiplied by a code, which can be distinguished from codes used by other users on the receiver side. As the data of all users is sent at the same time, the signals add up on the

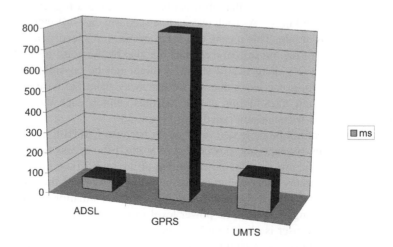

Figure 3.7   Round-trip delay time of UMTS (Release 99) compared to ADSL and GPRS

transmission path to the base station. The base station can then use the inverse mathematical approach that was used by the terminal as the base station knows the code of each user.

This principle can also be described within certain boundaries with the following analogies:

- Communication during a lecture: usually there is only one person speaking at a time while there are many persons in the room that are just listening. The bandwidth of the 'transmission channel' is high as it is only used by a single person. At the same time, however, the whispering of the students is creating a slight background noise, which has no impact on the transmission (of the speaker) due to its low volume.
- Communication during a party: there are again many persons in a room but this time they all talk with each other. Although the conversations add up in the air the human ear is still able to distinguish the different conversations from each other. Other conversations are filtered out by the ear as unwanted background noise. The more people speak at the same time, the higher the perceived background noise for the listeners. In order to be understood the speakers have to reduce their talking speed. Another method for talkers could also be to increase their volume to be able to be heard over the background noise. This, however means, that the background noise for others increases substantially.
- Communication in a disco: in this scenario, the background noise, i.e. the music, is very loud and communication is no longer possible.

These scenarios are analogous to a UMTS system as follows: if there are only few users that communicate with the base station at the same time, each user will experience only low interference on the transmission channel. Therefore, the transmission power can be quite low and the base station is still able to distinguish the signal from other sources. This also means that the available bandwidth per user is high and can be used if necessary to increase the transmission speed. If data is sent faster, the signal power needs to be increased to get a more favorable signal-to-noise ratio. As only few users are using the transmission channel in this scenario, increasing the transmission speed is no problem as all others are able to compensate.

If many users communicate with a base station at the same time, all users will experience a high background noise. This means that all users have to send at a higher power in order to overcome the background noise. As each user in this scenario can still increase the power level the system remains stable. This means that the transmission speed is not only limited by the 5 MHz bandwidth of the transmission channel but also by the noise generated by other users of the cell. While the system is still stable, it might not be possible to increase the data transmission speed for some users that are farther away from the base station as they cannot further increase their transmission power and thus cannot reach the signal-to-noise ratio required for a higher transmission speed. See Figure 3.8.

Transmission power cannot be increased indefinitely because UMTS terminals in Europe are limited to a maximum transmission power of 0.25 watt. If the access network could not continuously control and be aware of the power output of the mobile stations, a point would be reached at which too many users communicate with the system. As the signals of other users are perceived as noise from a single user's point of view a situation could occur when a mobile station can no longer increase its power level to get an acceptable signal-to-noise ratio. On the contrary, if a user is close to a base station and increases its power above the level commanded by the network, it could interfere with the signals of terminals, which are further away and thus weaker.

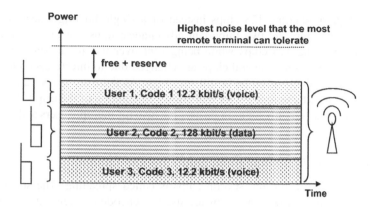

Figure 3.8 Simultaneous communication of several users with a base station in the uplink direction (axis not to scale and number of users per base station is higher in a real system)

From a mathematical point of view, CDMA works as follows.

The user data bits of the individual users are not transferred directly over the air interface but are first multiplied with a vector, which for example has a length of 128. The elements of the resulting vector are called chips. A vector with a length of 128 has the same number of chips. Instead of transmitting a single bit over the air interface, 128 chips are transmitted. This is called 'spreading', as more information, in this example 128 times more, is sent over the air interface compared to the transmission of the single bit. On the receiver side the multiplication can be reversed and the 128 chips are used to deduce if the sent bit represents a 0 or 1. Figure 3.9 shows the mathematical operations for two mobile stations that transmit data to a single receiver (base station).

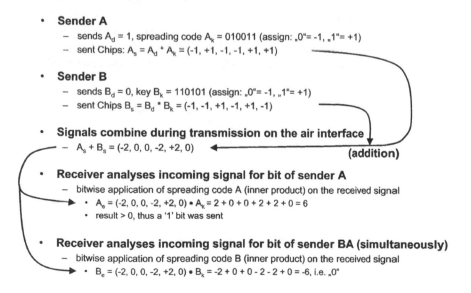

Figure 3.9 Simultaneous conversation of two users with a single base station and spreading of the data stream

The disadvantage of sending 128 chips instead of a single bit might seem quite severe but on the other hand there are two important advantages: transmission errors that change the values of some of the 128 chips while being sent over the air interface can easily be detected and corrected. Even if several chips are changed due to interference the probability of correctly identifying the original bit is still very high. As there are many 128-chip vectors, each user can be assigned a unique vector that allows calculation of the original bit out of the chips at the receiver side not only for a single user but also for multiple users at the same time.

3.3.1 Spreading Factor, Chip Rate, and Process Gain

The process of encoding a bit into several chips is called spreading. The spreading factor for this operation defines how many chips are used to encode a single bit. The speed with which the chips are transferred over the UMTS air interface is called the chip rate and is 3.84 MChips/s independent of the spreading factor.

As the chip rate is constant, increasing the spreading factor for a user means that his data rate decreases. Besides a higher robustness against errors there are a number of other advantages of a higher spreading factor: the longer the code, the more codes exist that are orthogonal to each other. This means that more users can simultaneously use the transmission channel than compared to a system in which only shorter spreading factors are used. As more users generate more noise, it is likely that the error rate increases at the receiver side. However, as more chips are used per bit, a higher error rate can be accepted than for a smaller spreading factor. This in turn means that a lower signal-to-noise ratio is required for a proper reception and thus the transmission power can be reduced if the number of users in a cell is low. As less power is required for a slower transmission, it can also be said that a higher spreading factor increases the gain of the spreading process (processing gain). See Figure 3.10.

If shorter codes are used, i.e. fewer chips per bit, the transmission speed per user increases. However, there are two disadvantages: due to the shorter codes, fewer people can communicate with a single base station at the same time. With a code length of eight (spreading factor 8), which corresponds to a user data rate of 384 kbit/s in the downlink direction, only eight users can communicate at this speed. With a code length of 256 on the other hand, 256 users

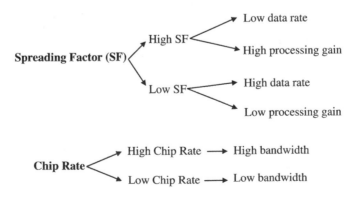

Figure 3.10 Relation between spreading factor, chip rate, processing gain, and available bandwidth per user

can communicate at the same time with the base station although the transmission speed is a lot slower. Due to the shorter spreading code, the processing gain also decreases. This means that the power level of each user has to increase in order to minimize transmission errors.

3.3.2 The OVSF Code Tree

The UMTS air interface uses a constant chip rate of 3.84 MChips/s. If the spreading factor is also constant, all users of a cell have to communicate with the network at the same speed. This is not desired because a single cell has to support many users with many different applications simultaneously. While some users may want to simply make voice calls, which require only a small bandwidth, other users at the same time might want to place video calls, watch some mobile TV (video streaming), or start a web-surfing session. All these services require much higher bandwidths and thus using the same spreading factor for all connections is not practical.

The solution to this problem is called orthogonal variable spreading factors (OVSF). While in the previous mathematical representation the spreading factors of both users were of the same length, it is possible to assign different code lengths to different users at the same time with the following approach.

As the codes of different lengths also have to be orthogonal to each other, the codes need to fulfill the following condition as shown in Figure 3.11: in the simplest case (C1,1), the vector is one dimensional. On the next level with two chips, four vectors are possible of

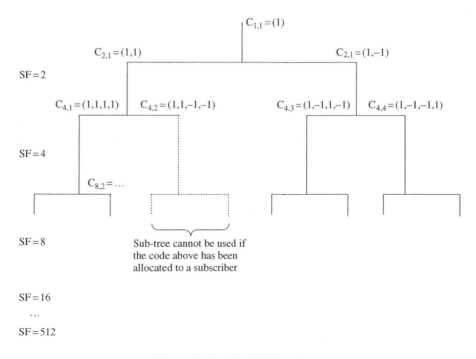

Figure 3.11 The OVSF code tree

which two are orthogonal to each other (C2,1 and C2,2). On the third level with four chips, there are 16 possible vector combinations and four that are orthogonal to each other. The tree which continues to grow for SF 8, 16, 32, etc., shows that the higher the spreading factor, the more subscribers can communicate with a cell at the same time.

If a terminal, for example, uses a spreading factor of eight, all longer codes of the same branch can no longer be used. This is due to the fact that the codes below are not orthogonal to the code on the higher level. As the tree offers seven other SF 8 spreading factors, it is still possible for other users to code with higher spreading factors from one of the other vertical branches of the code tree. It is up to the network to decide how many codes are used from each level of the tree. Thus the network has the ability to react dynamically to different usage scenarios.

Table 3.1 shows the spreading factors in the downlink direction (from the Node-B to the terminal) as they are used in a real system. The raw data rate is the number of bits transferred per second. The user data rate results from the raw data rate after removal of extra bits, which are used for channel coding that is necessary for error detection and correction, signaling data, and channel control.

3.3.3 Scrambling in the Uplink and Downlink Directions

By using OVSF codes, the data rate can be adapted for each user individually while still being able to differentiate the data streams with different speeds. Some of the OVSF codes are quite uniform. C(256,1) for example only comprises '1' chips. This creates a problem further down the processing chain because the modulation of long sequences that never change their value would result into a very uneven spectral distribution. To counter this effect the chip stream that results from the spreading process is scrambled. This is done by multiplying the chip stream as shown in Figure 3.12 with a pseudo random code called the scrambling code. The chip rate of 3.84 MChips/s is not changed by the process.

In the downlink direction the scrambling code is also used to enable the terminal to differentiate between base stations. This is necessary as all base stations of a network transmit on the same frequency. In some cases mobile operators have bought a license for more than a single UMTS frequency. However, this was done to increase the capacity in densely populated areas and not to make it easier for mobile stations to distinguish between different base stations. The use of a unique scrambling code per base station is also necessary to

Table 3.1 Spreading factors and data rates

Spreading factor (downlink)	Raw data rate (kbit/s)	User data rate (kbit/s)	Application
8	960	384	Packet data
16	480	128	Packet data
32	240	64	Packet data and video telephony
64	120	32	Packet data
128	60	12.2	Voice and packet data
256	30	5.15	Voice
512	15	1.7	Signaling, SMS, location update...

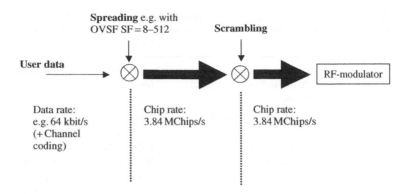

Figure 3.12 Spreading and scrambling

allow a base station to use the complete code tree instead of sharing it with the neighboring cells. This means that in the downlink direction, capacity is mainly limited by the number of available codes from the code tree as well as the interference of other base stations as experienced by the user equipment.

In the uplink direction each terminal is assigned its own scrambling code. Therefore, each terminal could theoretically use all codes of the code tree. This means that in the uplink direction the system is not limited by the number of codes but by the maximum transmitting power of the user equipment and by the interference that is created by other terminals in the current and neighboring cells.

Another reason for using a unique scrambling code per terminal in the uplink direction is the signal propagation delays. As different users have different distances to a base station the signals take different amounts of time to arrive. In the GSM radio network this was solved by controlling the timing advance (see Section 1.7.4). The use of a timing advance, however, is not possible in the UMTS radio network due to the soft handover state (see Section 3.7.1) in which the user equipment communicates with several base stations at the same time. As the user equipment is at a different distance to each base station it communicates with simultaneously, it is not possible to synchronize the mobile station to all base stations due to the different signal propagation delays. Therefore, if no scrambling code was used the mathematical equation shown in Figure 3.9 would no longer work as the chips of the different senders would be out of phase with each other and the result of the equation would change. See Table 3.2.

3.3.4 UMTS Frequency and Cell Planning

As all cells in a UMTS radio network can use the same frequency, the frequency plan is greatly simplified compared to a GSM radio access network. While it was of paramount importance in a GSM system to ensure that neighboring cells use different frequencies, it is quite the reverse in UMTS as all neighboring stations use the same frequency. This is possible due to the CDMA characteristics, which were described in the previous sections. While a thorough and dynamic frequency plan is indispensable for GSM, no frequency adaptations are necessary for new UMTS cells. If a new cell is installed in an area that

Table 3.2 Spreading and scrambling in the uplink and downlink directions

	Downlink	Uplink
Spreading	Addressing of different users Controls the individual data rate for each user	Controls the individual data rate for each user
Scrambling	Ensures consistent spectral distribution Used by the mobile terminal to differentiate base stations	Ensures consistent spectral distribution Differentiates users Removes the need for a timing advance by preserving orthogonal nature of the codes Necessary for soft handover

is already covered by other cells in order to increase the bandwidth the most important task in a UMTS network is to decrease the transmission power of the neighboring cells.

In GSM and UMTS radio networks alike it is necessary to properly define and manage the relationships between the neighboring cells. Incorrectly defined neighboring cells are not immediately visible but create difficulties for handovers (see Section 3.7.1) and cell reselections (see Section 3.7.2) of moving subscribers later on. Properly executed cell changes and handovers also improve the overall capacity of the system as they minimize interference of mobiles that stay in cells that are no longer suitable for them.

3.3.5 The Near-Far Effect and Cell Breathing

As all users transmit on the same frequency, interference is the most limiting factor for the UMTS radio network. The following two phenomena are a direct result of the interference problem.

In order to keep the interference at a minimum it is important to have a precise and fast power control. Users that are further away from the base station have to send with more power than those closer to the base station as the signal gets weaker the further it has to travel. This is called the near-far effect. Even small changes of the position of the user like moving from a free line of sight to a base station behind a wall or tree has a huge influence on the necessary transmission power. The importance of efficient power control for UMTS is also shown by the fact that the network can instruct each handset 1500 times per second to adapt its transmission power. A beneficial side effect of this for the mobile station is an increased operating time, which is very important for most devices as the battery capacity is quite limited.

Note: GSM also controls the transmission power of the handsets. The control cycle, however, is in the order of a second as interference in GSM is less critical then in UMTS. Therefore, power control is mostly beneficial to increase the operating time of the user equipment.

The dependency on low interference for each user also creates another unwanted side effect. Let us assume the following situation:

1. There is a high number of users in the coverage area of a base station and the users are dispersed at various distances from the center of the cell.
2. Because of interference the most distant user needs to transmit at the highest possible power.
3. An additional user who is located at a medium range from the center of the cell tries to establish a connection to the network for a data transfer.

In this situation the following things can happen: if the network accepts the connection request the interference level for all users will rise in the cell. All users thus have to increase their transmission power accordingly. The user at the border of the cell, however, already transmits at its maximum power and thus can no longer increase the power level. As a result his signal cannot be correctly decoded and the connection is broken. Seen from outside the system this means that the geographical area the cell can cover is reduced as the most distant user cannot communicate with the cell. The phenomenon is called cell breathing due to the fact that the cell expands and shrinks like a human lung, which increases and decreases its size while breathing. See Figure 3.13.

To avoid this effect the network constantly controls the signal-to-noise ratio of all active users. By actively controlling the transmission power of each user the network is aware of the impact an additional user would have on the overall situation of the cell. Therefore, the network has the possibility of rejecting a new user to protect the ongoing sessions.

In order to preserve all ongoing connections and additionally allow a new user to enter the system it is also possible to use a different strategy. The goal of this strategy is to

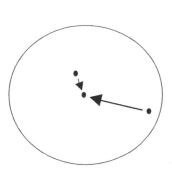

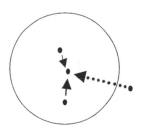

Two subscribers of a cell one of them close to the cell edge sending with its maximum possible power level

A third subscriber would like to communicate in the cell. This poses a problem for the second subscriber as he can't increase the power output to counter the additional interference

Figure 3.13 Cell breathing

reduce the interference to a level that allows all users including the prospective new one to communicate. This can be done in a number of ways. One way is to assign longer spreading codes to already established channels. As has been shown in Section 3.3.2, it is possible for terminals to reduce their transmission power by using longer spreading codes. This in turn reduces the interference for all other users. The disadvantage of using longer spreading codes is of course a reduction in the maximum transmission speed for some users. As not all connections might be impacted, there are again a number of possibilities for the selection process. Users could for example be assigned to different user classes. Changing spreading factors could then be done only for users of a lower user class who pay less for their subscription than others. It can also be imagined that the network already starts a congestion defense mechanism at a certain load threshold before the system gets into an overload situation. Once the threshold is reached, the network could then for example only assign short spreading factors to users with a higher priority subscription while the system load is above the threshold.

Besides cell breathing there are further interference scenarios. As already mentioned, it is necessary to increase the transmission power if the spreading factor is decreased in order to ensure a proper reception. Therefore, the maximum distance a user can be from the center of the cell also depends on the spreading factor. If a user roams between two cells it is possible that the current spreading factor would not allow data to be transferred as reliably as before due to the interference encountered at the cell edge. A lower spreading factor, however, would still allow a reliable data transfer. How this and similar scenarios at cell edges are resolved depends on the vendor's equipment and the parameter settings of the operator. As in other areas, the UMTS standard does not dictate a specific solution to these issues. Therefore, network vendors that have implemented clever solutions can gain a competitive advantage.

3.3.6 Advantages of the UMTS Radio Network Compared to GSM

While in the previous paragraphs the basic properties and methods of the UMTS W-CDMA air interface have been introduced, the following paragraph describes how this new air interface overcomes the limitations of GSM/GPRS.

One of the main reasons for the long delay times of GPRS are the constant reassignments of resources for bursty data traffic. UMTS solves this issue by assigning a dedicated channel not only for voice calls but also for packet data connections. The channel remains dedicated to the user even if there is no data transferred for some time. A downside of this approach, however, is that the spreading code is not available to other users. Due to the fact that only control information is sent over the established channel during times of inactivity, the interference level for other users decreases. As a result, only a little of the overall capacity of a cell is lost by keeping the spreading code assigned to a dormant user for some time. In a well-implemented network, from a subscriber's point of view, the spreading code should only be freed up for use by someone else if the session remains dormant for a prolonged amount of time. Once the system decides to reassign the code to someone else, it can assign a higher spreading factor to the dormant user of which a greater number exists per cell. If the user resumes data transmission, there is no delay as a dedicated channel still exists. If required, the bandwidth for the user can be increased again quite quickly by assigning a code with a shorter spreading factor. The subscriber, however, does not have to wait for this as in the meantime data transfer is possible over the existing channel. If the user remains dormant for an even longer time, the network might then go ahead and remove all resources on the air interface without cutting the logical connection. This saves further resources and also has a positive effect on the overall operating time of a terminal as it consumes less energy if the channel is released. The disadvantage of this approach is a longer reaction time once the user wants to resume the data transfer. In the uplink direction, the same methods are applied. It should be noted though that while the user is assigned a code, the mobile station will be constantly transmitting in the uplink direction. The transmission power will be lower while no user data is sent but the mobile station keeps sending power control and signal quality measurement results to the network. This is why it is beneficial to move the user into the Cell-FACH (forward access channel) state after a longer period of inactivity. In this state, no control information is sent from the user equipment to the network and no dedicated channel is assigned to the connection. The different connection states are described in more detail in Section 3.5.4. Furthermore, an analysis of how operational networks handle the code and state management of a packet call can be found in Section 3.9.2.

The assignment of dedicated channels for both circuit- and packet-switched connections in UMTS has a big advantage for mobile users compared to GPRS. In the GPRS network, the mobile station is solely responsible for performing a cell change. Once the cell has been changed, the mobile first needs to listen to the broadcast channel before the connection to the network can be re-established. In a practical environment, a cell change thus interrupts an ongoing data transmission for about one to three seconds. A handover, which is controlled by the network and thus results in no or only a minimal interruption of the data transmission, has not been foreseen for GPRS. Thus, GPRS users frequently experience interruptions of the data transmission during cell changes while traveling in cars or trains. With UMTS, however, there are no interruptions of an ongoing data transfer when changing cells due to a process called 'soft handover', which makes data transfers while on the move much more efficient. Furthermore, applications like voice over IP or video telephony on the move are thus also possible as they no longer experience interruptions during cell changes.

Another problem of GSM is the historical dimensioning of the transmission channel for narrow band voice telephony. This limitation was overcome for GPRS by combining several timeslots for the time of the data transfer. The maximum possible data rate, however, is still limited by the overall capacity of the 200 kHz carrier. For UMTS, high bandwidth applications were taken into consideration for the overall system design from the beginning. Due to this, a maximum data transfer rate of 384 kbit/s can be achieved with spreading factor eight in the downlink direction. In the uplink direction, data rates of 64 and 128 kbit/s can be reached, which are not as fast as in the downlink direction mainly due to the lower transmission power and omni-directional antenna design. These speeds are suitable for fast web surfing as well as for applications like voice over IP and video telephony.

UMTS also enables circuit-switched 64 kbit/s data connections in the up- and downlink directions. This speed is equal to an ISDN connection in the fixed-line network and is mostly used for video telephony between UMTS users.

UMTS can also react very flexibly to the current signal quality of the user. If the user moves away from the center of the cell, the network can react by increasing the spreading factor of the connection. This reduces the maximum transmission speed of the channel, which is usually preferred compared to losing the connection entirely.

The UMTS network is also able to react very flexibly to changing load conditions on the air interface. If the overall interference reaches an upper limit, or if a cell runs out of available codes due to a high number of users in the cell, the network can react and assign longer spreading factors to new or ongoing connections.

3.4 UMTS Channel Structure on the Air Interface

3.4.1 User Plane and Control Plane

GSM, UMTS, and other modern fixed and wireless communication systems differentiate between two kinds of data flows. In UMTS, these are also separated into two different planes. Data flowing in the user plane is data that is directly and transparently exchanged between the users of a connection like voice data or IP packets. The control plane is responsible for all signaling data that is exchanged between the users and the network. The control plane is thus used for signaling data to exchange messages for call establishment or messages, e.g. for a location update. Figure 3.14 shows the separation of user and control plane as well as some examples for protocols that are used in the different planes.

3.4.2 Common and Dedicated Channels

Both user plane data and control plane data are transferred over the UMTS air interface in so-called 'channels'. Three different kinds of channels exist.

Dedicated channels: these channels transfer data for a single user. A dedicated channel is used for example for a voice connection, for IP packets between the user and the network or a location update message.

The counterpart to a dedicated channel is a common channel. Data transferred in common channels is destined for all users of a cell. An example of this type of channel is the broadcast

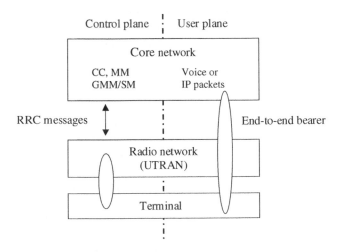

Figure 3.14   User and control planes

channel which transmits general information about the network for all users of a cell such as to which network the cell belongs to, current state of the network, etc. Common channels can also be used by several devices for the transfer of user data. In such a case, each device filters out its packets from the stream broadcast over the common channel and only forwards these to higher layers of the protocol stack.

Very similar to common channels are shared channels. These channels are not monitored by all devices but only by those which have been instructed by the network to do so. An example of such a channel is the high speed downlink shared channel of HSDPA (see Section 3.10).

3.4.3 Logical, Transport, and Physical Channels

In order to separate the physical properties of the air interface from the logical data transmission, the UMTS design introduces three different channel layers. Figure 3.15 shows the channels on different layers in the downlink direction while Figure 3.16 does the same for the uplink channels.

Logical Channels

The topmost channel layer is formed by the logical channels. Logical channels are used to separate different kinds of data flows that have to be transferred over the air interface.

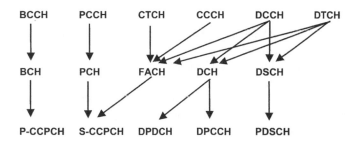

Figure 3.15 Logical, transport, and physical channels in the downlink direction

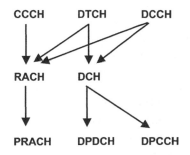

Figure 3.16 Logical, transport, and physical channels in the uplink direction

The channels contain no information on how the data is later transmitted over the air. The UMTS standards define the following logical channels:

- The BCCH (broadcast control channel): this channel is monitored by all terminals in idle state to receive general system information from the network. Information distributed via this channel for example includes how the network can be accessed, which codes are used by the neighboring cells, the LAC, the cell ID, and many other parameters. The parameters are further grouped into system information block (SIB) messages to help the terminal decode the information and to save air interface bandwidth. A detailed description of the messages and parameters can be found in 3GPP 25.331, chapter 10.2.48.8 [1].
- The PCCH (paging control channel): this channel is used to inform users of incoming calls or SMS messages. Paging messages are also used for packet-switched calls if new data arrives from the network once all physical resources (channels) for a subscriber have been released due to a long period of inactivity. If the terminal receives a paging message it has to first report its current serving cell to the network. The network will then re-establish a logical RRC connection with the terminal and the data waiting in the network is then delivered to the terminal.
- The CCCH (common control channel): this channel is used for all messages from and to individual terminals (bi-directional) that want to establish a new connection with the network. This is necessary for example if a user wants to make a phone call, send an SMS, or to establish a channel for packet-switched data transmission.
- The DCCH (dedicated control channel): while the three channels described above are common channels observed by many terminals in the cell, a DCCH only transports data for a single subscriber. A DCCH is used for example to transport messages for the mobility management (MM) and call control (CC) protocols for circuit-switched services, packet mobility management (PMM) and session management (SM) messages for packet-switched services from and to the MSC and SGSN. These protocols are described in more detail in Sections 3.6 and 3.7.
- The DTCH (dedicated traffic channel): this channel is used for user data transfer between the network and a single user. User data can for example be a digitized voice signal or IP packets of a packet-switched connection. If a dedicated logical channel carries a voice call, it is mandatory to map this channel to a dedicated physical channel. If the dedicated logical channel carries data of a packet-switched connection, however, it is also possible to map the dedicated logical connection onto a common or shared physical channel. As can be seen in Figure 3.15 it is thus possible to map a DTCH not only on a dedicated transport and physical channel but also on a common/shared channel.
  Note: In UMTS Release 99 most IP packet-switched connections will always be carried over a dedicated physical channel. By using this approach individual user speeds of up to 384 kbit/s in the downlink direction are possible. Common/shared physical channels (e.g. the FACH which is introduced below) are only used if the user has been inactive for a long time or if the user only sends or receives small amounts of data infrequently.
- The CTCH (common traffic channel): this channel is used for cell broadcast information. In GSM, the same mechanism is used, for example, by Vodafone in Germany to inform subscribers of fixed-line phone network area codes which are used around the current cell that can be called from the mobile phone for a cheaper tariff.

Transport Channels

Transport channels prepare downlink data frames for transmission over the air interface by splitting them into smaller parts, which are encapsulated into RLC/MAC frames that are more suitable for the transmission over the air interface. The RLC/MAC header, which is put in front of each frame, contains among other things the following information:

- length of the frame (10, 20, 40, or 80 ms);
- type of integrity checking mechanism (CRC checksum);
- channel coding format for error detection and correction;
- rate matching if the speed of the physical channel and the layers above do not match;
- control information for detection of discontinuous transmission (DTX) if the other end has no data to send at a particular time.

All of these properties are combined into a so-called transport format. The actual channel coding, however, is only performed on the physical layer on the Node-B. This is very important as channel coding includes the addition of error detection and correction bits to the data stream, which can be a huge overhead. In Chapter 1, for example, the half-rate convolutional decoder for channel coding was introduced, which practically doubles the data rate. UMTS also makes use of this channel coder and further introduces a number of additional ones.

Logical channels are mapped to the following transport channels:

- The BCH (broadcast channel): transport channel variant of the logical BCCH.
- The DCH (dedicated channel): this transport channel combines data from the logical DTCH and the logical DCCH. The channel exists in both the uplink and downlink directions as data is exchanged in both directions.
- The PCH (paging channel): transport channel variant of the logical PCCH.
- The RACH (random access channel): The bi-directional logical CCCH is called RACH on the transport layer in the uplink direction. This channel is used by terminals to send RRC connection request messages to the network if they wish to establish a dedicated connection with the network (e.g. to establish a voice call). Furthermore, the channel is used by terminals to send user packet data (in Cell-FACH state, see Section 3.5.4) if no dedicated channel exists between the terminal and the network. It should be noted, however, that this channel is only suitable for small amounts of data.
- The FACH (forward access channel): this channel is used by the network to send RRC connection setup messages to terminals which have indicated via the RACH that they wish to establish a connection with the network. The message contains information for the terminal on how to access the network. If the network has assigned a dedicated channel, the message contains, for example, which spreading codes will be used in the uplink and downlink directions. The FACH can also be used by the network to send user data to a terminal if no dedicated channel has been allocated for a data transfer. The terminal is then in the Cell-FACH state, which is further described in Section 3.5.4. In the uplink direction data is transferred via the RACH.

Physical Channels

Physical channels are responsible for offering a physical transmission medium for one or more transport channels. They are also responsible for channel coding, i.e. the addition of redundancy and error detection bits to the data stream.

The intermediate products between transport channels and physical channels are called composite coded transport channels (CCTrCh) and are a combination of several transport channels which are subsequently transmitted over one or more physical channels. This intermediate step was introduced because it is not only possible to map several transport channels onto a single physical channel (e.g. the PCH and FACH on the S-CCPCH) but it is also possible to map several physical channels onto a single transport channel (e.g. the DPDCH and DPCCH onto the DCH).

The following physical channels are used in a cell:

- The P-CCPCH (primary common control physical channel): this channel is used for distributing broadcast information in a cell.
- The S-CCPCH (secondary common control physical channel): this channel is used to broadcast the PCH and FACH. As the spreading factor for this channel is variable, data rates from a few kbit/s up to several 100 kbit/s can be achieved by using spreading factors between 4 and 256. This has been done as the FACH can not only transport channel assignment messages to terminals, but it can also be used to transport user data for terminals (Cell-FACH state). Due to the unpredictability of the amount of user traffic, the operator thus has a tool to adjust the channel bandwidth depending on the traffic situation.
- The PRACH (physical random access channel): the physical implementation of the random access channel.
- The AICH (acquisition indication channel): this channel is not shown in the channel overview figures as there is no mapping of it to a transport channel. The channel is used exclusively together with the PRACH during the connection establishment of a terminal with the network. More about this channel and the process of establishing a connection can be found in Section 3.4.5.
- The DPDCH (dedicated physical data channel): this channel is the physical counterpart of a dedicated channel to a single terminal. The channel combines user data and signaling messages from (packet) mobility management, call control, and session management.
- The DPCCH (dedicated physical control channel): this channel is used in addition to a DPDCH in both the uplink and downlink directions. It contains layer 1 information such as transmit power control (TPC) bits for adjusting the transmission power. Furthermore, the channel is also used to transmit so-called pilot bits. These bits always have the same value and can thus be used by the receiver to generate a channel estimation, which is used to decode the remaining bits of the DPCCH and the DPDCH. More information about the DPCCH can be found in 3GPP TS 25.211 chapter 5.2.1 [2].

While the separation of channels in GSM into logical and physical channels is still quite easy to understand, the UMTS concept of logical, transport, and physical channels and the mappings between them is somewhat difficult to understand at first. Therefore, the following list summarizes the different kinds of channels and their main tasks:

- **Logical channels**: these channels describe different flows of information like user data and signaling data. Logical channels contain no information about the characteristics of the transmission channel.
- **Transport channels**: these channels prepare data packets that are received from logical channels for transmission over the air interface. Furthermore, this layer defines which channel coding schemes (e.g. error correction methods) are to be applied on the physical layer.
- **Physical channels**: these channels describe how data from transport channels is sent over the air interface and apply channel coding and decoding to the incoming data streams.

In order to get an impression of which way the different channels are used, the next section shows how the channels are used for two different procedures:

3.4.4 Example: Network Search

When a terminal is switched on, one part of the start-up procedure is the search for available networks. Once a suitable network has been found, the terminal performs an attach procedure. Afterwards the terminal is known to the network and ready to accept incoming calls, short messages, etc. When the users switches the terminal off, the current information about the network (e.g. the frequency, scrambling code, and cell ID of the current cell) is saved to the SIM card. This enables the terminal to skip most activities required for the network search once it is powered on and thus substantially reduces the time it takes to find and attach to the network again. In this example, it is assumed that the terminal has no or only invalid information about the last used cell when it is powered on. This can be the case if the SIM card is used for the first time or if the cell for which information was stored on the SIM card is not found.

As in all communication systems it is also necessary in UMTS to synchronize the terminals with the network. Without correct synchronization it is not possible to send an RRC connection request message at the correct time or to detect the beginning of an incoming data frame. Therefore, the terminal's first task after power on is to synchronize to the cells of the networks around it. This is done by searching all frequency bands assigned to UMTS for primary synchronization channels (P-SCH). As can be seen in Figure 3.17, a UMTS data frame consists of 15 slots in which 2560 chips per slot are usually transported. On the P-SCH only the first 256 chips per slot are sent and all base stations use the same code. If several signals (originating from several base stations) are detected by the mobile at different times due to the different distances of the terminal to the various cells, the terminal synchronizes to the timing of the burst with the best signal quality.

Once a P-SCH is found the terminal is synchronized to the beginning of a slot. In the next step the terminal then has to synchronize itself with the beginning of a frame. To do this the terminal will search for the secondary synchronization channel (S-SCH). Again only 256 chips per slot are sent on this channel. However, on this channel each slot has a different chip pattern. As the patterns and the order of the patterns are known, the terminal is able to determine the slot that will contain the beginning of a frame.

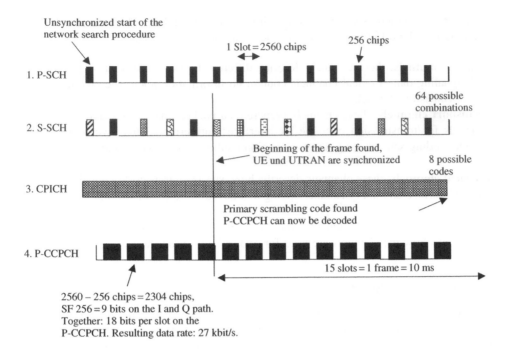

Figure 3.17 Network search after the terminal is switched on

If an operator only has a license for a single channel, all cells of the network operator send on the same frequency. The only way to distinguish them from each other is by using a different scrambling code for each cell. The scrambling code is used to encode all downlink channels of a cell including the P-CCPCH, which contains the system broadcast information. The next step of the process is therefore to determine the primary scrambling code of the selected cell. The first part of this process was already started with the correct identification of the S-SCH and the chip pattern. Altogether, 64 different S-SCH chip patterns are specified in the standard. This means that in theory the terminal could distinguish up to 64 individual cells at its current location. In an operational network, however, it is very unlikely that the terminal would receive more then a few cells at a time. In order to determine the primary scrambling code the terminal then decodes the common pilot channel (CPICH), which broadcasts another known chip pattern. Eight possible scrambling codes are assigned to each of the 64 chip patterns, which can be found on the S-SCH. In order to find out which code is used by the cell out of the eight scrambling codes for all other channels, the terminal now applies each of the eight possible codes on the scrambled chip sequence and compares the result to the chip pattern that is expected to be broadcast on the CPICH. As only one of the scrambling codes will yield the correct chip pattern the terminal can stop the procedure as soon as it has found the correct one.

Once the primary scrambling code has been found by using the CPICH the terminal can now read the system information of the cell which is broadcast via the P-CCPCH. The P-CCPCH is always encoded with spreading code C256,1 with a spreading factor of 256 which is easy to find by the mobile even under difficult radio conditions. Only after having deciphered the information which is broadcast on this channel is the mobile aware of which network the cell belongs to. The following list shows some parameters that are broadcast on the P-CCPCH:

- The identity of the network the cell belongs to (MCC/MNC), location area (LAC) and cell-ID.
- Cell access restrictions, i.e. which groups of subscribers are allowed to communicate with the cell. Usually all subscribers are allowed to communicate with a cell. Only under certain conditions will the network operator choose to temporarily restrict access to parts of the network for some subscribers. This can help during catastrophic events to allow important users of the network like police, doctors, etc. to communicate with facilities such as hospitals etc. Without access restrictions, cells quickly overload during such events as the number of call attempts of normal users increase dramatically and can thus delay important calls.
- Primary scrambling codes and frequencies of neighboring cells. As described above the frequencies of the other cells in the area are usually the same as the frequency of the current cell. Only in areas of very high usage might operators deploy cells in other frequency bands to increase the overall available bandwidth. Both scrambling codes and frequencies of neighboring cells are needed by the mobile to be able to easily find and measure the reception quality of other cells while they are in idle mode for cell reselection purposes.
- Frequency information of neighboring GSM cells. This information is used by the mobile to be able to reselect a GSM cell if the signal quality of the current cell deteriorates and no suitable neighboring UMTS cell can be received.
- Parameters that influence the cell reselection algorithm. This way the network is able to instruct the terminal to prefer some cells over others.
- Maximum transmission power the mobile is allowed to use when sending a message on the RACH.
- Information about the configuration of the PRACH and S-CCPCH which transport the RACH and FACH respectively. This is necessary because some parameters, such as the spreading factor, are variable in order to allow the operator to control the bandwidth of these channels. This is quite important as they do not only transport signaling information but can also transport user data as will be further described below.

If the cell belongs to the network the terminal wants to attach to, the next step in the process is to connect to the network by performing a circuit-switched location update, or a GPRS attach, or both. These procedures use the higher protocol layers of the mobility management and packet mobility management respectively which are also used in GSM and GPRS. For UMTS both protocol stacks were only slightly adapted. Further information on these procedures can be found in Sections 1.8.1 and 2.7.1.

3.4.5 Example: Initial Network Access Procedure

If the terminal is in idle state and wants to establish a connection with the network, it has to perform an initial network access procedure. This may be done for the following reasons:

- To perform a location update.
- For a mobile originated call.

- To react to a paging message.
- To start a data session (PDP context activation).
- To access the network during an ongoing data session for which the physical air interface connection was released by the network due to long inactivity.

For all the above scenarios, the terminal needs to access the network to request a connection over which further signaling messages can be exchanged. As can be seen in Figure 3.18, the terminal starts the initial network access procedure by sending several preambles with a length of 4096 chips. The time required to transmit the 4096 chips is exactly one millisecond. If the terminal receives no answer from the network it increases the transmission power and repeats the request. The terminal keeps increasing the transmission power for the preambles until a response is received or the maximum transmission power and number of retries have been reached without a response. This is necessary as the terminal does not know which transmission power level is sufficient to access the network. Thus the power level is very low at the beginning which on the one hand creates only low interference for other subscribers in

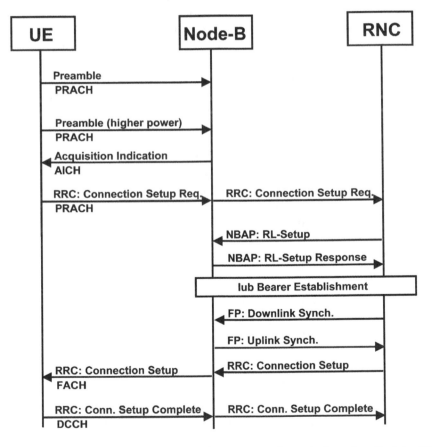

Figure 3.18 Initial network access procedure (RRC connection setup) as described in 3GPP TS 25.931 [3]

the cell but on the other hand does not guarantee success. To allow the network to answer, the preambles are spaced three slots apart. Once the preamble is received correctly, the network then answers on the acquisition indication channel (AICH). If the terminal receives the message correctly, it is then aware of the transmission power to use and proceeds by sending a 10 or 20 millisecond frame on the PRACH which contains an RRC connection request message. As the spreading factor of the PRACHs is variable the message can contain between 9 and 75 bytes of information.

In order to avoid collisions of different terminals, the PRACH is divided into 15 slots. Furthermore, there are 16 different codes for the preamble. Thus, it is very unlikely that two terminals use the same slot with the same code at the same time. Nevertheless, if this happens the connection request will fail and has to be repeated by the terminals as their requests cancel out each other. Once the RNC has received the RRC connection request message as shown in Figure 3.18, it will allocate the required channels in the radio network and on the air interface. There are two possibilities the RNC can choose from:

- The RNC can use a DCH for the connection as shown in Figure 3.18. Thus, the terminal changes into the Cell-DCH RRC state (see also Section 3.5.4). Using a dedicated channel is a good choice for the network if the connection request message indicates that the terminal wishes to establish a user data connection (voice or packet-switched data).
- The RNC can also decide to continue to use the RACH and FACH for the subsequent exchange of messages. The terminal will thus change into the Cell-FACH state. The decision to use a shared channel instead of a dedicated channel can be made, for example, if the connection request message indicates to the network that the channel is only required for signaling purposes, such as performing a location update.

After choosing a suitable channel and reserving the necessary resources, the RNC replies with an RRC connection setup message to the terminal via the FACH. If a dedicated channel was established for the connection, the terminal will switch to the new channel and return an RRC connection setup complete message.

After this common procedure, the terminal can then proceed to establish a higher layer connection between itself and the core network to request the establishment of a phone call or data connection (PDP context). A number of these scenarios are shown in Section 3.7.

3.4.6 The Uu Protocol Stack

UMTS uses channels on the air interface as shown in the previous section. As on any interface, several protocol layers are used for encapsulating the data and sending it to the correct recipient. In UMTS Release 99 the RNC is responsible for all protocol layers except for the physical layer which is handled in the Node-B. The only exception to this rule is the BCCH, which is under the control of the Node-B. This is because the BCCH only broadcasts static information that does not have to be repeatedly sent from the RNC to the Node-B.

As is shown in Figure 3.19, higher layer PDUs (packet data units) are delivered to the RNC from the core network. This can be user data like IP packets or voice frames, as well as control plane messages of the MM, CM, PMM, or SM subsystems.

If the PDUs contain IP user data frames, the PDCP (packet data convergence protocol) can optionally compress the IP header. The compression algorithm used by UMTS is described

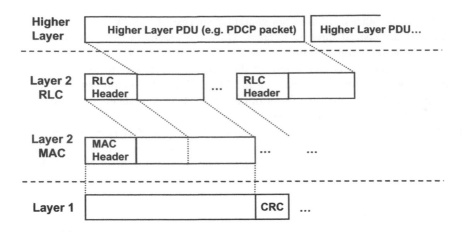

Figure 3.19 Preparation of user data frames for air interface (Uu) transmission

in RFC 2507 [4]. Depending on the size of the transmitted IP frames, header compression can substantially increase the transmission speed. Small frames in particular benefit from this as the IP header requires a proportionally oversized part of the frame.

The RLC (radio link control) layer is aware of the physical properties of the air interface and splits the packets it receives from higher layers for transmission over the air interface. This procedure is called segmentation and is required because PDCP frames that contain IP frames can be of variable size and can even be over 1000 bytes long. Frames on the air interface, however, are usually much smaller and are always of the same length. The length of those frames is determined by the spreading factor, the transmission time interval (TTI, 10–80 ms), and the applied coding scheme.

Just like GSM and GPRS, the UMTS radio network has been designed to send only small frames over the air interface. This has the advantage that in the case of packet loss or corruption only a few bytes have to be retransmitted. Depending on the spreading factor and thus the speed of the connection, the frame sizes vary. For a 384 kbit/s bearer with a TTI of 10 milliseconds for example, each data frame contains 480 bytes of user data. For a 64 kbit/s bearer with a TTI of 20 milliseconds, a frame contains only 160 bytes. For a voice call with a TTI of 20 milliseconds and a data rate of 12.2 kbit/s, a frame contains only 30 bytes.

If RLC frames are smaller than a frame on the air interface it is also possible to concatenate several RLC frames for a single TTI. In the event that there is not enough data arriving from higher layers to fill an air interface frame, padding is used to fill the frame. Instead of padding the frame it is also possible to use the remaining bits for RLC control messages.

Depending on the kind of user data one of three different RLC modes is used:

- The RLC transparent mode is used primarily for the transmission of circuit-switched voice channels and for the information that is broadcast on the BCCH and the PCCH. As the length of voice frames does not vary and as they are sent in a predefined format every 20 milliseconds, padding is also not necessary. Therefore, no adaptation or control functionality is required on the RLC layer, hence the use of the RLC transparent mode.
- The RLC non-acknowledged mode offers segmentation and concatenation of higher layer frames as described above. Furthermore, this mode allows marking of the beginning and

end of layer 3 user data frames. Thus, it is possible to always completely fill an air interface frame regardless of the higher layer frames. As no acknowledgment for RLC frames is required in this mode, frames that are not received correctly or lost cannot be recovered on this layer.

- The third mode is the RLC acknowledged mode, which is mostly used to transfer IP frames. In addition to the services offered by the non-acknowledged mode this mode offers flow control and automatic retransmission of erroneous or missing blocks. Similar to TCP, a window scheme is used to acknowledge the correct reception of a block. By using an acknowledgement window it is not necessary to wait for a reply for every transmitted block. Instead, further blocks can be transmitted up to the maximum window size. Up to this time the receiver has the possibility of acknowledging frames which in turn advances the window. If a block was lost, the acknowledgement bit in the window will not be set which automatically triggers a retransmission. The advantage of this method is that the data flow in general is not interrupted by a transmission error. The RLC window size can be set between 1 and 2^{12} frames and is negotiated between the terminal and RNC. This flexibility is the result of the experience gained with GPRS. There, the window size was static, it offered only enough acknowledgment bits for 64 frames. In GPRS, this proved to be problematic especially for coding schemes three and four during phases of increased block error rates (BLER) which lead to interrupted data flows as frames cannot be retransmitted quickly enough to advance the acknowledgment window.

Once the RLC layer has segmented the frames for the transmission over the air interface and has added any necessary control information, the medium access control (MAC) layer performs the following operations:

- Selection of a suitable transport channel: as shown in Figure 3.15, logical channels can be mapped onto different transport channels. User data of a DTCHcan for example be transferred either on a DCH or on the FACH. The selection of the transport channel can be changed by the network at any time during the connection in order to increase or decrease the speed of the connection.
- Multiplexing of data on common and shared channels: the FACH cannot only be used to transport RRC messages for different users but can also carry user data frames. The MAC layer is responsible for mapping all logical channels selected on a single transport channel and for adding a MAC header. The header describes, among other things, for which subscriber the MAC frame is intended. This part of the MAC layer is called MAC c/sh (common/shared).
- For dedicated channels the MAC layer is also responsible for multiplexing several data streams on a single transport channel. As can be seen in Figure 3.15, several logical user data channels (DTCH) and the logical signaling channel (DCCH) of a user are mapped onto a single transport channel. This permits the system to send user data and signaling information of the MM (mobility management), PMM (packet mobility management), CC (call control), and SM (session management) subsystems in parallel. This part of the MAC layer is called the MAC-d (dedicated).

Before the frames are forwarded to the physical layer, the MAC layer includes additional information in the header to inform the physical layer which transport format it should select for transmission of the frames over the air interface. This so-called transport format set (TFS) describes the combination of data rate, the TTI of the frame, and which channel coding and puncturing scheme to use.

For most channels, all layers described before are implemented in the RNC. The only exception is the physical layer, which is implemented in the Node-B. The Node-B therefore is responsible for the following tasks.

In order not to send the required overhead for error detection and correction over the Iub interface, channel coding is performed in the Node-B. This is possible as the header of each frame contains a TFS field that describes which channel encoder and puncturing scheme is to be used. UMTS uses the half-rate convolutional decoder already known from GSM as well as a new 1/3 rate and turbocode coder for very robust error correction. These coders double or even triple the number of bits. It should be noted that puncturing is used to remove some of the redundancy again before the transmission to adapt the data to the fixed frame sizes of the air interface. Afterwards, the physical layer performs the spreading of the original data stream by converting the bits into chips, which are then transferred over the air interface.

Finally, the modulator converts the digital information into an analog signal, which is sent over the air interface. Quadrature phase shift keying (QPSK) modulation is used for the UMTS Release 99 air interface which transmits two chips per transmission step. This is done in the Node-B in the downlink direction by sending one chip over the complex I-path and a second chip over the complex Q-path. As each path uses a fixed transmission rate of 3.84 MChips/s, the total data rate of the transmission is 2×3.84 MChips/s. The DPDCH and the DPCCH, which only use a small percentage of the frame especially for low spreading factors, are thus time multiplexed in the downlink direction as shown in Figure 3.20.

For the uplink direction, which is the direction from the terminal to the network, a slightly different approach was chosen. As in the downlink direction, QPSK modulation is used. Instead of multiplexing user and signaling data over both the I- and Q-path, user data is only sent on the I-path in the uplink (see Figure 3.21). The Q-path is used exclusively for the

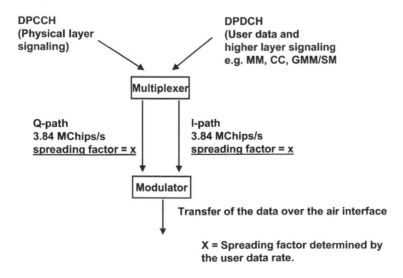

Figure 3.20 User data transmission in downlink direction via the complex I- and Q-path

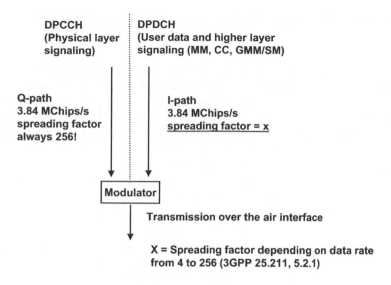

Figure 3.21   User data transmission via the I-path only

transmission of the DPCCH, which carries layer 1 messages for power control (see 3GPP 25.211, 5.2.1 [2]). Thus, only one path is used for the transmission of user data in the uplink direction. This means that for an equal bandwidth in the uplink and downlink directions, the spreading factor in the uplink direction is only half that of the downlink direction.

Note: The DPCCH is only used to transmit layer 1 signaling for power control. Control and signaling information of the MM, PMM, CC, and SM subsystems, which are exchanged between the terminal and the MSC or SGSN, are not transferred over the DPCCH but use the logical DCCH. This channel is sent together with the logical DTCH (user data) in the DPDCH transport channel (see Figures 3.14, 3.19 and 3.20).

The decision to use only the I-path for user data in the uplink direction was made for the following reason: while a dedicated channel has been assigned, there will be times in which no user data has to be transferred in the uplink direction. This is the case during a voice call for example while the user is not talking. During packet calls it also happens quite frequently that no IP packets have to be sent in the uplink for some time. Thus switching off the transmitter during that time could save battery capacity. The disadvantage of completely switching off the uplink transmission path, however, is the fact that the interference caused by this can be heard for example in radio receivers which are close by. This can be observed with GSM mobile phones, which use only some time slots on the air interface and thus have to frequently activate and deactivate the transmitter. In UMTS, only the transmission on the I-path is stopped while the DPCCH on the Q-path continues to be transmitted. This is necessary, as power control and signal quality information need to be sent even if no user data is transferred in order to maintain the channel. The transmission power is thus only reduced and not completely switched off. The typical interference of GSM mobile phones in radio receivers that are close to the device thus cannot be observed with a UMTS terminal.

3.5 The UMTS Terrestrial Radio Access Network (UTRAN)

3.5.1 Node-B, Iub Interface, NBAP, and FP

The base station, called Node-B in the 3GPP standards, is responsible for all functions required for sending and receiving data over the air interface. This includes, as shown in Section 3.4, channel coding, spreading and de-spreading of outgoing and incoming frames as well as modulation. Furthermore, the Node-B is also responsible for the power control of all connections. The Node-B only receives a transmission quality target from the RNC for each connection and then it decides if it is necessary to increase or decrease the transmission power in both the uplink and downlink directions to meet the target.

The size and capacity of a Node-B are variable. For locations with a high amount of data and voice calls, the Node-B is used in a sectorized configuration. This means that the 360-degrees coverage area of a Node-B is divided into several independent cells, each covering a certain area. Each cell has its own cell-ID, scrambling code, and OVSF tree. Each cell also uses its own directional antennas which cover either 180 degrees (two-sector configuration), 120 degrees (three-sector configuration), or only 90 degrees (four-sector configuration) in very densely populated areas. The capacity of the Iub interface, which connects the Node-B to an RNC, depends mainly on the number of sectors of the Node-B. While GSM only uses some of the 64 kbit/s timeslots on an E-1 link to the base station, UMTS base stations require a much higher bandwidth. In order to deliver high data rates, a Node-B needs to be connected to the RNC by at least one E-1 connection (2 Mbit/s). If a Node-B serves several sectors, multiple E-1 links are required in order to support the available data rates of the air interface.

For regions with low voice and data traffic, a Node-B with only a single cell is used. In order to decrease costs, transmission power is increased to cover a larger omnidirectional geographical area. From the outside, it is not always possible to distinguish between different configurations, as even a single cell can use a sectorized antenna installation. In the downlink direction, the signal is sent over all antennas and thus the total capacity of the base station is identical to a configuration which uses an omnidirectional antenna. In uplink direction, a sectorized antenna installation has the big advantage that the signal coming from the subscriber can be much better received due to the much higher antenna gain of a sectorized antenna.

For very dense traffic areas such as streets in a downtown area, a Node-B microcell can be an alternative to a sectorized configuration. A microcell is usually only equipped with a single antenna which covers a very small area such as several hundred meters of a street. As the necessary transmission power for such a small coverage area is very low, most network vendors have specialized micro Node-Bs with very compact dimensions. Usually, such micro Node-Bs are not much bigger than a PC workstation.

For the exchange of control and configuration messages on the Iub interface, the Node-B application part (NBAP) is used between the RNC and the Node-B. It has the following tasks:

- cell configuration;
- common channel management;
- dedicated channel management such as the establishment of a new connection to a subscriber;

- forwarding of signal and interference measurement values of common and dedicated channels to the RNC;
- control of the compressed mode which is further explained in Section 3.7.1.

User data is exchanged between the RNC and the Node-Bs via the frame protocol (FP), which has been standardized for dedicated channels in 3GPP 25.427 [5]. It is responsible for the correct transmission and reception of user data over the Iub interface and transports user data frames in a format that the Node-B can directly transform into a Uu (air interface) frame. This is done by evaluating the TFI which is part of every FP frame. The TFI among other things instructs the Node-B to use a certain frame length (e.g. 10 milliseconds) and which channel coding algorithm to apply.

The frame protocol is also used for the synchronization of the user data connection between the RNC and the Node-B. This is especially important for the data transfer in the downlink direction, as the Node-B has to send an air interface frame every 10, 20, 40, or 80 milliseconds to the terminal. In order not to waste resources on the air interface and to minimize the delay it is necessary that all Iub frames arrive at the Node-B in time. To ensure this, the RNC and Node-B exchange synchronization information at the setup of each connection and also when the synchronization of a channel has been lost.

Finally, FP frames are also used to forward quality estimates from the Node-B to the RNC. These help the RNC during a soft handover state of a dedicated connection to decide which Node-B has delivered the best data frame for the connection. This topic is further discussed in Secton 3.7.1.

3.5.2 The RNC, Iu, Iub, and Iur Interfaces, RANAP and RNSAP

The heart of the UMTS radio network is the RNC. As can be seen in Figures 3.22 and 3.23, all interfaces of the radio network are terminated by the RNC.

In the direction of the mobile subscriber the Iub interface is used to connect several dozen Node-Bs to an RNC. For the first few years after the initial deployment of the network, most Node-Bs will be connected to the RNC via one or more dedicated 2 Mbit/s E-1 connections

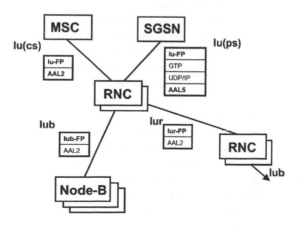

Figure 3.22 RNC protocols and interfaces for user data (user plane)

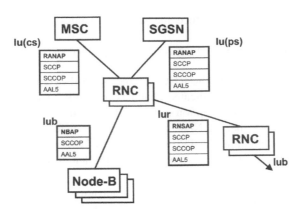

Figure 3.23 RNC protocols and interfaces used for signaling (control plane)

either via fixed-line or microwave links. The number of links used per Node-B mainly depends on the number of sectors and number of frequencies used. It is foreseeable that over time, and especially with the introduction of HSDPA, which is further discussed at the end of this chapter, bandwidth requirements will increase. Thus, Node-Bs will more and more be connected via optical or microwave STM-1 links (155 Mbit/s) in the future, which will use ATM or alternatively IP as a transport protocol. While ATM is already used for the 2 Mbit/s E-1 connections between the Node-B and RNC since the initial deployment of the network, using IP poses a greater challenge due to the non-real-time behavior of the protocol. The benefit of using IP instead of ATM for this interface, however, could be a substantial cost reduction in many cases, as IP connections are widely available for a much lower price than ATM connections, especially in metropolitan areas. As the cost of transmission links in the UTRAN make up a large portion of the overall operating expenditure (OPEX) of a wireless network, this reduction can have a big impact on the total operating cost of the network.

The RNC is connected to the core network via the Iu interface. As has been shown in Figure 3.3, UMTS Release 99 continues to use independent circuit-switched and a packet-switched networks for the following purposes.

For voice and video telephony, the circuit-switched core network, which is already known from GSM, continues to be used. The mobile switching center (MSC) therefore remains the bridge between the core and access network. Due to the new functionalities offered by UMTS such as video telephony, a number of adaptations were necessary on the interface which connects the MSC to the radio network. While in GSM, the transcoding and rate adaptation unit (TRAU) was logically part of the radio network, it was decided to put this functionality into the core network for UMTS. This was done due to the fact that even in GSM, the TRAU is physically located near the MSC to save transmission resources as described in Section 1.7.5. The interface between the MSC/TRAU and RNC has been named Iu(cs), which indicates that this interface connects the radio network to the circuit-switched part of the core network. The Iu(cs) interface therefore corresponds to the GSM A-interface and reuses many functionalities on the higher layers for mobility management and call control.

The BSSMAP protocol, which is used on the GSM A-interface, has been enhanced and modified for UMTS and renamed the radio access network application part (RANAP). In the standards, RANAP, which is described in 3GPP TS 25.413 [6], forms the basis for mobility management, call control, and the session management. Furthermore, RANAP is used by the MSC and SGSN for requesting the establishment and clearing of radio bearers (RABs) by the RNC.

In order to support RANAP with an already existing GSM MSC, a software update and new hardware for the TRAU is necessary. As the A-interface is still part of the MSC, a single switching center can be used with both the UTRAN (via the Iu(cs) interface) and the GSM radio network (via the A-interface).

With GSM and the A-interface, the MSC can only handle 12.2 kbit/s circuit-switched connections for voice calls and 9.6 or 14.4 kbit/s channels for data calls. With UMTS and the Iu(cs) interface, the MSC is also able to establish 64 kbit/s circuit-switched connections to the RNC, which equals the speed of an ISDN B-channel. This functionality is mainly used for video telephony. By using optimized video and speech compression algorithms, which are part of the H.323M (mobile) standard, this bearer ideally fulfills the tough requirements of this service concerning guaranteed bandwidth, delay time, and optimized handling of handovers in the UTRAN. The Iu(cs) interface is, like all other interfaces in the UTRAN, based on ATM and can thus use a number of different transmission technologies. In operational networks, STM-1 (155 Mbit/s) connections are commonly used.

All packet-switched services, which in most cases require a connection to the Internet, are routed to and from the core network via the Iu(ps) interface. The functionality of this interface corresponds to the GSM/GPRS Gb interface that has been described in Chapter 2. Some network vendors offer SGSNs that support both the Gb and Iu(ps) interface in a single node, which allows the use of only a single SGSN in a region to connect both types of radio network to the packet-switched core network. If this functionality is not supported, two SGSNs are necessary to cover the two radio networks of a region. As the GGSN in most cases will be able to handle both GPRS and UMTS PDP contexts, this is not a big disadvantage.

Similarly to the Iu(cs) interface, the higher layer signaling protocols were reused for UMTS and only slightly enhanced for the new capabilities of the radio network. On lower layers, however, ATM and IP are used instead of the old frame relay protocol.

The handling of user data has also changed significantly for the SGSN with UMTS. In the GSM/GPRS system, the SGSN is responsible for processing incoming GTP packets from the GGSN and converting them into a BSSGP frame for transmission to the correct PCU and vice versa. In UMTS, this is no longer necessary as the SGSN can forward the GTP packets arriving from the GGSN directly to the RNC via an IP connection and can send the GTP packets it receives from the RNC to the GGSN. The UMTS SGSN is thus no longer aware in which cell a subscriber is currently located. This change was made mainly for two reasons:

- The SGSN has been logically separated from the radio network and its cell-based architecture. It merely needs to forward GTP packets to the RNC, which then processes the packets and decides to which cell(s) to forward them. This change is especially important for the implementation of the soft handover mechanism, which is further described in Section 3.7.1, as the packet can be sent to a subscriber via several Node-Bs simultaneously. This complexity, however, is concealed from the SGSN as it is a pure radio network issue which is outside the scope of a core network node. As a consequence, a UMTS SGSN is only aware of the current serving RNC (S-RNC) of a subscriber.

- Using GTP and IP on the Iu(ps) interface on top of the ATM transport layer substantially simplifies the protocol stack when compared to GSM/GPRS. The use of GTP and IP via ATM and ATM adaptation layer (AAL) 5 is also shown in Figure 3.22.

The SGSN is still responsible for the mobility and session management (GMM/SM) of the subscribers as described in Chapter 2. Only a few changes were made to the protocol to address the specific needs of UMTS. One was made to allow the SGSN to request the setup of a radio bearer when a PDP context is established. This concept is not known in GSM/GPRS as 2G subscribers do not have any dedicated resources on the air interface. As has been shown in Chapter 2, GPRS users are only assigned a certain number of timeslots for a short time, which are shared or immediately reused for other subscribers once there is no more data to transmit.

In the Release 99 UTRAN a different concept is used: here, the RNC can assign a dedicated radio bearer (RAB) for a packet-switched connection in a very similar way as for circuit-switched voice calls. On the physical layer this means that the user gets its own DPDCH and DPCCH for the packet connection. The bandwidth of the channel remains assigned to the subscriber even if not fully used for some time. When no data has to be sent in the downlink direction, discontinuous transmission (DTX) is used as described in Section 3.5.4. This reduces interference in the cell and helps the terminal to save energy. The RNC can select from different spreading factors during the setup of the connection to establish bearers with a guaranteed bandwidth of 8, 32, 64, 128, or 384 kbit/s. Later on, the RNC can then change the bandwidth at any time by assigning a different spreading factor to the connection which is useful, for example, if the provided bandwidth is not sufficient or not fully used by the subscriber for some time. As the standard is very flexible in this regard, different network vendors have implemented different strategies for the radio resource management. Section 3.9.2 thus compares the radio resource management strategies of two operational networks.

Packet-switched data and a circuit-switched voice or video call can be transmitted over a single dedicated radio bearer. Thus, one of the big limitations of GSM/GPRS, which exists in most networks to this date, has been resolved. In order to transmit several data streams for a user simultaneously, the RNC has to be able to modify a radio bearer at any time. If a circuit-switched voice call is added to an already existing packet-switched data connection, the RNC modifies the RAB to accommodate both streams. It is of course also possible to add a packet-switched data session to an ongoing circuit-switched voice call. It should be noted that during an ongoing circuit-switched voice connection the speed for the packet-switch connection is limited to 64 or 128 kbit/s. In practice this means that in the case where a 384 kbit/s bearer is established for a packet call and the user wants to place a phone call, the bandwidth for the packet data is reduced to be able to carry the voice traffic over the same bearer as well.

Additionally a dedicated channel has the following advantages:

- A cell change can be controlled by the network. Thus no interruptions in the data flow occur when changing a cell as is the case with GPRS. Together with the much higher data rates offered by the UTRAN it is possible to also hand over services like video streaming and IP video telephony without any impact to the service.
- No resources need to be assigned for sending data in the up- or downlink direction if a dedicated channel is used. Therefore, data can be sent or received with only minimal

delay, which noticeably improves the web-surfing experience compared to GPRS. There, long delay has a noticeable impact on the time it takes for the browser to get enough data to be able to display the first parts of the page.

- If a subscriber does not use the dedicated channel for some time, there are several possibilities to modify the radio bearer. One possibility is to assign a longer spreading code to the connection. This reduces the maximum speed of the user but allows the network to reuse the short spreading code for other users. If the subscriber resumes data transmission, the radio bearer is still established and thus no delay is experienced. The network can then quickly modify the bearer and again assign a shorter spreading code.

A second option for packet-switched data transfer over the air interface is to send data to the user via the FACH. For data in the uplink direction, the RACH is used. As these are shared channels, they are not exclusively assigned to a single subscriber but are potentially used by many terminals simultaneously. This is an interesting concept as the primary role of those channels is to carry signaling information for radio bearer establishments. As the capacity of those channels is quite limited and also has to be shared, the use of common channels only makes sense for small amounts of data or as a fallback if no data has been transferred over a dedicated connection for some time. Another disadvantage of using common channels is that the terminal cannot use the soft handover procedure when moving into a new cell and thus no seamless mobility is ensured (see Section 3.7.1). Therefore whenever the network detects that the amount of data transferred to or from a terminal increases again, a dedicated connection is quickly re-established.

Finally, there is a third option to transfer packet-switched data. UMTS Release 99 also defines a downlink shared channel (DSCH), which is a similar to using the FACH. Compared to the FACH, however, the DSCH is only used for packet data and not for signaling and can be used with variable spreading codes to adapt the channel bandwidth to the current traffic load. Furthermore, power control is used to control the power output of the terminals, which reduces interference in the cell. While most operational networks today use both the dedicated channels as well as the FACH for data transfer, the use of a DSCH has not become very popular. This is probably due to the fact that it was soon realized that a dedicated 384 kbit/s bearer would not be sufficient in the long run to satisfy user demand. This triggered the standardization of an enhancement of the DSCH concept, which is now referred to as high speed downlink packet access (HSDPA). As HSDPA introduces a number of new concepts, which lead away from the dedicated channel approach of UMTS Release 99, this topic is covered in more detail in Section 3.10.

Independent of whether a dedicated, common, or shared channel is assigned at the request of the SGSN during the PDP context activation, the bandwidth of the established connection depends on a number of factors. Important factors for dedicated radio bearers are for example the current load of a cell and the reception conditions of the terminal at its current location. Furthermore, the number of available spreading codes and distance of the terminal to the Node-B also influence the length of the spreading code. If the interference level in the cell is already very high, only the assignment of a long spreading code might ensure reliable communication, which of course limits the data rate.

The terminal can also influence the assignment of radio resources during establishment of a PDP context. By using optional parameters of the 'at+cgdcont' command (see Section 2.9) the application can ask the network to establish a connection for a certain quality of service

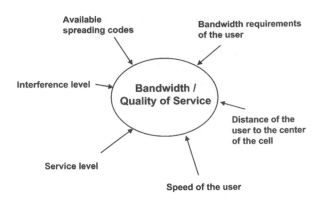

Figure 3.24 Factors influencing the quality of service and the maximum bandwidth of a connection

(QoS) level. The QoS describes properties for a new connection like the minimal acceptable data rate or the maximum delay time allowed that the network has to guarantee throughout the duration of the connection. It is also possible to use different access point names (APNs) to let the network automatically assign the correct QoS settings to a connection. The HLR therefore stores a QoS profile for each user that defines which APNs a user is allowed to use and which QoS level a user is allowed to request for a new connection. See Figure 3.24.

The assignment of resources on the air interface can also be influenced by the service level that is assigned to a user. This allows a network operator to allocate privileges to some users, who for example pay a higher subscription fee each month and expect higher data rates to be reserved for them. Another way for network vendors to use the subscription level concept is to allocate users with a higher service level during cell congestion situations and assign them shorter spreading codes while others only receive a longer code. In practice, it can be observed that most operators do not make use of this functionality today.

The Iur interface completes the overview of the UTRAN interfaces for this section. This interface connects RNCs with each other in order to be able to perform the soft handover procedure with the Node-Bs that are connected to different RNCs. Further information about this topic can be found in Section 3.7.1. Furthermore, the Iur interface allows the maintenance of a packet-switched connection that is currently in the Cell-FACH, Cell-PCH or URA-PCH state if the functionality is implemented in the network. More about the different connection states can be found in Section 3.5.4. The protocol responsible for these tasks is called the radio network subsystem application part (RNSAP).

3.5.3 Adaptive Multi Rate (AMR) Codec for Voice Calls

For UMTS it was decided to use the AMR codec for voice encoding, which has been introduced as an optional voice codec in Chapter 1 for GSM. As AMR is mandatory for UMTS networks, it is used in all UMTS networks while only a few GSM networks make use of its advanced possibilities. GSM networks mostly still use the full rate (FR) or enhanced full rate (EFR) codecs that transcode a 64 kbit/s voice data stream of the core network into a 12 kbit/s data stream for the radio network. This rate adaptation is necessary due to the limited resources on the air interface. At the beginning of a voice call the network and terminal agree on a codec, which is then used for the entire duration of the call. The base station adds error correction and detection bits to the incoming data stream which result in an overall data rate of about 22 kbit/s. UMTS on the other hand took advantage of the increase in processing power in both terminal and network nodes which allows the use of even more efficient codecs that require an even lower bit rate than the FR or EFR codecs. With AMR the codec is no longer negotiated at the establishment of a voice call the system can now change the codec every 20 milliseconds. As the name adaptive multi rate already suggests this functionality is quite useful in adapting to a number of changes that can occur during the lifetime of a call.

If the reception quality deteriorates during a call, the network could decide to use a voice codec with a lower bit rate. If the spreading factor of the connection is not changed, more bits of the bearer can then be used to add additional redundancy. A lower bit rate codec naturally lowers the quality of the voice transmission, which is, however, still better than a reduced voice quality due to an increased error rate. If the reception quality increases again during the connection, AMR returns to a higher bit rate codec and decreases the number of redundancy bits again.

Another application of AMR could be to increase the number of possible simultaneous calls in a cell during cell congestion. In this case a higher spreading factor is used for a connection which only allows the lower bit rate AMR codes to be used. This reduces the voice quality for the subscriber but increases the number of simultaneous voice calls.

Table 3.3 gives an overview of the different AMR codecs that have been standardized in 3GPP TS 26.071 [7]. While UMTS terminals have to support all bit rates it is optional for the network.

3.5.4 Radio Resource Control (RRC) States

The activity of a subscriber determines in which way data is transferred over the air interface between the terminal and the network. In UMTS, a terminal can therefore be in one of five radio resource control (RRC) states (see Figure 3.25).

Idle state: in this state a terminal is attached to the network but does not have a physical or logical connection with the radio network. This means that the user is neither involved in a voice call nor in a data transfer. From the packet-switched core network point of view the subscriber might still have an active PDP context (i.e. an IP address) even if no radio

Table 3.3 AMR codecs and bit rates

Codec mode	Bit rate
AMR_12.20	12.20 kbit/s (GSM EFR)
AMR_10.20	10.20 kbit/s
AMR_7.95	7.95 kbit/s
AMR_7.40	7.40 kbit/s (IS-641)
AMR_6.70	6.70 kbit/s (PDC-EFR)
AMR_5.90	5.90 kbit/s
AMR_5.15	5.15 kbit/s
AMR_4.75	4.75 kbit/s

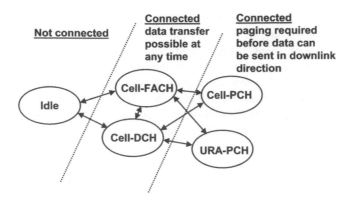

Figure 3.25 Radio resource control (RRC) states

resources are assigned at the moment. Due to the user's period of inactivity the radio network has decided to release the radio connection. This means that if the user wants to send some data again (e.g. request a new web page) the terminal needs to establish a new radio bearer in either the Cell-DCH or Cell-FACH state, which are both described below.

The Cell-DCH RRC state is used similarly to the GSM dedicated mode for circuit-switched voice calls. While in this state, a physical connection is established between the terminal and the network. In the UTRAN this means that the terminal has been assigned its own spreading code in the downlink direction and its own spreading and scrambling codes in the uplink direction.

The Cell-DCH state is also used for packet-switched connections. At first, this contradicts the packet-switched approach. The advantage of packet-switched connections is normally the fact that resources are only dedicated to a logical connection while they are needed to transfer data. In the Cell-DCH state, however, resources are not immediately freed once there is no more data to be transferred. The CDMA system offers an elegant solution to this. If a subscriber does not send or receive data for some time, only control information is sent over the established channel. Other subscribers benefit indirectly from this due to the reduced overall interference level of the cell during such periods. If new data arrives, which has to be sent over the air interface, no new resources have to be assigned as the dedicated channel is still established. Once data is sent again the interference level increases for other subscribers in the cell. This effect can be observed for example during web-surfing sessions. For this application the page requests of different subscribers are statistically multiplexed which thus helps to counter the effect of the dedicated bearer approach without the negative effects of resource assignments as is the case in a GSM/GPRS radio network. See Figure 3.26.

By using signal measurements of the terminal and the Node-B it is possible to control the power level of each terminal in a cell, which is a task that is shared between the Node-B and the RNC. By using the downlink of the PDCCH the network is able to instruct the terminal to adapt its transmission power to the current conditions 1500 times a second. The rate at which power control is performed also shows the importance for UMTS, as interference is the major limiting factor of the number of connections that can be established simultaneously in a cell.

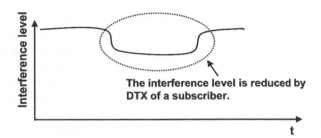

Figure 3.26 Discontinuous transmission (DTX) on a dedicated channel reduces the interference for other subscribers

While in the Cell-DCH state, the mobile continuously measures the reception quality of neighboring cells and reports the results to the network. Based on these values the RNC can then decide to start a handover procedure when required. While the GSM radio network uses a static reporting interval, a much more flexible approach was selected for UMTS. In the first step the RNC instructs the terminal, similar to the GSM approach, to send periodic measurement reports. The measurement interval itself is now flexible and can be set by the network between 0.25 and 64 seconds. Furthermore, the network can also instruct the terminal to send measurement reports only if certain conditions are met. This way, it is possible to send measurement reports for neighboring cells to the network only if the measurement values reach a certain threshold. This removes some signaling overhead, which can be used to send more user data over the bearer instead. Another advantage of this method for the RNC is the fact that it has to process fewer messages for each connection compared to periodical measurement reports.

Depending on the requirements of the data to be sent, different properties can be assigned to a dedicated channel. One property for example is the length of the spreading code, which affects the maximum bandwidth available for user data. Therefore, depending on the length of the spreading code, data rates in the range of a few kilobits per second up to several hundred kilobits per second are possible for a single connection (see Section 3.3.2).

While the Idle and Cell-DCH RRC states are mandatory for the network all other states like the Cell-FACH, Cell-PCH, and URA-PCH states, which are further described below, are optional. The Cell-FACH state is mainly used when only a small amount of data needs to be transferred to or from a subscriber. In this mode, the subscriber does not get a dedicated channel but uses the FACH to receive data. As described in Section 3.4.5, the FACH's primary task is to carry RRC connection setup messages for subscribers that have requested access to the network via the RACH. If the Cell-FACH state is implemented in the network the channel can also be used to send user data or signaling messages from the MSC and SGSN to the terminals. The FACH is a 'common channel' as it is not exclusively assigned to a single user. Therefore, the MAC header of each FACH data frame has to contain a destination ID, which consists of the S-RNTI (serving-radio network temporary ID) which was assigned to a terminal during connection establishment, and the ID of the S-RNC. The terminals therefore have to inspect the header of each FACH data frame and only forward those frames to higher layers of the protocol stack that contain the terminal's ID. The approach of Cell-FACH RRC state is thus similar to Ethernet (802.11) and GSM/GPRS for packet-switched data transmission. If data is received in the downlink direction, no resources have to be assigned and the data can be sent to the subscriber more or less quickly depending on the current traffic load of the FACH. As several subscribers share the same channel, the network cannot ensure a certain data rate and constant delay times for any terminal in Cell-FACH state. Furthermore, it should be noted that the FACH usually uses a high spreading factor which limits the total available bandwidth for subscribers on this channel. See Figure 3.27.

Compared to the Cell-DCH state in which the mobility of the subscriber is controlled by the network no such control has been foreseen for the Cell-FACH state. In the Cell-FACH state the terminal is responsible for changing cells which is therefore called cell update instead of handover. As the network does not control the cell update it is also not possible to ensure an uninterrupted data transfer during the procedure. Due to these reasons the Cell-FACH RRC state is not suited for real-time or streaming applications. For bursty and low-speed data transmission such as WAP browsing, the Cell-FACH state is an alternative to the establishment of a dedicated bearer. As the displays of mobile devices are usually quite small the amount of data that has to be transferred for a WAP page is usually also quite small. Therefore, a dedicated transmission channel is not strictly necessary. In operational networks, it can be observed that a dedicated channel is established even for WAP browsing but is released again very quickly after the data transfer. More about the use of the different RRC states in operational networks can be found in Section 3.9.2.

The Cell-FACH state is also suitable for the transmission of mobility management and packet mobility management signaling messages between the terminal and the MSC or SGSN. As the terminal already indicates the reason for initiating the connection to the network in the RRC connection setup message, the network can flexibly decide if a dedicated channel is to be used for the requested connection or not. If the Cell-FACH channel is implemented in the network and used for signaling exchanges, no dedicated channel has to be assigned for example for a location update procedure.

If the terminal is in Cell-FACH state, uplink data frames are sent via the RACH whose primary task is to forward RRC connection setup request messages. As has been shown in Section 3.4.5, access to the RACH is a time intensive procedure which causes some delay before the actual data frame can be sent. This is another reason why the Cell-FACH state is not suited for real-time applications.

There are two possibilities for a terminal to change to the Cell-FACH state. As already discussed, the network can decide during the RRC connection setup phase to use the FACH

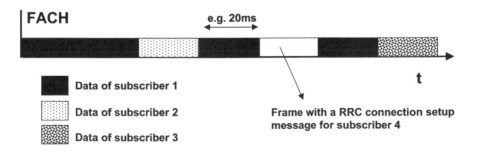

Figure 3.27   Data of different subscribers is time multiplexed on the FACH

for MM/PMM signaling or user data traffic. Furthermore, it is possible to enter the Cell-FACH state from the Cell-DCH state. The RNC can decide to modify the radio bearer this way if, for example, no data has been sent or received by the terminal for some time. The spreading code which is thus released can then immediately be used for another subscriber. Furthermore, a fallback to the Cell-FACH state reduces the power consumption of the terminal. As long as only small amounts of data are exchanged, the Cell-FACH state is usually maintained. If the data volume increases again, the network can immediately establish a new dedicated bearer and instruct the terminal to enter Cell-DCH state to be able to transfer data more quickly.

The optional Cell-PCH (cell-paging channel) RRC state and the URA-PCH (UTRAN registration area – paging channel) RRC state can be used to reduce the power consumption of the terminal even further during extended times of inactivity. Similar to the Idle state, no resources are assigned to the terminal. If data arrives for a subscriber from the network, the terminal needs to be paged first. The terminal then answers the paging request with an RRC connection request message which allows the RNC to establish a new connection. Depending on the decision of the RNC, the terminal then either changes to the Cell-FACH or Cell-DCH state.

As the name Cell-PCH already indicates, the subscriber is only paged in a single cell if new data from the core network arrives. This means that the mobile station has to send a cell update message to the RNC whenever it selects a new cell. In the URA-PCH state, the mobile only informs the RNC whenever it enters a new UTRAN registration area (URA). Consequently the paging message needs to be sent to all cells of the URA in case of incoming data (see Section 3.7.3).

The difference between the Cell-PCH and URA-PCH state compared to the Idle state is that the network and terminal still maintain a logical connection. As the RRC states are managed by the RNC, the SGSN as a core network component has no information on the RRC state of the terminal. Therefore, the SGSN simply forwards all incoming data packets from the GGSN to the RNC regardless of the current state of the mobile. If the mobile is currently in either the Cell-PCH or the URA-PCH state the RNC needs to buffer the packets, page the terminal, wait for an answer, and then establish a physical connection to the terminal again. If the terminal is in the Cell-DCH or Cell-FACH state the RNC can directly forward any incoming packets. The distinction between a logical and physical connection has been made in order to separate the connection between the terminal and core network (SGSN and MSC) on the one hand and the connection between the terminal and the radio network (RNC) on the other hand. The advantage of this concept is the decoupling of the MSC and SGSN from the properties and functionality of the radio network. Thus, it is possible to evolve the radio network and core network independently from each other.

In an operational network the difference between the Idle, Cell-PCH, and URA-PCH is very small from a user point of view. Both the power consumption of the terminal as well as the resumption of a data transfer are only slightly different. Therefore, it is questionable if the Cell-PCH and URA-PCH states will ever be implemented. At the time this book was published, only the Idle state, the Cell-DCH state, and the Cell-FACH state were used in operational networks.

As described in Chapter 2, the GSM/GPRS SGSN is aware of the state of a terminal as the Idle, Ready, and Standby states as well as the Ready timer is administered by the SGSN. Thus, a core network component performs tasks of the radio network such as cell

Table 3.4   RNC and SGSN states

	RNC state	SGSN state
Idle	Not connected	Not connected
Cell–DCH	Connected, data is sent via the DCH or HS-DSCH	Connected
Cell-FACH	Connected, incoming data is sent immediately via the FACH (common channel)	Connected
Cell-PCH	Connected, but subscriber has to be paged and needs to reply before data can be forwarded. Once the answer to the paging has been received the subscriber is put in either the Cell-FACH or Cell-DCH state	Connected
URA-PCH	Same as Cell-PCH. Furthermore, the network only needs to be informed of a cell change if the terminal is moved into a cell which is part of a different UTRAN registration area	Connected

updates. On the one hand this has the advantage that the SGSN is aware of the cell in which a subscriber is currently located, which can be used for supplementary location-dependent functionalities. The advantage of implementing the UMTS state management in the RNC is the distribution of this task on several RNCs and thus a reduction of the signaling load of the SGSN as well as a clear separation between core network and radio access network responsibilities. See Table 3.4.

3.6 Core Network Mobility Management

From the point of view of the MSC and the SGSN, the terminal can be in one of the following mobility management (MM) or packet mobility management (PMM) states.

The MSC knows the following MM states:

- MM detached: the terminal is switched off and the current location of the subscriber is unknown. Incoming calls for the subscriber cannot be forwarded to the subscriber and are either rejected or forwarded to another destination if the call forward unreachable (CFU) supplementary service is activated.
- MM idle: the terminal is powered on and has successfully attached to the MSC (see Attach procedure). The subscriber can at any time start an outgoing call. For incoming calls, the terminal is paged in its current location area.
- MM connected: the terminal and MSC have an active signaling and communication connection. Furthermore, the connection is used for a voice or a video call. From the point of view of the RNC, the subscriber is in the Cell-DCH RRC state as this is the only bearer that supports circuit-switched connections.

The SGSN implements the following PMM states:

- PMM detached: the terminal is switched off and the location of the subscriber is unknown to the SGSN. Furthermore, the terminal cannot have an active PDP context, i.e. no IP address is currently assigned to the subscriber.
- PMM connected: the terminal and the SGSN have an active signaling and communication connection. The PMM connected state is only maintained while the subscriber has an active PDP context, which effectively means that the GGSN has assigned an IP address for the connection. In this state, the SGSN simply forwards all incoming data packets to the serving-RNC (S-RNC). In contrast to GSM/GPRS the UMTS SGSN is only aware of the S-RNC for the subscriber and not of the current cell. This is due to the desired separation of radio network and core network functionality and also to the soft handover mechanism (see Section 3.7). The SGSN is also not aware of the current RRC state of the terminal. Depending on the QoS profile, the network load, the current data transfer activity, and the required bandwidth, the terminal can be either in Cell-DCH, Cell-FACH, Cell-PCH or URA-PCH state.
- PMM idle: in this state, the terminal is attached to the network but no logical signaling connection is established with the SGSN. This can be the case for example if no PDP context is active for the subscriber. Furthermore, the RNC has the possibility to modify the RRC state of a connection at any time. This means that the RNC, for example, can decide after a period of inactivity of the connection to set the terminal into the RRC Idle state. As the RNC no longer controls the mobility of the subscriber it requests the SGSN to set the connection into PMM Idle state as well. Therefore, even though the subscriber no longer has a logical connection to either the RNC or the SGSN, the PDP context remains active and the subscriber can keep the assigned IP address. For the SGSN, this means that if new data arrives for the subscriber from the GGSN, a new signaling and user data connection has to be established before the data can be forwarded to the terminal.

3.7 Radio Network Mobility Management

Depending on the MM state of the core network, the radio network can be in a number of different RRC states. How the mobility management is handled in the radio network depends on the respective state. Table 3.5 gives an overview of the MM and PMM states in the core network and the corresponding RRC states in the radio network.

3.7.1 Mobility Management in the Cell-DCH State

For services like voice or video communication it is very important that no or only a very short interruption of the data stream occurs during a cell change. For these services, only the Cell-DCH state can be used. In this state the network constantly controls the quality of the connection and is able to redirect the connection to other cells if the subscriber is moving. This procedure is called handover or handoff. In UMTS a number of different handover variants have been defined.

Hard handover as shown in Figure 3.28: this kind of handover is very similar to the GSM handover. By receiving measurement results from the terminal of the active connection and measurement results of the signal strength of the broadcast channel of the neighboring cells,

Table 3.5 Core network and radio network states

MM states and possible RRC states	MM idle	MM connected	PMM idle	PMM connected
Idle	X		X	
Cell-DCH		X		X
Cell-FACH				X
Cell-PCH				X
URA-PCH				X

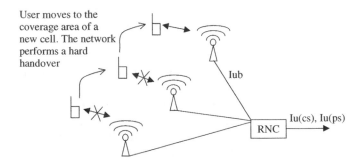

User moves to the coverage area of a new cell. The network performs a hard handover

Iub

Iu(cs), Iu(ps)

RNC

Figure 3.28 UMTS hard handover

the RNC is able to recognize if a neighboring cell is more suitable for the connection. In order to redirect the call into the new cell a number of preparatory measures have to be performed in the network before the handover is executed. This includes for example the reservation of resources on the Iub interface and if necessary also on the Iur interface. The procedure is similar to the resource reservation of a new connection.

Once the new connection is in place the terminal receives a command over the still established connection to change into the new cell. The handover command contains, among other parameters, the frequency of the new cell and the new channelization and scrambling code to be used. The terminal then suspends the current connection and attempts to establish a connection in the new cell. The interruption of the data stream during this operation is usually quite short and takes about 100 milliseconds on average, as the network is already prepared for the new connection. Once the terminal is connected to the new cell the user data traffic can resume immediately. This kind of handover is called UMTS hard handover as the connection is shortly interrupted during the process.

Soft Handover: with this kind of handover, user data traffic is not interrupted at any time during the procedure. Based on signal quality measurements of the current and neighboring cells, the RNC can decide to set the terminal into soft handover state. All data from and to the terminal will then be sent and received not only over a single cell but also over two or even more cells simultaneously. All cells that are part of the communication are put into the so-called active set of the connection. If a radio connection of a cell in the active set deteriorates, it is removed from the connection. Thus it is ensured that despite the cell change, the terminal never losses contact to the network. The active set can contain up to six cells at the same time although in operational networks no more than two or three cells are used at a time. Figure 3.29 shows a soft handover situation with three cells.

The soft handover procedure has a number of advantages over the hard handover described before. As no interruption of the user data traffic occurs during the handover procedure the overall connection quality increases. As the soft handover procedure can be initiated while the signal quality of the current cell is still acceptable the possibility of a sudden loss of the connection is reduced.

Furthermore, the transmission power and thus the energy consumption of the terminal can be reduced in some situations as shown in Figure 3.30. In this scenario, the subscriber first roams into an area in which it has a good coverage by cell 1. As the subscriber moves, there are times when buildings or other obstacles are in the way of the optimal transmission path to cell 1. As a consequence, the terminal needs to increase its transmission power.

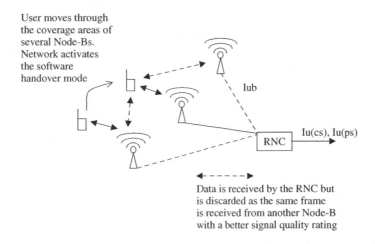

Figure 3.29 Connections to a terminal during a soft handover procedure with three cells

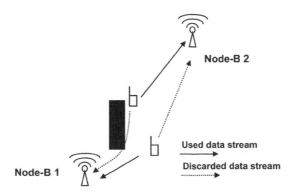

Figure 3.30 Soft handover reduces energy consumption of the mobile due to lower transmission power

If the terminal is in soft handover state, however, cell 2 still receives a good signal from the terminal and can thus compensate for the deterioration of the transmission path to cell 1. As a consequence, the terminal is not instructed to increase the transmission power. This does not mean, however, that the connection to cell 1 is released immediately, as the network speculates on an improvement of the signal conditions.

As the radio path to cell 1 is not released, the RNC receives the subscriber's data frames from both cell 1 and cell 2 and can decide, based on the signal quality information included in both frames, that the frame received from cell 2 is to be forwarded into the core network. This decision is made for each frame, i.e. the RNC has to make a decision for every connection in handover state every 10, 20, 40, or 80 milliseconds depending on the size of the radio frame.

In the downlink direction, the terminal receives identical frames from cell 1 and cell 2. As the cells use different channelization and scrambling codes the terminal is able to separate the two data streams on the physical layer. This means that the terminal has to decode the data stream twice, which of course slightly increases the power consumption as more processing power is required. See Figure 3.31.

From the network point of view, the soft handover procedure has the following advantages: as the terminal uses less transmission power compared to a single cell scenario in order to be able to reach at least one of the cells in the active set, the interference is reduced in the uplink direction. This increases the capacity of the overall system, which in turn increases the number of subscribers that can be handled by a cell.

On the other hand, there are some disadvantages for the network as well: in the downlink direction, data has to be duplicated so it can be sent over two or even more cells. In the reverse direction, the RNC receives a copy of each frame from all cells of the active set. Thus, the capacity that has to be reserved for the subscriber on the different interfaces of the radio network is much higher than for a subscriber that only communicates with a single cell. Therefore, good network planning tries to ensure that there are no areas of the network in which more than three cells need to be used for the soft handover state.

A soft handover gets even more complicated if cells need to be involved that are not controlled by the S-RNC. In this case, a soft handover is only possible if the S-RNC is connected to the RNC that controls the cell in question. RNCs in that role are called the drift RNCs (D-RNC). Figure 3.32 shows a scenario that includes an S-RNC and a D-RNC. If a foreign cell needs to be included in the active set, the S-RNC has to establish a link to

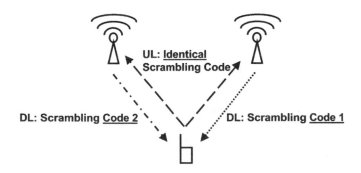

Figure 3.31 Use of scrambling codes while a terminal is in soft handover state

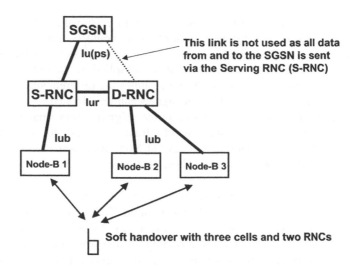

Figure 3.32   Soft handover with S-RNC and D-RNC

the D-RNC via the Iur interface. The D-RNC then reserves the necessary resources to its cell on the Iub interface and acknowledges the request. The S-RNC then in turn informs the terminal to include the new cell in its active set via an 'update active set' message. From this point onwards, all data arriving at the S-RNC from the core network will be forwarded via the Iub interface to the cells that are directly connected to the S-RNC and also via the Iur interface to all D-RNCs which control a cell of the active set. These in turn forward the data packets to the cells under their control. In the reverse direction, the S-RNC is the point of concentration for all uplink packets as the D-RNCs forward all incoming data packets for the connection to the S-RNC. It is then the task of the S-RNC to decide which of the packets to use based on the signal quality indications embedded in each frame.

A variation of the soft handover is the so-called softer handover, which is used when two or more cells of the same Node-B are part of the active set. For the network, the softer handover has the advantage that no additional resources are necessary on the Iub interface as the Node-B already decides which of the frames received from the terminal via the different cells to forward to the RNC. In the downlink direction, the point of distribution for the data frames is also the Node-B, i.e. it duplicates the frames it receives from the RNC for all cells which are part of the active set of a connection.

One of the most important parameters of the GSM air interface is the timing advance. Terminals that are further away from the base station have to start sending their frames earlier compared to terminals that are closer to the base station due to the time it takes the signal to reach the base station. This is called timing advance control. In UMTS controlling the timing advance is not possible. This is due to the fact that while a terminal is in soft handover state, all Node-Bs of the active set receive the same data stream from the terminal. The distance of the terminal to each Node-B is different and thus each Node-B receives the data stream at a slightly different time. For the terminal, it is not possible to control this by starting to send data earlier, as it only sends one data stream in the uplink direction for all Node-Bs. Fortunately, it is not necessary to control the timing advance in UMTS because all active subscribers are transmitting simultaneously. As no time slots are used, no collisions can occur between the different subscribers. In order to ensure the orthogonal nature of the channelization codes of the different subscribers it would be necessary, however, to receive the data streams of all terminals synchronously. As this is not possible, an additional scrambling code is used for each subscriber that is multiplied by the data that has already been treated with the channelization code. This decouples the different subscribers and thus a time difference in the arrival of the different signals can be tolerated.

The time difference of the multiple copies of a user's signal is very small compared to the length of a frame. While the transmission time of a frame is 10, 20, 40, or 80 milliseconds, the delay experienced on the air interface of several Node-Bs is less then 0.1 milliseconds even if the distances vary by 30 kilometers. Thus, the timing difference of the frames on the Iub interface is negligible.

If a subscriber continues to move away from the cell in which the radio bearer was initially established, there will be a point at which not a single Node-B of the S-RNC is part of the transmission chain. Figure 3.33 shows such a scenario. As this state is a waste of radio network resources, the S-RNC can request a routing change from the MSC and the SGSN on the Iu(cs)/Iu(ps) interface. This procedure is called a serving radio network subsystem (SRNS) relocation request. If the core network components agree to perform the change, the D-RNC becomes the new serving RNC and the resources on the Iur Interface can be released.

An SRNS relocation is also necessary if a handover needs to be performed due to degrading radio conditions and no Iur connection is available between two RNCs. In this case it is not the optimization of radio network resources that triggers the procedure but the need to maintain the radio bearer. Therefore not only is an SRNS relocation necessary but also a hard handover into the new cell, as a soft handover is not possible due to the missing Iur interface.

When the first GSM networks were built at the beginning of the 1990s, many earlier generation networks already covered most parts of the country. The number of users was

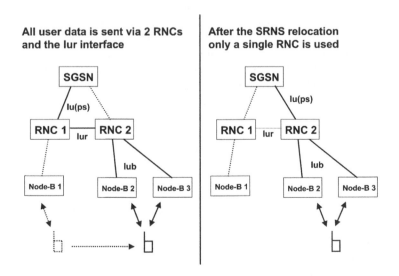

Figure 3.33   SRNS relocation procedure

very small and it was not immediately necessary to reach the same coverage area with GSM as well. When the first UMTS networks became operational, the situation had changed completely. Due to the enormous success of GSM, most people in Europe already possess a mobile phone. As network deployment is a lengthy and costly process it was therefore not possible to ensure the same countrywide coverage for UMTS right from the start. Therefore, it was necessary to ensure a seamless integration of UMTS into the already existing GSM infrastructure. For the design of UMTS mobile phones this meant that right from the beginning the phone also had to support GSM and GPRS. Thus, while a user roams in an area covered by UMTS, both voice calls and packet data are handled by the UMTS network. If the user roams into an area which is only covered by a 2G network, the mobile phone would automatically switch over to GSM and packet-switched connections would use the GPRS network. In order not to interrupt ongoing voice or data calls, the UMTS standards also include procedures to allow handing over an active connection to a 2G network (Figure 3.34). This handover procedure is called intersystem handover.

In UMTS there are a number of different possibilities to perform an intersystem handover.

The first intersystem handover method is the blind intersystem handover. For this case, the RNC is aware of GSM neighboring cells for certain UMTS cells. In the event of severe signal quality degradation, the RNC reports to the MSC or SGSN that a handover into a 2G cell is necessary. The procedure is called a 'blind handover', as no measurement reports of the GSM cell are available for the handover decision.

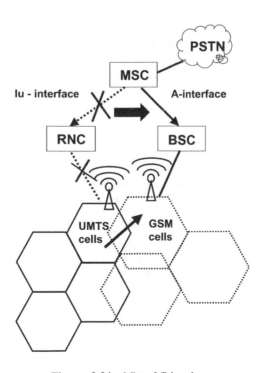

Figure 3.34   3G to 2G handover

The advantage of this procedure, of course, is simple implementation in the network and in the terminals. However, there are a number of problems linked to a blind intersystem handover:

- The network has no information if the GSM cell can be received by the terminal.
- The terminal and the target GSM cell are not synchronized. This considerably increases the time it takes for the terminal to contact the new cell once the handover command has been issued by the network. For the user this means that during a voice call he might notice a short interruption of the voice path.
- If a UMTS cell has several GSM neighboring cells, as shown in Figure 3.35, the RNC cannot make a good decision into which cell to hand over the subscriber. Thus, such a network layout should be avoided. In practice, however, this is often not possible.

In order to improve the success rate and quality of intersystem handovers, the UMTS standards also contain a controlled intersystem handover procedure. To perform a controlled handover, UMTS cells at the border of the coverage area inform terminals about both UMTS and GSM neighboring cells. A terminal can thus measure the signal quality of neighboring cells of both systems during an active connection. As described before, there are several ways to report the measurement values to the RNC. The RNC in turn can then decide to request an intersystem handover from the core network based on current signal conditions rather than purely guessing that a certain GSM cell is suitable for the handover.

Performing neighboring cell signal strength measurements is quite easy for UMTS cells as they usually use the same frequency as the current serving cell. The terminal thus merely applies the primary codes of neighboring cells on the received signal in order to get signal strength indications for them. For the terminal this means that it has to perform some additional computing tasks during an ongoing session. For neighboring GSM cells, the process is somewhat more complicated as they send on different frequencies and thus cannot be received simultaneously with the UMTS cells of the active set. The same problem occurs when signal quality measurements need to be made for UMTS cells that operate on a different frequency in order to increase the capacity of the radio network. The only

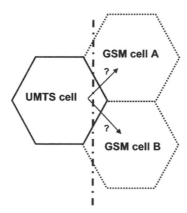

Figure 3.35 A UMTS cell with several GSM neighboring cells presents a problem for blind intersystem handovers

way for the terminal to perform measurements for such cells therefore is to stop sending and receiving frames in a predefined pattern in order to perform measurements on other frequencies. This mode of operation is referred to as compressed mode and is activated by the RNC if necessary in the terminal and all cells of the active set of a connection. The standard defines three possibilities for implementing compressed mode. While network vendors can choose which of the options described below they want to implement, the support of all options is required in the terminal:

- Reduction of the spreading factor: for this option, the spreading factor is reduced for some frames. Thus, more data can be transmitted during those periods that increase the speed of the connection. This allows injecting short transmission gaps without reducing the overall speed of the connection for inter-frequency measurement purposes. As the spreading factor changes, the transmission power has to be increased during these times in order to ensure an acceptable error rate.
- Puncturing: after the channel coder has added error correction and error detection bits to the original data stream some of them are removed again in order to have time for inter-frequency measurements. To keep the error rate of the radio bearer within acceptable limits, the transmission power has to be increased.
- Reduction of the number of user data bits per frame: as fewer bits are sent per frame, the transmission power does not have to be increased in this method. The disadvantage is the reduced user data rate while being in compressed mode.

The goal of the measurements while in compressed mode is to be able to successfully decode the frequency correction channel (FCCH) and the synch channel (SCH) of the surrounding GSM cells. For further information on these channels see Section 1.7.3.

Figure 3.36 shows how an intersystem handover from UMTS to GSM is performed. The procedure starts on the UTRAN side just like a normal inter-MSC handover by the RNC sending an SRNS relocation request. As the SRNS relocation is not known in GSM, the 3G MSC uses a standard 2G prepare handover message to initiate the communication with the 2G MSC. Thus, for the 2G MSC, the handover looks like a normal GSM to GSM handover and is treated accordingly.

3.7.2 Mobility Management in Idle State

While in Idle state, the terminal is passive, i.e. no data is sent or received. Nevertheless, there are a number of tasks that have to be performed periodically by the terminal.

In order to be able to respond to incoming voice calls, short messages, MMS messages etc., the paging channel (PCH) is monitored. If a paging message is received that contains the subscribers IMSI or TMSI, the terminal reacts and establishes a connection with the network. As the monitoring of the paging channel consumes some power, subscribers are split into a number of groups based on their IMSI (paging group). Paging messages for a subscriber of each group are then only broadcast at certain intervals. Thus, a terminal does not have to listen for incoming paging messages all the time but only at a certain interval. At all other times, the receiver can be deactivated and thus battery capacity can be saved. A slight disadvantage of this approach is, however, that the paging procedure takes a little bit longer than if the paging channel was constantly monitored by the terminal.

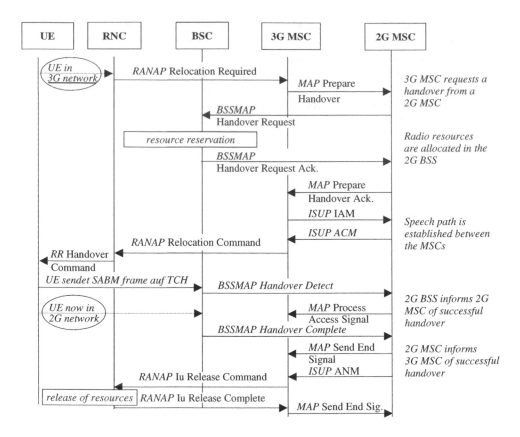

Figure 3.36   3G–2G intersystem hard handover message flow

In the event that the subscriber has an active PDP context while the terminal is in Idle state, the network will also need to send a paging message in the case of an incoming IP frame. Such a frame could for example originate from a messaging application. When the terminal receives a paging message for such an event, it has to re-establish a logical connection with the network before the IP frame can be forwarded.

In Idle state, the terminal is responsible for mobility management, i.e. changing to a more suitable cell when the user is moving. As the network is not involved in the decision-making process, the procedure is called cell reselection.

While in Idle state, no physical or logical connection exists between the radio network and the terminal. Thus, it is necessary to re-establish a physical connection over the air interface if data needs to be transported again. For the circuit-switched part of the network the RRC Idle state therefore implies that no voice connection is established. For the SGSN on the other hand the situation is different. A PDP context can still be established in Idle state, even though no data can be sent or received. To transfer data again, the terminal needs to re-establish the connection and the network then either establishes a DCH or uses the FACH for the data exchange. In practice, it can be observed that the time it takes to re-establish a channel is about 2.5 to 3 seconds. Therefore, the mobile should only be put into Idle state after a prolonged period of inactivity as this delay has a negative impact on the quality of experience of the user, for example during a web-browsing session. Instead of an instantaneous reaction to the user clicking on a link, the user notices an undesirably delay before the new page is presented.

While a terminal is in Idle state, the core network is not aware of the current location of the subscriber. The MSC is only aware of the subscriber's current location area. A location area usually consists of several dozen cells and therefore it is necessary to page the subscriber for incoming calls. This is done via a paging message that is broadcast on the paging channel in all cells of the location area. This concept has been adopted from GSM without modifications and is described in more detail in Section 1.8.1.

From the point of view of the SGSN, the same concept is used if an IP packet has to be delivered while the terminal is in Idle state. For the packet-switched part of the network the cells are divided into routing areas (RA). A routing area is a subset of a location area but most operators use only a single routing area per location area. Similar to the location area, the routing area concept was adopted from the 2G network concept without modification and is described in more detail in Section 2.8.1.

In the event that the terminal moves to a new cell that is part of a different location or routing area, a location or a routing area update has to be performed. This is done by establishing a signaling connection which prompts the RNC to set the state of the terminal to Cell-DCH or Cell-FACH. Afterwards, the location or routing area update is performed transparently over the established connection with the MSC and the SGSN. Once the updates are performed, the terminal returns to Idle state.

3.7.3 Mobility Management in Other States

While in Cell-FACH, Cell-PCH, or URA-PCH state, the terminal is responsible for mobility management and thus for cell changes. The big difference between these states and the Idle state is that a logical connection exists between the terminal and the radio network. If these states are implemented in the network, they are used while a data connection is in a dormant state (see Section 3.5.4). Depending on the state, the terminal has to perform the following tasks after a cell change.

In Cell-FACH state the terminal can exchange data with the network at any time. If the terminal performs a cell change it has to inform the network straight away via a cell update message. Afterwards, all data is exchanged via the new cell. If the new cell is connected to a different RNC, the cell update message will be forwarded to the serving RNC of the subscriber via the Iur interface. As the terminal has a logical connection to the network, no location or routing area update is necessary if the new cell is in a different area. This means that the core network is not informed that the subscriber has moved to a new location or routing area. This is, however, not necessary as the S-RNC will forward any incoming data over the Iur interface via the D-RNC to the subscriber. In practice, changing the cell in Cell-FACH state results in a short interruption of the connection which is tolerable as this state is not used for real-time or streaming services.

If the new serving cell is connected to an RNC that does not have an Iur interface to the S-RNC of the subscriber, the cell update will fail. As the new RNC cannot inform the

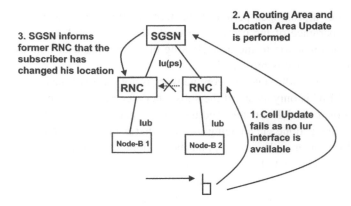

Figure 3.37 Cell change in PMM connected state to a cell that cannot communicate with the S-RNC

S-RNC of the new location of the subscriber it will reset the connection and the terminal automatically defaults to Idle state. In order to resume data transmission, the terminal then performs a location update with the MSC and SGSN as shown in Figure 3.37.

As the SGSN detects during the location and routing area update that there is still a logical connection to a different RNC, it sends a message to the previous RNC that the subscriber is no longer under its control. Thus, it is ensured that all resources that are no longer needed to maintain the connection are released.

From the mobility management point of view, the Cell-PCH is almost identical to the Cell-FACH state. The only difference is that no data can be transmitted to the terminal in the Cell-PCH state. If data is received for the terminal while being in the Cell-PCH state, the RNC needs to page the terminal first. Once the terminal responds, the network can then put the terminal in the Cell-DCH or Cell-FACH state and the data transfer can resume.

In the event of an even longer period of inactivity of a PDP context, the radio network can set the terminal to URA-PCH state. A cell update message thus only has to be sent to the network if the subscriber roams into a new UTRAN registration area (URA). The URA is a new concept that has been introduced with UMTS. It refines a location area as shown in Figure 3.38.

The core network is not aware of URAs. Furthermore, even single cells have been abstracted into so-called service areas. This is in contrast to a GSM/GPRS network, where the MSC and SGSN are aware of the location area and even the cell ID in which the terminal is located during an active connection. In UMTS, the location area does not contain single cells but one or more service areas. It is possible to assign only a single cell to a service area to be able to better pinpoint the location of a terminal in the core network. By this abstraction it was possible to clearly separate the location principles of the core network which is aware of location areas, routing areas and services areas, and the radio network which deals with URAs and single cells. Core network and radio network are thus logically decoupled. The mapping between the location principles of core and radio network is done at the interface between the two networks, the RNC.

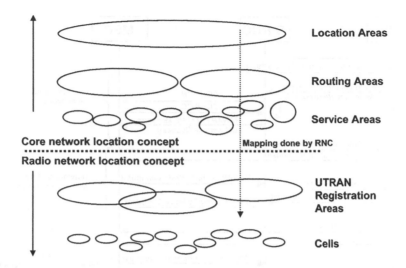

Figure 3.38   Location concepts of radio and core network

3.8 UMTS CS and PS Call Establishment

In order to establish a circuit-switched or packet-switched connection, the terminal has to contact the network and request the establishment of a session. The establishment of the user data bearer is then performed in several phases.

As a first step, the terminal needs to perform an RRC connection setup procedure as shown in Figure 3.39 to establish a signaling connection. The procedure itself was introduced in Section 3.4.5 and Figure 3.18. The goal of the RRC connection setup is to establish a temporary radio channel that can be used for signaling between the terminal, the RNC, and a core network node. The RNC can choose either to assign a dedicated channel (Cell-DCH state) or to use the FACH (Cell-FACH state) for the subsequent exchange of messages.

If a circuit-switched connection is established as shown in Figure 3.39, the terminal sends a CM service request DTAP message (see Section 1.4.2) over the established signaling connection to the RNC that transparently forwards the message to the MSC. DTAP messages are exchanged between the RNC and the MSC via the connection-oriented SCCP protocol (see Section 1.4.1). Therefore, the RNC has to establish a new SCCP connection before the message can be forwarded.

Once the MSC has received the CM service request message, it verifies the identity of the subscriber via the attached TMSI or IMSI. This is done in a challenge and response procedure similar to GSM. In addition to the terminal authentication already known from GSM, a UMTS network has to authenticate itself to the user to protect against air interface eavesdropping with a false base station. Once the authentication procedure has been performed, the MSC activates the ciphering of the radio channel by issuing a security mode command. Optionally, the MSC afterwards assigns a new TMSI to the subscriber which, however, is not shown in Figure 3.39 for clarity.

After successful authentication and activation of the encrypted radio channel, the terminal then proceeds to inform the MSC of the exact reason of the connection request. The call control (CC) setup message contains among other things the telephone number (MSISDN)

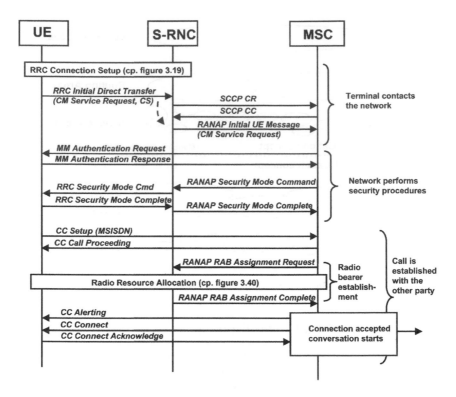

Figure 3.39 Messaging for a mobile originated voice call (MOC)

of the destination. If the MSC approves the request, it returns a call proceeding message to the terminal and starts two additional procedures simultaneously.

At this point, only a signaling connection exists between the terminal and the radio network, which is not suitable for a voice call. Thus, the MSC requests the establishment of a speech path from the RNC via an RAB assignment request message. The RNC proceeds by reserving the required bandwidth on the Iub interface and instructs the Node-B to allocate the necessary resources on the air interface. Furthermore, the RNC also establishes a bearer for the speech path on the Iu(cs) interface to the MSC. As a dedicated radio connection was already established for the signaling in our example, it is only modified by the radio resource allocation procedure (radio link reconfiguration). The reconfiguration includes for example the allocation of a new spreading code as the voice bearer requires a higher bandwidth connection than a slow signaling connection. If the RNC has performed the signaling via the FACH (Cell-FACH state), it is necessary at this point to establish a dedicated channel and to move the terminal over to a dedicated connection. Figure 3.40 shows the necessary messages for this step of the call establishment.

Simultaneous to the establishment of the resources for the traffic channel in the radio network, the MSC tries to establish the connection to the called party. This is done for example via ISUP signaling to the gateway MSC for a fixed-line destination as described in Section 1.4. If the destination is reachable, the MSC informs the caller by sending call control 'altering' and 'connect' messages.

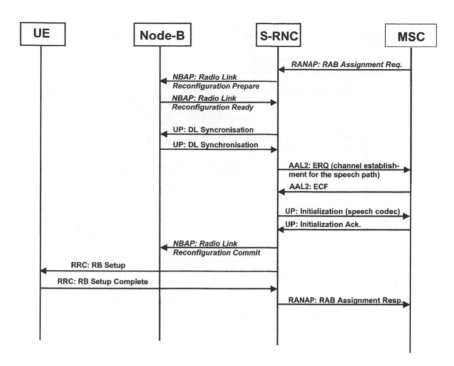

Figure 3.40 Radio resource allocation for a voice traffic channel

The establishment of a packet-switched connection is also called packet data protocol (PDP) context activation (see Figure 3.41). From the users' point of view, the activation of a PDP context means getting an IP address in order to be able to communicate with the Internet or another IP network. Further background information on the PDP context activation can be found in Chapter 2. As shown for a voice call in the previous example the establishment of a packet-switched connection also starts with an RRC connection setup procedure.

Once the signaling connection has been established successfully, the terminal continues the process by sending an 'activate PDP context request' message via the RNC to the SGSN. As shown in the previous example, this triggers the authentication of the subscriber and activation of the air interface encryption. Once encryption is in place, the SGSN continues the process by establishing a tunnel to the GGSN which in turn assigns an IP address to the user. Furthermore, the SGSN requests the establishment of a suitable bearer from the RNC based on the QoS parameters (e.g. minimal bandwidth, latency, etc.) for the new connection which have been given to the SGSN at the beginning of the procedure in the 'activate PDP context request' message. However, these values can be modified by the SGSN or GGSN in the event that the user is not subscribed to the requested QoS or if the connection requires a different QoS setting. The establishment of the radio bearer is done in the same way as shown in Figure 3.39 for a circuit-switched channel. However, as the bearer for a packet-switched connection uses other QoS attributes, the parameters inside the messages will be different.

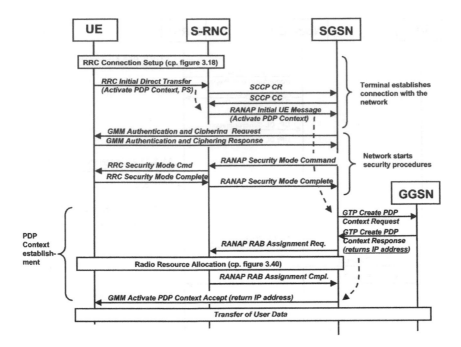

Figure 3.41 PDP context activation

3.9 UMTS Release 99 Performance

As shown in this chapter, UMTS Release 99 is a rich standard leaving both network manufacturers and operators a great deal of choice in the implementation of the functionality and operation of the network. The following section shows how some of these features are used in an operational network and how they influence the overall user experience of the network. The information presented here was gathered by using a standard terminal with special network monitoring software to be able to observe the parameters presented below during an ongoing packet data session. Furthermore, the tests were performed in a number of different operational networks to show how operators have chosen to run their network and how these choices influence the user experience.

3.9.1 Data Rates, Delay, and Applications

As has been shown in Section 3.3.2 the maximum data rate that can be achieved in the downlink direction with a UMTS Release 99 network is 384 kbit/s. This translates into a maximum user data speed of about 40–45 kbyte/s which is about seven times faster than a fixed-line dial-up modem connection and comparable to entry-level DSL speeds. While not being quite as fast as current DSL offerings for users close to the DSL access multiplexer, this kind of transmission speed is easily capable of offering a fast web access experience to the user similar to a DSL connection. For browsing ordinary web pages the speed difference to fixed-line high-speed Internet access is hardly noticeable because standard web pages are usually less then 100 kb in size and are thus transferred very quickly over a 384 kbit/s bearer. However, for large file transfers of PDF documents for example, the speed difference is noticeable, but UMTS still delivers a good user experience even for this application.

In the event of high network load or bad network coverage, the network automatically assigns a lower bandwidth bearer of 128 kbit/s or even less to the user. However, a speed reduction due to a high air interface load has not been observed during performance evaluation tests, which stretched over several weeks. While some networks reduced the bearer speed during bad reception conditions, most networks were able to maintain a high data rate even under unfavorable conditions.

Delay is another crucial factor in delivering good user experience for applications like web browsing (see Section 3.9.4). With delay times ranging from 160–200 ms in Cell-DCH state depending on the supplier of the radio network, UMTS is fast enough to be used as a bearer for voice over IP applications like IMS or Skype. While Skype was initially developed for fixed-line networks, its efficient codec, which requires about 40 kbit/s of bandwidth in each direction, requires less bandwidth then other fixed-line voice over IP systems like SIP (session initiation protocol) while at the same time providing a speech quality that far exceeds fixed-line or mobile AMR speech quality. However, even an efficient VoIP system exceeds by far the resource requirements of a circuit-switched voice call in the radio network for which only a 12.2 kbit/s bearer is needed. For a VoIP call on the other hand, a bearer of at least 64 kbit/s is required. While a circuit-switched bearer for voice has been specifically optimized for fast mobility environments, the same cannot be said for 64 kbit/s packet-switched bearers. Thus, VoIP sessions might not behave as well in fast mobile environments as current circuit-switched voice calls. Current operational networks assign an even higher bandwidth bearer for a VoIP call as will be shown in Section 3.9.2 and do not throttle back to a more conservative spreading code even if the data rate does not increase above 40 kbit/s for a considerable amount of time. Therefore, while it is possible to use VoIP in UMTS Release 99 networks with high speech quality for stationary users, it also wastes a lot of resources on the air interface and operators are not keen to see these kinds of applications being used in their networks. HSDPA will change the situation somewhat as radio resources are assigned in a more efficient way especially for streaming services that do not make use of the entire bandwidth of a dedicated bearer. Furthermore, streaming applications also have a rather negative impact on the total number of users a cell can service simultaneously as is further described in Section 3.9.4.

Audio and video streaming are also becoming more and more popular with high-speed Internet users today. UMTS is capable of reliably streaming audio and video content with a speed of up to 40–45 kbyte/s. However, a prolonged heavy use of a high bit rate bearer for a single user severely reduces the total number of users a cell can support simultaneously. This is also further described in Section 3.9.3. Therefore, audio and video streaming is another application for which HSDPA is required in the future in order to meet the bandwidth requirements of a high number of users with a reasonable number of cell sites.

3.9.2 Radio Resource Management Example

For bursty applications, today's operational UMTS networks are especially optimized to only assign a dedicated bearer for the duration of a web page download, for example, and to put the terminal into Cell-FACH state as quickly as possible. This is done in order to support as many users as possible in a single cell. Table 3.6 shows how two network operators have set their air interface parameters and timers and how they perform their radio resource management.

As can be seen in the first line of the table, both networks quickly release the bearer after a PDP context has been assigned if no data is sent. Operator-1 does not use the Cell-FACH state and thus the RNC frees all resources on the radio interface after an inactivity timer expires. Therefore the logical connection to the terminal is released and the mobile is set to Idle state. Operator-2 uses the Cell-FACH state and downgrades the connection from Cell-DCH state to Cell-FACH state after the inactivity timer expires. Only after another 30 seconds of inactivity is the logical connection released as well and the mobile is put into Idle state.

Line two of the table shows the behavior of the network in a low traffic situation, which was simulated by constantly sending ping requests to the network and waiting for the

Table 3.6 Comparison of radio resource management of two UMTS networks

	Operator-1 with UTRAN equipment from vendor A	Operator-2 with UTRAN equipment from vendor B
Initial state after PDP context activation and idle time	Idle-PCH	Idle-PCH after about 30 seconds of Cell-FACH
Ping traffic state	Cell-DCH	Cell-FACH if Cell-FACH before, Cell-DCH if startup from Idle-PCH, no fallback to Cell-FACH
First ping RTD to first available hop	>4 seconds	2600 ms from Idle-PCH 160 ms in Cell-DCH 183–207 ms in Cell-FACH
Subsequent ping RTD to first available hop	210–230 ms to 265–290 ms depending on spreading factor of the bearer	160 ms in Cell-DCH 183–207 ms in Cell-FACH
Downgrade state/timer with constant background ping	Remains in DCH but bearer downgrade	Remains in initial state Cell-FACH or Cell-DCH
Downgrade state/timer after stop of ping	Idle-PCH, <10 seconds	<3–4 seconds (Cell-FACH)
DCH downgrade after web page download complete with no further traffic/timer	Idle-PCH, <10 seconds	Cell-FACH / <5 seconds
Timer to final state after web page download without further traffic	<10 seconds to Idle-PCH	Idle-PCH after about 30 seconds of Cell-FACH

reply. While operator-1 immediately assigns a dedicated channel, operator-2 only assigns a dedicated channel if the subscriber has previously been in Idle state. If the connection of the subscriber was already in Cell-FACH state, it remains unchanged.

An important factor for the user experience is the time that is required to serve a user's request for a web page for example once the connection has been dormant for some time e.g. because the user has been reading the current page. As operator-1 quickly releases the logical connection, a new request triggers the establishment of a new radio bearer, which is a time-intensive process. Therefore, it takes over four seconds before the request can be served. Operator-2 only puts the connection in Idle state after over half a minute and is thus able to quickly process the request over the Cell-FACH channel. The delay therefore is only 183 to 207 milliseconds and a page can be displayed almost instantly. Even while the first parts of the page are transferred over the air interface the network reacts to the increased traffic and establishes a dedicated connection for the terminal again. Once in place, the network sends a command to the terminal and it seamlessly changes to the higher bandwidth channel to download the remainder of the page. The user notices a big difference between the two radio resource management approaches especially during a web-browsing session. While the web page is almost instantly delivered in operator-2's network, there is a noticeable delay of over four seconds in operator-1's network.

To be able to optimize the use of radio resources, a network should also be capable of reacting to a decrease in traffic of a subscriber and reducing the bandwidth of the connection. Operator-1 has implemented such a scheme and it can be observed by sending a constant stream of ping requests to the network after a web page has been loaded that after several seconds the round-trip delay times increase slightly. This is caused by the network assigning a longer spreading code for the dedicated connection and thus reduces the bandwidth which implicitly increases the round-trip delay time as data packets cannot be sent as quickly as before. Operator-2 on the other hand does not reduce the bandwidth of a connection unless there is no traffic observed for some time. Therefore, a continuous stream of small ping messages is enough to keep a 384 kbit/s bearer. While for a simple web-surfing session such behavior is not a problem, a cell will become congested very quickly if several subscribers use their connection for streaming or interactive applications, like voice over IP, which only require a fraction of the bandwidth of the connection.

The three rows at the bottom of Table 3.6 show which way the two operators have set the parameters regarding the downgrade of the connection. As operator-1 does not use the Cell-FACH channel for user traffic, all three rows indicate a release of the physical bearer and release of the logical connection after only 10 seconds of inactivity resulting in a poor user experience during a web-browsing session as described below in Section 3.9.3. Operator-2 has implemented a two-step mechanism and downgrades the dedicated connection by putting the subscriber on the FACH common channel after no data packets have been observed by the network for about four seconds. The network only takes the next step and releases the logical connection by putting the terminal into Idle state if the connection has been completely dormant for more than 30 seconds.

An additional radio resource management approach used by operator-2 which is not shown in Table 3.6 is the slow increase of the bearer bandwidth after the transition to Cell-DCH state from Idle or Cell-FACH state. Instead of assigning a high bandwidth bearer right from the start, it can be observed that initially only a 64 kbit/s bearer is assigned which is quickly increased to 128 kbit/s and 384 kbit/s when the network detects a large number of packets in its transmit buffer. This way, the network saves resources especially for WAP surfing sessions which are mostly used to download only small web pages for which a high-speed bearer is not required.

From the users' point of view the behavior of the network of operator-2 results in a much better web-browsing experience. While network 1 downsizes the bandwidth of dedicated bearer once it detects a decrease in the flow of traffic, network 2 does just the opposite and only uses a limited bandwidth when entering Cell-DCH state that is quickly increased if demanded by incoming traffic. For optimal use of the available resources of a cell, both methods should be combined for the management of a dedicated connection. While not done widely today, it is very likely that operators will adapt their radio resource management schemes over time to the increasing use of the network.

3.9.3 UMTS Web Browsing Experience

As has been shown in Section 2.12.1 for (E)GPRS, the delay caused by the network has a considerable impact on the user experience in a web-browsing session. When requesting a new page, the following delays are experienced before the page can be downloaded and displayed.

- The URL has to be converted into the IP address of the web server that hosts the requested page. This is done via a DNS (domain name service) query that causes a delay in the order of the time it takes to send one IP frame to a host in the Internet and waiting for the reply. This delay is also called the round-trip delay time.
- Once the IP address of the server has been established, the web browser needs to establish a TCP connection. This is done via a three-way handshake. During the handshake, the client sends a synchronization TCP frame to the server which is answered by a synchronization-ack frame. This is in turn acknowledged by the client by sending an acknowledgment frame. As three frames are sent before the connection is established, the whole operation causes a delay of 1.5 times the round-trip delay of the connection. As the first frame containing user data is sent right after the acknowledgment frame, however, the time is reduced to approximately a single round-trip delay time.
- Only after the TCP connection has been established can the first frame be sent to the web server, which usually contains the actual request of the web browser. The server then analyzes the request and sends back a frame that contains the beginning of the requested web page. As the request (e.g. 300–500 bytes) and the first response frame (1400 bytes) are quite large, the network requires somewhat more time to transfer those packets then the simple round-trip delay time.

As has been shown in Section 3.9.2, the round-trip delay times of the UMTS radio access network are in the order of 160–250 ms for the Cell-FACH and Cell-DCH states and only exceeds several seconds if the terminal is in Idle state and a radio bearer has to be established first. As most operational networks configure their networks in a similar way as operator-2 in Table 3.6, it rarely happens that the connection is in Idle state. Therefore, assuming an average round-trip delay time of about 200 milliseconds for Cell-FACH state and 160 ms for Cell-DCH state, the time it takes between requesting a page and the browser showing the first parts of the page can be roughly calculated as follows:

$$\text{Total delay (UMTS Release 99)} = \text{Delay DNS query (Cell-FACH)}$$
$$+ \text{Delay TCP establish (Cell-DCH/FACH)}$$
$$+ \text{Delay request/response}$$
$$= 200\,\text{ms} + 160\,\text{ms} + 250\,\text{ms} = 610\,\text{ms}$$

The request/delay response time is slightly larger than the normal round-trip delay time due to the bigger packet size of the first web server response packet which already contains a part of the web page and usually has a size of about 1460 bytes. Figure 3.42 shows the timing of a sample request for a web page in the same way as was already presented in Section 2.12.1 for EGPRS, again without the initial DNS query which is not part of the TCP connection establishment for the actual web page.

3.9.4 Number of Simultaneous Users per Cell

Calculating the number of simultaneous users a cell can handle is very difficult for a several reasons. First, a cell handles many different radio bearers at the same time for different

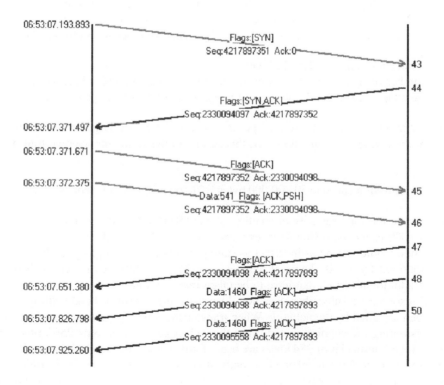

Figure 3.42 IP packet flow and delay times during the download of a web page

purposes. Bearers for voice calls with 12.2 kbit/s require only little bandwidth while bearers for high-speed Internet access with 384 kbit/s quickly saturate a cell. Another factor for Internet connections is the usage profile for subscribers that describe how much data they transfer in a given time frame. Also the way in which data is transferred influences the total capacity of the cell. As has been shown in Section 3.9.2, some radio network implementations do not reduce the bearer capacity when the user only needs a part of the available bandwidth that was once assigned while streaming audio or being engaged in a voice over IP call. Even though discontinuous transmission is used while no data is transferred, it nevertheless reduces the capacity of the cell that could otherwise be made available to other subscribers.

Under operational conditions a single cell can handle around three simultaneous 384 kbit/s bearers in order to have enough capacity for lower bit rate data bearers and voice calls and in order not to create too much interference for neighboring cells. If three users in a cell use their high-speed bearer for video and audio streaming over a longer period of time then the cell would be saturated. Thus the maximum number of simultaneous high-speed users would be only three. Most subscribers, however, use their Internet connection mainly for web browsing and only rarely transfer bigger files like pictures, PDF documents or anti-virus program updates. Voice over IP applications used by some subscribers slightly change the average user profile. However, an intelligent radio network resource management can adapt the assigned bandwidth of the bearer for the relatively low constant data rate of 4 kbyte/s of real-time data traffic flow in the up- and downlink directions and thus free up capacity for other subscribers in the cell.

In order to roughly calculate a more realistic number of high-speed Internet subscribers per cell, the following assumptions are made. These are supported by the author's observations of the traffic generated while using UMTS as a high speed Internet access for his daily work:

- Total connection time per day: 12 hours.
- Total number of bytes transmitted during that time by using applications like email with large file attachments, web browsing and voice over IP: 50 Megabytes.

At first, the total volume transmitted per day seems to be very large. When dividing the volume by the connection time, however, this results in the following average transmission speed:

Average transmission speed $= 50.000.000$ bytes$/12$ h$/60$ minutes$/60$ seconds $= 1.15$ kbyte/s

Compared to the total capacity of a cell of about 1500 kbyte/s, this number suggests that more than 1000 users could simultaneously use a single cell for high-speed Internet access. However, that would just be as incorrect as stating that only three high-speed Internet users can be supported by a cell. The real number is in between these extremes as there are a number of additional factors that have to be taken into account.

Wasted resources: a subscriber does not always use the complete bandwidth of a bearer that has been assigned by the network. When transferring a web page for example, there are many times during the ongoing download of the web page in which only 250 kbit/s or even less of the total bandwidth of 384 kbit/s are used. Furthermore, the network keeps the bearer assigned to the user for a number of seconds after the web page transfer is finished and only releases the bearer after the expiry of an inactivity timer. Studies like the one presented in [8] use network simulations to explore the impact of different inactivity timer settings to the overall capacity of a cell. Small inactivity timers increase the overall capacity of a cell but decrease the user experience due to increased page request times, while a higher value has the opposite effect. If it is assumed for this example that the resource efficiency is only one-third of the total capacity of the cell, the number of concurrently supported users of a cell is reduced from 1000 down to 333.

Busy hour: in operational networks, so-called busy hours can be observed in which the majority of users are much more active then during other times of the day. If it is therefore assumed that the average use is five times higher during busy hour, a cell would still be able to support $333/5 = 66$ users.

If one finally takes into account that a single base station is usually composed of three cells, the total number of high-speed Internet subscribers that can be supported by a single base station is about 200. If this number is used to calculate the amount of revenue an operator can generate per base station by offering high-speed Internet access at a price of 30 euros per month, as offered in some countries, the following financial calculation can be made:

$$\text{Revenue per cell: 30 euros subscription fee per subscriber per month}$$

$$\times\, 200 \text{ subscribers} \times 12 \text{ months} = 72,000 \text{ euros/year}$$

Over the average lifetime of a base station of 10 years this would result in revenue of 720,000 euros. On top of this, the cell is also used for other services such as voice and video calls, low-speed WAP browsing, SMS, MMS, etc., which generate additional revenue.

An important factor that has not been part of the calculations above is the tariffs for which an operator decides to offer high-speed Internet access to users. Depending on the price itself and its relation to end-user prices for other high-speed Internet access technologies like DSL or WiMAX, the operator has a tool to influence the total number of users which subscribe to high-speed UMTS Internet access and the amount of traffic generated by them. By limiting the amount of traffic included in the tariff or even by setting a limit on a daily basis, it is possible to keep users from excessively using streaming applications and thus prevent the network from saturating. However, such an approach reduces the attractiveness of the offer compared to other technologies that might not have such restrictions and should therefore be used carefully.

3.10 UMTS Release 5: High-Speed Downlink Packet Access (HSDPA)

As shown in Section 3.1, UMTS is an evolving system and the standards are constantly extended to take advantage of the latest technical advances to enhance the user experience. With the high-speed downlink packet access (HSDPA), which is introduced in Release 5 of the 3GPP specifications, the UMTS standard was enhanced to be able to deliver much higher data rates per cell and per user than before. With data rates between 500 kbit/s and 3.6 Mbit/s per user, HSDPA allows network operators to compete head on with other high-speed Internet access technologies such as DSL.

Important standards documents that were newly created or enhanced for HSDPA are the overall system description Stage 2 in 3GPP TS 25.308 [9], the physical layer description TR 25.858 [10], physical layer procedures in TS 25.214 [11], Iub and Iur interface enhancements in TR 25.877 [12], RRC extensions in TS 25.331 [1], and signaling procedure examples in TS 25.931 [3].

3.10.1 HSDPA Channels

As shown in Figures 3.43 and 3.44, HSDPA combines the concepts of dedicated and shared channels. For user data in the downlink direction one or more high-speed physical downlink shared channels (HS-PDSCH) are used which are shared among several users. Thus, it is possible to send data to several subscribers simultaneously or to increase the transmission speed for a single subscriber by bundling several HS-PDSCH which each use a different code.

Each HS-PDSCH uses a spreading factor length of 16 which means that in theory up to 15 simultaneous HS-PDSCH channels can be used in a single cell. In operational networks, however, 5 to 10 HS-PDSCH are used per cell as other services like voice or packet-switched data transfers for Release 99 terminals also require bandwidth. Furthermore, some services like video telephony will continue to use Release 99 dedicated channels for some time to come, as only these channels guarantee a stable bandwidth, delay, and handover behavior during the lifetime of the connection. These parameters are not guaranteed for an HSDPA connection as the bandwidth available to a user depends on the current signal quality and the number of simultaneous users of the current cell. HSDPA thus sacrifices the concept of a dedicated channel with a guaranteed bandwidth for a substantially increased bandwidth. For many applications like web surfing or the transfer of big files or emails with file attachments, this is very beneficial and, most of the time, not disadvantageous.

The assignment of timeslots on high-speed downlink shared channels (HS-DSCH) to a user is done via several simultaneous broadcast high-speed shared control channels (HS-SCCH) which use a spreading factor length of 128. A terminal has to be able to receive and decode at least four of those channels simultaneously. Thus, it is possible to inform many users at the same time about which of the HS-PDSCH channels data is sent for them in the next timeslot.

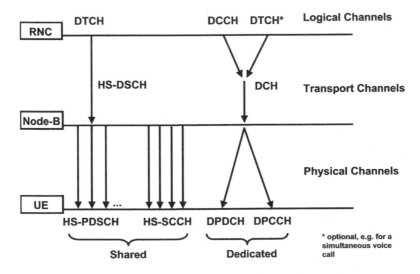

Figure 3.43 Simplified HSDPA channel overview in the downlink direction

Apart from shared channels, an HSDPA connection furthermore uses a number of dedicated channels per subscriber:

- A dedicated physical control channel (DPCCH) in the uplink direction with a spreading factor of 256 for HSDPA control information such as acknowledgments and retransmission requests for bad frames as well as for transmitting signal quality information. This channel uses its own channelization code and is not transmitted with other channels by using time or IQ-multiplexing.
- A dedicated control channel (DCCH) for RRC messages in the uplink and downlink directions between the RNC and the terminal, which is used for tasks like mobility management which is for example necessary for cell changes.
- A dedicated traffic channel (DTCH) for IP user data in the uplink direction, as HSDPA only uses shared channels in the downlink direction. The uplink bearer has a bandwidth of 64, 128, or optionally 384 kbit/s.
- Optionally, an additional dedicated traffic channel (DTCH) is used in the uplink and downlink directions if a circuit-switched connection is established during an HSDPA connection. The channel can have a bandwidth of up to 64 kbit/s.

3.10.2 Shorter Delay Times and Hybrid ARQ (HARQ)

Apart from offering increased bandwidth to individual users and increasing the capacity of a cell in general, another goal of HSDPA was to reduce the round-trip delay (RTD) time for both stationary and mobile users. HSDPA further reduces the RTD times experienced with Release 99 dedicated channels of 160–200 milliseconds to about 70 milliseconds. This is important for applications like web browsing, as shown in Section 3.9.1 for UMTS Release 99 and also in Section 2.3.4 for EDGE, that require several frame round trips for the DNS query and establishment of the TCP connections before the content of the web page is sent to the user. In order to reduce the round-trip time, the air interface block size has been reduced to 2 milliseconds. This is quite small compared to the block sizes of dedicated channels of at least 10 milliseconds.

Due to the frequently changing signal conditions experienced when the user is for example in a car or train, or even while walking in the street, transmission errors will frequently occur. Due to error detection mechanisms and retransmission of faulty blocks, no packet loss is experienced on higher layers. However, every retransmission increases the overall delay of the connection. Higher layer protocols like TCP for example react very sensitively to changing delay times and interpret them as congestion. In order to minimize this effect, HSDPA adds an error detection and correction mechanism on the MAC layer in addition to the mechanisms that already exist on the RLC layer. This mechanism is directly implemented in the Node-B and is called hybrid ARQ (HARQ). Together with a block size of 2 milliseconds instead of at least 10 milliseconds for dedicated channels, an incorrect or missing block can be retransmitted by the Node-B in less then 10 milliseconds. This is a substantial enhancement over Release 99 dedicated channels as they only use a retransmission scheme on the RLC layer, which needs at least 80 to 100 milliseconds for the detection and retransmission of a faulty RLC frame.

Compared to other error detection and correction schemes, which are used for example on the TCP layer, HARQ does not use an acknowledgment scheme based on a sliding window mechanism but sends an acknowledgment or error indication for every single frame. This mechanism is called stop and wait (SAW). Figure 3.45 shows how a frame is transmitted in the downlink direction, which the receiver cannot decode correctly. The receiver therefore sends an error indication to the Node-B, which then in turn retransmits the frame. In detail, the process works as follows.

Before the transmission of a frame, the Node-B informs the terminal of the pending transmission on the HS-SCCH. Each HS-SCCH frame thus contains the following information:

- ID of the terminal for which a frame is sent in one or more HS-PDSCH channels in the next frame.
- Channelization codes of the HS-PDSCH channels that are assigned to a terminal in the next frame.
- Transport format and resource indicator (channel coding information).
- Modulation format (QPSK or 16QAM).
- HARQ process number (see below).
- If the block contains new data or is used for retransmission, which redundancy version (RV) is used (see below).

Each frame on the HS-SCCH is split into three slots. The information in the control frame is arranged in a way that the terminal has all information necessary to receive the frame once it has received the first two of the three slots. Thus, the network does not wait till the complete control frame is sent but already starts sending the user data on the HS-PDSCH once the terminal has received the first two slots of the control frame. This means that the shared control channel and the downlink shared channels are sent with a time shift of one slot. After the reception of a user data frame, the terminal has exactly 5 milliseconds to decode the frame and to check if it was received correctly. If the frame was sent correctly

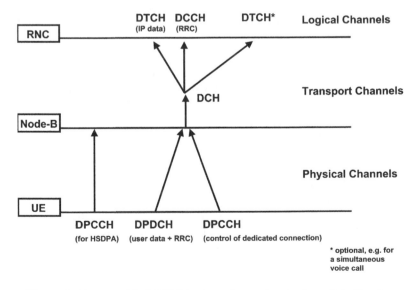

Figure 3.44   Simplified HSDPA channel overview in the uplink direction

the terminal sends an acknowledge (ACK) in the uplink direction. If the terminal is not able to decode the packet correctly, a not acknowledge (NACK) is sent. In order to save further time, the uplink control channel is also slightly time shifted against the downlink shared channel. This allows the network to quickly retransmit a frame. See Figure 3.45.

As HARQ can only transmit a frame once the previous frame has been acknowledged, the terminal must be able to handle up to eight simultaneous HARQ processes. Thus it is ensured that the data flow is not interrupted by a problem with a single frame. As higher layers of the protocol stack expect the data in the right order, the data stream can only be forwarded once a frame has been received correctly. Therefore, the terminal has to have a buffer to store frames of other HARQ processes that need to be reassembled with other frames that have not yet been received correctly.

For the network there are two possibilities for retransmitting a frame. If the incremental redundancy method is used, the network uses error correction information that was punctured out after channel coding to make the data fit into the MAC frame. Puncturing is a method that is already used in UMTS Release 99, GPRS and EDGE, and further information can be found in Section 2.3.4. If a frame needs to be retransmitted, the network sends different redundancy bits and the frame is thus said to have a different redundancy version (RV) of the data. By combining the two frames, the overall redundancy is increased on the receiving side and the chance that the frame can be decoded correctly increases. If the frame still cannot be decoded, there is still enough redundancy information left which has not yet been sent to assemble a third version of the frame.

The second retransmission method is called chase combining and retransmits a frame with the same redundancy version as before. Instead of combining the two frames on the MAC layer, this method combines the signal energy of the two frames on the physical layer before attempting to decode the frame again. The method which is used for retransmission

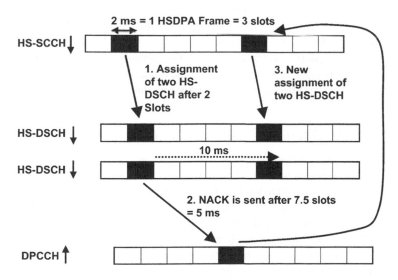

Figure 3.45 Detection and report of a missing frame with immediate retransmission within 10 milliseconds

is controlled by the network. However, the mobile can indicate to the network during bearer establishment which of the two methods it supports.

3.10.3 Node-B Scheduling

The HS-DSCH channels have been designed in a way so that different channels can be assigned to different users at the same time. The network then decides for each frame which channels to assign to which users. As shown before, the HS-SCCH channels are then used to inform the terminals on which channels to listen for their data. This task is called scheduling. In order to quickly react to changing radio conditions of each subscriber, the scheduling for HSDPA has not been implemented on the RNC as for other channels but directly on the Node-B. This can also be seen in Figure 3.43 as the HS-SCCH originate from the Node-B. This means that for HSDPA yet another task that was previously located in the RNC for dedicated channels has been moved to the border node of the network. Therefore the scheduler can for example react very quickly to deteriorating radio conditions (fading) of a terminal. Instead of sending frames to a terminal while it is in a deep fading situation and thus most likely unable to receive the frame correctly, the scheduler can use the frames during this time for other terminals. This helps to increase the total bandwidth available in the cell as fewer frames have to be used for retransmission of bad or missing blocks. Studies like [13] and [14] have shown that a scheduler that takes channel conditions into consideration can increase the overall cell capacity by about 30% for stationary users. Apart from the signal quality of the radio link to the user, the scheduling is also influenced by other factors like the priority of the user. As with many other functionalities, the standard does not say which factors should influence the scheduling in which way and thus a good scheduling implementation of a vendor can lead to an advantage.

As the RNC has no direct influence on the resource assignment for a subscriber, it is also not aware how quickly data can be sent. Therefore, a flow control mechanism is required on the Iub interface between the RNC and the Node-B. For this reason, the Node-B has a data buffer for each user priority from which the scheduler then takes the data to be transmitted over the air interface. In order for the RNC to find out how much space is left in those buffers, it can send a capacity request message to the Node-B which reports to the RNC the available buffer sizes using a capacity allocation message. It should be noted that a Node-B does not administer a data buffer per user but only one data buffer per user priority.

3.10.4 Adaptive Modulation, Coding, and Transmission Rates

In order to reach the highest possible transmission rate during favorable transmission conditions a new modulation scheme has been introduced with HSDPA. The new modulation scheme allows the system to transfer four bits per transmission step. As 16 values can be coded in four bits (2^4), the modulation scheme is called 16QAM. Under ideal conditions this increases the total capacity of a cell by a factor of two while maintaining the channel bandwidth of 5 MHz. Under less favorable radio conditions, HSDPA uses the existing QPSK modulation which transmits two bits per transmission step.

In addition to changing the modulation scheme the network can also alter the coding scheme and the number of simultaneously used HS-DSCH channels for a terminal on a per frame basis. This behavior is influenced by the channel quality index (CQI) which is frequently reported by the terminal. The CQI has a range from 1 (very bad) to 31 (very good) and tells the network how many redundancy bits are required to keep the block error rate (BLER) below 10%. For a real network this means that under less favorable conditions more bits per frame are used for error detection and correction. This reduces the transmission speed but ensures that a stable connection between network and terminal is maintained. As modulation and coding is controlled on a per user basis, bad radio conditions for one user have no negative effects for other users in the cell to which the same HS-DSCHs are assigned for data transmission.

By adapting the modulation and coding schemes it is also possible to keep the power needed for the HSDPA channels at a constant level. The strategy of HSDPA to use a constant power level of for example 40% of the total transmission power available in a cell and instead adapt the transmission speeds of the users is quite different from the strategy of Release 99 dedicated channels. Here the bandwidth of a connection is stable while the transmission power is adapted depending on the user's changing signal quality. Only if the power level can no longer be increased to ensure a stable connection, does the network take action and increase the spreading factor to reduce the bandwidth of the connection.

The capabilities of the terminal also influence the maximum data rate of a connection. The standard defines a number of different device categories which are listed in 3GPP TS 25.306 [15]. Table 3.7 shows some of these categories and their properties.

With a Category 6 terminal that supports both QPSK and 16QAM, the following maximum transmission speed can be reached:

$$7298 \text{ bits per TTI (which are distributed over five HS-PDSCH channels)}$$

$$\text{every 2 milliseconds} = (1/0.002) \times 7298 = 3.6 \text{ Mbit/s}$$

This corresponds to a speed of 720 kbit/s per channel with a spreading factor of 16. Compared to a Release 99 dedicated channel with 384 kbit/s, which uses a spreading factor of eight the transmission is four times faster. This is achieved by using the 16QAM modulation instead of QPSK which doubles the maximum speed and by reducing the number of error detection and correction bits while signal conditions are favorable. Thus, an HSDPA cell is in theory about four times as fast as a Release 99 cell. In operational networks the speed increase will not be as high because not all terminals will be close to the cell and thus many will only be able to use QPSK and a higher number of redundancy bits in order to ensure a stable transmission.

Table 3.7   A selection of HSDPA terminal categories

HS-DSCH category	Maximum number of simultaneous HS-PDSCH	Minimum TTI interval	Maximum number of transport block bits per TTI
6	5	1	7298 (16QAM)
11	5	2	3630 (QPSK only)
12	5	1	3630 (QPSK only)

A Category 11 terminal on the other hand, which is limited to QPSK modulation, can only receive data in every second frame. Thus for this category of terminals the maximum speed is:

$$3630 \text{ bits every 4 milliseconds} = (1/0.004) \times 3630 = 900 \text{ kbit/s}$$

In the future there may also be devices that are capable of receiving more than five HS-DSCH channels simultaneously and can thus reach a maximum speed of 14.4 Mbit/s. However, this would consume all resources of a cell and could of course only be reached under ideal signal conditions.

As has been shown, there are many factors that influence how fast data can be sent to a terminal. The following list summarizes once again the main factors:

- signal quality;
- number of active HSDPA users in a cell;
- number of established channels for voice and video telephony in the cell;
- number of users that use a dedicated channel for data transmission in a cell;
- terminal category;
- bandwidth of the connection of the Node-B to the RNC;
- interference generated by neighboring cells;
- achievable throughput in other parts of the network, as high data rates cannot be sustained by all web servers or other end points.

It should be noted at this point that the transmission speeds that can be reached with HSDPA also have an impact on other parts of the terminal. Apart from increased processing power of the terminal in general, the interface to a remote terminal like a notebook needs to be capable of handling data at these speeds. The maximum transmission rate of Bluetooth up to version 1.2, for example, is just 700 kbit/s (see Chapter 6). This is not sufficient by far for HSDPA data rates and thus new devices should also support the Bluetooth enhanced data rate extension in order not to become the bottleneck of the end-to-end connection.

3.10.5 Establishment and Release of an HSDPA Connection

To establish an HSDPA connection, an additional dedicated channel (DCH) is required in order to be able to send data in the uplink direction as well. If the network detects that the terminal is HSDPA capable during the establishment of a dedicated channel for packet data transfer, it automatically allocates the necessary resources during the setup of the connection as shown in Figure 3.46.

To establish an HSDPA connection, the S-RNC informs the Node-B that a new connection is required and the Node-B will configure the HS-PDSCH accordingly. In a further step, the RNC then reserves the necessary resources on the Iub interface between itself and the Node-B. Once this is also done, the network is ready for the high-speed data transfer and informs the terminal via a RRC radio bearer reconfiguration message that data will now be sent on the HS-DSCH. Once data is received by the RNC from the SGSN, flow control information is exchanged between the Node-B and the RNC as described in Section 3.10.3. Thus it is ensured that the data buffer of the Node-B is not flooded as the RNC has no

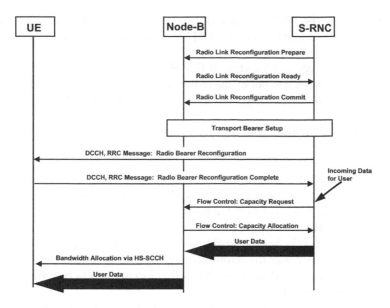

Figure 3.46 Establishment of an HSDPA connection to a terminal

direct information on or control of how fast the incoming data can be sent to the terminal. When the Node-B receives data for a user, it is then the task of the HSDPA scheduler in the Node-B to allocate resources on the air interface and to inform the user's terminal via the shared control channels whenever it sends data on one or more HS-PDSCHs.

While the terminal is in HSDPA reception mode, it has to constantly monitor all assigned HS-SCCH channels and also has to maintain the necessary dedicated channels. This of course results in higher power consumption which is acceptable while data is transferred. If no data is transferred for some time, this state is quite unfavorable as the power consumption remains high and thus the runtime of the terminal decreases. This state is also not ideal for the network as bandwidth on the air interface is wasted for the dedicated channel of the HSDPA connection. Thus, the network can decide to release the HSDPA connection after a period of time and put the subscriber into the Cell-FACH state (see Section 3.5.4). In this state the terminal can still send and receive data, but the bandwidth is very small. Nevertheless, this is quite acceptable as an HSDPA connection can be reestablished again very quickly when required.

3.10.6 HSDPA Mobility Management

HSDPA has been designed for both stationary and mobile users. Therefore, it is necessary to maintain the connection while the user is moving from cell to cell. For this reason, the terminal keeps a so-called active set for the dedicated channel of the HSDPA connection which is required for the soft handover mechanism as described in Section 3.7.1. In contrast to a pure dedicated connection, the terminal only receives its data over one of the Node-Bs of the active set. Based on the configuration of the network, the terminal then reports to the RNC if a different cell of the active or candidate sets would provide better signal quality than the current cell. The RNC can then decide to redirect the data stream to a different cell. As the concept is different from the UMTS soft handover, the standards refer to this operation as cell change procedure.

Compared to the cell update procedure of (E)GPRS, the cell change procedure of HSDPA is controlled by the network and not by the terminal. As the terminal is already synchronized with the new cell, a cell change only leads to a short interruption of the data transfer on the HS-PDSCHs.

Depending on the relationship between the old and the new cell, there are several different kinds of cell changes:

- Intra Node-B cell change: old and new cells are controlled by the same Node-B. This is the simplest version of the operation because data that is still available in the buffer of the Node-B can simply be sent over the new cell.
- Inter Node-B cell change: old and new cells belong to different Node-Bs. In this scenario the RNC has to instruct the new Node-B to allocate resources for the HSDPA connection. This is done in a similar way to establishing a new connection as shown in Figure 3.46. User data that is still buffered in the old Node-B is lost and has to be retransmitted by the RLC layer which is controlled in the RNC.
- Cell change with Iur interface: if the old and new cells are under the control of different RNCs, the HSDPA connection has to be established over the Iur interface.
- Cell change without Iur interface: if the old and new cells are under the control of different RNCs, which are not connected via the Iur interface, an SRNS relocation has to be performed which also involves core network components (SGSN and possibly also the MSC).
- Old and new cells use different frequencies (inter-frequency cell change): in this scenario additional steps are required in the terminal in order to find cells on different frequencies and to synchronize to them before data transmission can resume.
- New cell does not support HSDPA: in this scenario the network can allocate a dedicated channel in the new cell and then hand over the terminal to this channel in the new cell. The user will notice that the maximum speed of the connection decreases but the overall connection is maintained.
- Inter-RAT cell change: if the subscriber leaves the UMTS coverage area completely a handover from UMTS/HSDPA to GSM is performed. In the same way as the inter-frequency cell change described above, HSDPA connections can use a compressed mode similar to that of dedicated channels to allow the terminal to search for cells on other frequencies.

During all scenarios it is of course also possible that an additional voice or video call is established. This further complicates the cell change/handover as this connection also has to be maintained next to the data connection and handed over into a new cell.

3.11 UMTS Release 6: High-Speed Uplink Packet Access (HSUPA)

Due to the emergence of peer-to-peer applications like multimedia calls and video conferencing, the demand for uplink bandwidth is continually increasing. Other applications like sending emails with large file attachments or large MMS messages also benefit from high uplink data rates. However, UMTS uplink speeds have not been enhanced since Release 99. Thus, even with the introduction of HSDPA the uplink is still limited to 64–128 kbit/s and to 384 kbit/s in some networks under ideal conditions. The solution to satisfy the increasing demand in the uplink direction is introduced into the UMTS radio network with UMTS Release 6 and is called high-speed uplink packet access (HSUPA). HSUPA increases uplink user data rates up to theoretical 5.8 Mbit/s. When taking realistic radio conditions into account, the number of simultaneous users, terminal capabilities, etc., user speeds of 800 kbit/s and beyond can still be reached.

For the network, HSUPA has a number of benefits as well. For HSDPA (see Section 3.10), an uplink dedicated channel (DCH) is required for all terminals which are receiving data via the high-speed downlink shared channels for TCP acknowledgments and other user data. This is problematic for bursty applications as a dedicated channel in the uplink direction wastes uplink resources of the cell despite the terminal reducing its power output during periods in which no user data is sent. Nevertheless, HSUPA continues to use the dedicated concept of UMTS for the uplink by introducing an enhanced dedicated channel (E-DCH) functionality for the uplink only, which includes a number of enhancements to decrease the impact of bursty applications on the dedicated channel concept. In order to have both high-speed uplink and downlink performance, using an E-DCH introduced with HSUPA makes the most sense when combined with high-speed downlink shared channels which were introduced with HSDPA.

While a Release 99 dedicated channel (DCH) ensures a constant bandwidth and delay time for data packets with all its advantages and disadvantages discussed in previous chapters, the E-DCH trades in this concept for higher data rates. Thus, while still being a dedicated channel, an E-DCH does not necessarily guarantee a certain bandwidth to a user in the uplink direction. For many applications this is quite acceptable and allows the increase of the number of simultaneous users that can share the uplink resources of a cell. This is due to the fact that the network can control the uplink noise in a much more efficient way by dynamically adjusting the uplink bandwidth on a per subscriber basis in a cell to react to changing radio conditions and traffic load. This reduces the overall cost of the network by requiring fewer base stations for the same number of users which in turn can result in cheaper subscription costs.

The E-DCH concept also ensures full mobility for subscribers. However, the radio algorithms are clearly optimized to ensure the highest throughput for low speed or stationary use.

The main purpose of the E-DCH concept is to support streaming (e.g. mobile TV), interactive (e.g. web browsing) and background services (e.g. FTP). In order to ensure good performance for real-time applications like IMS video conferencing, the E-DCH enhancements also contain optional mechanisms to ensure a minimal bandwidth to a user. As these methods are optional, it can be expected that first implementations will focus on the basic E-DCH concept first and only add optional features later on. Despite this, voice and video over IP services can still be used over a non-optimized E-DCH without any problems as long as the cell's bandwidth is sufficient to satisfy the demand of all users currently transmitting data in a cell regardless of the application.

As the uplink bandwidth increases, the E-DCH approach also further reduces the round-trip delay times for applications like web surfing and interactive gaming as described in Section 3.9.3.

Finally, it is important to note that the E-DCH concept is backward compatible. Thus, a cell can support Release 99 terminals that were only designed for dedicated channels, HSDPA terminals that require a DCH in the uplink direction, and terminals which support a combination of HSDPA in the downlink and HSUPA in uplink.

As the E-DCH concept is an evolution of existing standards, it has triggered the creation of a number of new documents as well as the update of a number of existing specifications. Most notably, 3GPP TR 25.896 [16] was created to discuss the different options which were analyzed for HSUPA. Once consensus on the high-level architecture was reached, 3GPP TS 25.309 [17] was created to give a high-level overview of the selected solution. Among the specification documents that were extended are 3GPP TS 25.211 [2] which describes physical and transport channels and 3GPP TS 25.213 [18] which was extended to contain information about E-DCH spreading and modulation.

3.11.1 E-DCH Channel Structure

For the E-DCH concept a number of additional channels were introduced in both the uplink and downlink directions as shown in Figures 3.47 and 3.48. These are used in addition to existing channels, which are also shown in the figure below. For further explanation of these channels, see Section 3.4.3 for Release 99 channels and Section 3.10.1 for HSDPA.

As shown on the left side in Figure 3.47, HSUPA introduces a new transport channel which is called the enhanced-DCH (E-DCH). While still being a dedicated channel for a single user, the dedicated concept was adapted to use a number of features, which were already introduced with HSDPA for the downlink direction. Therefore, the following overview just

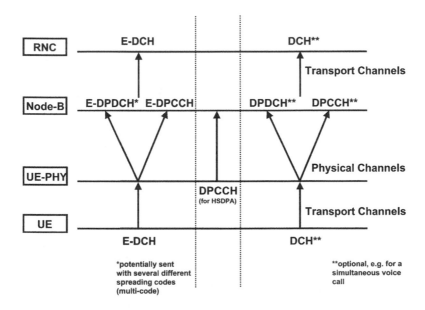

Figure 3.47 Transport and physical channels used for HSUPA

gives a short introduction to the feature and the changes required to address the needs of a dedicated channel:

- Node-B scheduling: while standard dedicated channels are managed by the RNC, E-DCH channels are managed by the Node-B. This allows a much quicker reaction to transmission errors which in turn decreases the overall round-trip delay time of the connection. Furthermore, the Node-B is able to react much more quickly to changing conditions of the radio environment and variations of user demands for uplink resources, which help to better utilize the limited bandwidth of the air interface.
- HARQ: instead of leaving the error detection and correction to the RLC layer, the E-DCH concept uses the hybrid automatic retransmission request (HARQ) scheme which is also used by HSDPA in the downlink direction. This way, errors can be detected on a per MAC-frame basis by the Node-B. For further details see Section 3.10.2 which describes the HARQ functionality of HSDPA in the downlink direction. While the principle of HARQ in the uplink direction is generally the same, it should be noted that the signaling of acknowledgments is done in a slightly different way due to the nature of the dedicated channel approach.
- Chase combining and incremental redundancy are used in a similar way for E-DCH as described in Section 3.10.2 for HSDPA in order to retransmit a frame when the HARQ mechanism reports a transmission error.

On the physical layer the E-DCH is split into two channels: the enhanced dedicated physical data channel (E-DPDCH) is the main transport channel and is used for user data (IP frames carried for RLC/MAC frames) and layer 3 RRC signaling between the terminal on the one side and the RNC on the other. As will be further shown below the spreading factor used for this channel is quite flexible and can be dynamically adapted from 64 to 2 depending on the current signal conditions and the amount of data the terminal wants to send. It is even possible to use several channelization codes at the same time to increase the overall speed. This concept is called a multi-code channel and is similar to the HSDPA concept of assigning frames on several downlink shared channels to a single terminal. As will be shown in more detail below the maximum number of simultaneous code channels has been limited to four per terminal with two channels being used with $SF = 2$ and the other two with $SF = 4$. In terms of frame length, 10 milliseconds are used for the E-DPDCH by default with 2 millisecond frames having been standardized as an optional feature for the terminal.

The enhanced dedicated physical control channel (E-DPCCH) on the other hand is used for physical layer control information. For each E-DPDCH frame a control frame is sent on the E-DPCCH to the Node-B which most importantly contains the seven-bit traffic format combination ID (TFCI). Only by analyzing the TFCI is the Node-B able to decode the MAC frame on the E-DPDCH as the terminal can choose the spreading factor and coding of the frame from a set given to it by the Node-B to adapt to the current signal conditions and uplink user data buffer state. Furthermore, each frame on the E-DPCCH contains a two-bit retransmission sequence number (RSN) to signal HARQ retransmissions and the redundancy version (see Section 3.10.2) of the frame. Finally, the control frame contains a so-called 'happy' bit to indicate to the network if the maximum bandwidth currently allocated to the terminal is sufficient or if the terminal would like the network to increase it. While the spreading factor of the physical data channel is variable, a constant spreading factor of 256 is used for the E-DPCCH.

A number of existing channels, which might also be used together with an E-DCH, is shown in the middle and on the right of Figure 3.47. Most of the time, an E-DCH is used together with HSDPA high-speed downlink shared channels which require a separate dedicated physical control channel (DPCCH) to send control information for downlink HARQ processes. In order to enable applications like voice and video telephony during an E-DCH session a mobile must also support simultaneous Release 99 dedicated data and control channels in the uplink. This is necessary because these applications require a fixed and constant bandwidth of 12.2 and 64 kbit/s, respectively. In total, an E-DCH capable terminal must therefore be able to simultaneously encode the data streams of at least five uplink channels. If multi-code operation for the E-DPDCH is used, up to eight code channels are used in uplink direction at once.

In the downlink direction, HSUPA additionally introduces two mandatory and one optional channel to the other already numerous channels that have to be monitored in downlink direction. Figure 3.48 shows all channels that a mobile station has to decode while having an E-DCH assigned in the uplink direction, HSDPA channels in the downlink direction and an additional dedicated channel for a simultaneous voice or video session via a circuit-switched bearer.

While HSUPA only carries user data in the uplink direction, a number of control channels in the downlink direction are nevertheless necessary. For the network to be able to return acknowledgments for received uplink data frames to the terminal, the enhanced HARQ information channel (E-HICH) is introduced. The E-HICH is a dedicated channel, which means that the network needs to assign a separate E-HICH to each terminal currently in E-DCH state.

In order to dynamically assign and remove bandwidth to and from individual users quickly, a shared channel called the enhanced access grant channel (E-AGCH) is used by the network

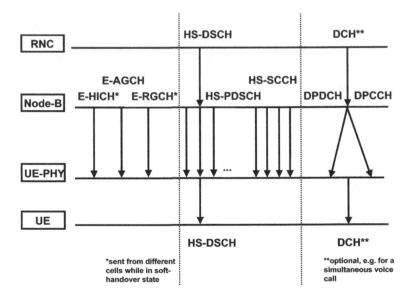

Figure 3.48 Simultaneous downlink channels for simultaneous HSUPA, HSDPA and dedicated channel use

that must be monitored by all terminals in a cell. A fixed spreading factor of 256 is used for this channel. Further details about how this channel is used to issue grants (bandwidth) to the individual terminals are given below in Section 3.11.3.

Finally, the network can also assign an enhanced relative grant channel (E-RGCH) to individual terminals to increase or decrease an initial grant which was given on the E-AGCH. The E-RGCH is again a dedicated channel which means that the network has to assign a separate E-RGCH to every active E-DCH terminal. The E-RGCH is optional, however, and depending on the solutions of the different network vendors there might be networks in which this channel is not used. If not used, only the E-AGCH is used to control uplink access to the network. Note that although all channels are called 'enhanced', none of these channels has a Release 99 predecessor.

Besides these three control channels, an E-DCH terminal must also be able to decode a number of additional downlink channels simultaneously. As HSUPA will normally be used together with HSDPA, the terminal also needs to be able to simultaneously decode the HS-DSCHs as well as up to four HS-SCCHs. If a voice or video call is established besides the high-speed packet session, the network will add another two channels in the downlink direction as shown in Figure 3.48 on the right-hand side. In total, an E-DCH mobile must therefore be capable of decoding 10–15 downlink channels at the same time. If the mobile is put into soft handover state by the network (see Section 3.7.1) the number of simultaneous channels increases even further as some of these channels are then broadcast via different cells of the terminal's active set.

3.11.2 The E-DCH Protocol Stack and Functionality

In order to reduce the complexity of the overall solution, the E-DCH concept introduces two new layers which are called the MAC-e and MAC-es. Both layers are below the existing MAC-d layer. As shown in Figure 3.49, higher layers are not affected by the enhancements and thus the required changes and enhancements for HSUPA in both the network and the terminals are minimized.

While on the terminal the MAC-e/es layers are combined, the functionality is split on the network side between the Node-B and the RNC. The lower layer MAC-e functionality is implemented on the Node-B in the network. It is responsible for scheduling, which is further described below, and the retransmission (HARQ) of faulty frames.

The MAC-es layer in the RNC is responsible for recombining frames received from different Node-Bs if an E-DCH connection is in soft handover state. Furthermore, the RNC is also responsible for setting up the E-DCH connection with the terminal at the beginning. This is not part of the MAC-es layer but part of the radio resource control (RRC) algorithm which has to be enhanced for HSUPA as well. As the RNC treats an E-DCH channel like a dedicated channel, the mobile station is in Cell-DCH state while an E-DCH is assigned. While scheduling of the data is part of the Node-B's job, overall control of the connection rests with the RNC. Thus, the RNC can decide to release the E-DCH to a terminal after some period of inactivity and put the terminal into Cell-FACH state. Therefore, HSUPA becomes part of the Cell-DCH state and thus part of the overall radio resource management as described in Section 3.5.4.

One of the reasons for enhancing the dedicated connection principle in order to increase uplink speeds instead of using a shared channel approach lies in the fact that this enables

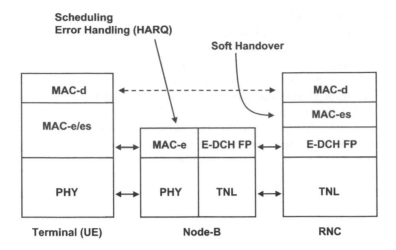

Figure 3.49 E-DCH protocol stack

the soft handover principle to be used in the uplink. This is not possible with a shared channel approach, which is used by HSDPA in the downlink, because cells would have to be synchronized to assign the same timeslots to a user. In practice, this would create a high signaling overhead in the network. By using dedicated channels the timing between the different terminals that use the same cells in soft handover state is no longer critical as they can send at the same time without being synchronized. The only issue arising from sending at the same time is the increased noise level in the cells. However, neighboring cells can minimize this by instructing mobiles in soft handover state to decrease their transmission power via the relative grant channel (E-RGCH) as further described below. Using soft handover in the uplink direction might prove to be very beneficial, as the mobile station's transmit power is much less than that of the Node-B. Furthermore, there is a higher probability that one of the cells can pick up the frame correctly and thus the terminal only has to retransmit a frame if all cells of the active set send a negative acknowledge for a frame. This in turn reduces the necessary transmission power on the terminal side and increases the overall capacity of the air interface. As soft handover for E-DCH has been defined as optional in the standards, most initial implementations, however, will most likely not make use of it.

Another advantage of the dedicated approach is that terminals do not have to be synchronized within a single cell and thus do not have to wait for their turn to send data. This further reduces the round-trip delay times.

3.11.3 E-DCH Scheduling

If the RNC wants to put a terminal into Cell-DCH state due to the establishment of a packet connection or due to renewed activity on a downgraded bearer (Cell-FACH state), it can establish an E-DCH instead of a DCH if the following criteria are fulfilled:

- The current cell is E-DCH capable.
- The terminal is E-DCH capable.

- The QoS requirements allow the use of an E-DCH. Some E-DCH implementations might require the use of a standard DCH instead of an E-DCH for packet connections that are established for real-time services like VoIP or packet-switched video calls. However, more advanced E-DCH implementations will be able to manage such connections over an E-DCH as well and still ensure a minimal bandwidth and constant delay time by using non-scheduled grants as described further below.

If the decision is made by the RNC to assign an E-DCH to the terminal, the bearer establishment or modification messaging is very similar to establishing a standard DCH. During the E-DCH establishment procedure, the RNC informs the terminal of the transport format combination set (TFCS) that can be used for the E-DCH. A TFCS is a list (set) of data rate combinations, coding schemes, and puncturing patterns for different transport channels that can be mapped on to the physical channel. In practice, at least two channels, a DTCH for user data, and a DCCH for RRC messages, are multiplexed over the same physical channel (E-DPDCH). This is done in the same way as for a standard dedicated channel. By using this list the terminal can later select a suitable transport format combination for each frame depending on how much data is currently waiting in the transmission buffer and the current signal conditions. By allowing the RNC to flexibly assign a TFC set to each connection it is possible to restrict the maximum speed on a per subscriber basis based on the subscription parameters. During the E-DCH setup procedure the terminal is also informed which of the cells of the active set will be the serving E-DCH cell. The serving cell is defined as being the cell over which the network later controls the bandwidth allocations to the terminal.

Once the E-DCH has been successfully established, the terminal has to request a bandwidth allocation from the Node-B. This is done by sending a message via the E-DCH even though no bandwidth has so far been allocated. The bandwidth request contains the following information for the Node-B:

- UE estimation of the available transmit power after subtracting the transmit power already necessary for the DPCCH and other currently active dedicated channels.
- Indication of the priority level of the highest priority logical channel currently established with the network for use via the E-DCH.
- Buffer status for the highest priority logical channel.
- Total buffer status (taking into account buffers for lower priority logical channels).

Once the Node-B receives the bandwidth request, it takes the terminal's information into account together with its own information about the current noise level, bandwidth requirements of other terminals in the cell, and the priority information for the subscriber it has received from the RNC when the E-DCH was initially established. The Node-B then issues an absolute grant, also called a scheduling grant, via the absolute grant channel (E-AGCH) which contains information about the maximum power ratio the mobile can use between the E-DPDCH and the E-DPCCH. As the mobile has to send the E-DPCCH with enough power to be correctly received at the Node-B, the maximum power ratio between the two channels implicitly limits the maximum power that can be used for the E-DPDCH. This in turn limits the number of choices the terminal can make from the TFC set that was initially assigned by the RNC. Therefore, as some TFCs can no longer be selected, the overall speed in the uplink direction is implicitly limited.

Furthermore, an absolute grant can be addressed to a single terminal only or to several terminals simultaneously. If the network wants to address several terminals at once, it has to issue the same enhanced radio network temporary ID (E-RNTI) to all group members when their E-DCH is established. This approach minimizes signaling when the network wants to schedule terminals in the code domain.

Another way to dynamically increase or decrease a grant given to a terminal or a group of terminals is the use of relative grants, which are issued via the optional relative grant channel (E-RGCH). These grants are called relative grants because they can increase or decrease the current power level of the mobile step by step with an interval of one TTI or slower. Thus, the network is quickly able to control the power level and therefore implicitly the speed of the connection every 2 or 10 milliseconds. Relative grants can also be used by all cells of the active set. This allows cells to influence the noise level of E-DCH connections currently controlled by another cell in order to protect themselves from too much noise being generated in neighboring cells. This means that the terminal needs to be able to decode the E-RGCH of all cells of the active set. As shown in Figure 3.50, each cell of the active set can assume one of three roles:

- One of the cells of the active set is the serving E-DCH cell from which the mobile receives absolute grants via the E-AGCH (cell 4 in Figure 3.50). The serving E-DCH cell can furthermore instruct the terminal to increase, hold, or decrease its power via commands on the E-RGCH.
- The serving E-DCH cell and all other cells of the Node-B which are part of the active set of a connection (cell 3 and 4 in Figure 3.50) are part of the serving radio link set. The commands sent over the E-RGCH of these cells are identical and thus the terminal can combine the signals for decoding.
- All other cells of the active set are part of the non-serving radio link set (cell 1, 2, and 5 in Figure 3.50). The terminal has to decode all E-RGCHs of these cells separately. Cells in the non-serving RLS can only send hold or down commands.

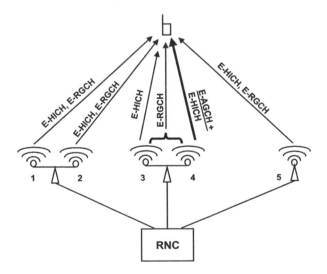

Figure 3.50 Serving E-DCH cell, serving RLS, and non-serving RLS

If an 'up' command is received from the serving RLS, the terminal is allowed to increase its transmission power only if at the same time no 'down' command is received by one or more cells of the non-serving RLS. In other words, if a 'down' command is received by the terminal from any of the cells, the terminal has to immediately decrease its power output. Therefore only the serving E-DCH is able to increase or decrease the power output of the mobile via the relative grant channels while all other cells of the non-serving RLS are only permitted to decrease the power level.

It should be noted that in a real environment it is unlikely that the five cells as shown in Figure 3.50 are part of the active set of a connection, as the benefit of the soft handover would be eaten up by the excessive use of air interface and Iub link resources. Thus in a normal environment, it is the goal of radio engineering to have two or at most three cells in the active set of a connection in soft handover state.

As has been shown, the Node-B has quite a number of different pieces of information to base its scheduling decision on. The standard, however, does not describe how these pieces of information are used to ensure a certain QoS level for the different connections and leaves it to the network vendors to implement their own algorithms for this purpose. Again, the standards encourage competition between different vendors, which unfortunately increases the overall complexity of the solution.

In order to enable the use of the E-DCH concept for real-time applications like voice and video over IP, the standard contains an optional scheduling method which is called a non-scheduled grant. If the RNC decides that a certain constant bandwidth and delay time is required for an uplink connection, it can instruct the Node-B to reserve a sufficiently large power margin for the required bandwidth. The terminal is then free to send data at this speed to the Node-B without prior bandwidth requests. If such E-DCH connections are used, which is again implementation dependent, the Node-B has to ensure that even peaks of scheduled E-DCH connections do not endanger the correct reception of the non-scheduled transmissions.

3.11.4 E-DCH Mobility

Very high E-DCH data rates can only be achieved for stationary or low mobility scenarios due to the use of low spreading factors and few redundancy bits. Nevertheless, the E-DCH concept uses a number of features to enable high data rates in high-speed mobility scenarios.

Early E-DCH implementations might only make use of a single serving cell, i.e. no macro diversity (soft handover) is used. For mobility this means that in between cells the maximum possible speed achievable might not be ideal as the terminal does not have enough power to use low spreading factors and coding rates. When the RNC then decides to use a better suited cell as serving E-DCH cell, a short interruption of the data traffic in the uplink direction will occur as the mobile first has to establish a new E-DCH channel in the new serving cell.

More advanced implementations will make use of macro diversity (soft handover) as shown in Figure 3.50. This means that the uplink data is received by several cells which forward the received frames to the RNC. Each cell can then indicate to the terminal if the frame has been received correctly and thus the frame only has to be repeated if none of the cells were able to decode the frame correctly. This is especially beneficial for mobility scenarios in which reception levels change quickly due to obstacles suddenly appearing in between the terminal and one of the cells of the active set as shown earlier in Figure 3.30.

Furthermore, the use of soft handover ensures that no interruptions in the uplink occur while the user is moving through the network with the terminal.

Inter-frequency and inter-RAT (radio access technology) handovers have also been enhanced for HSUPA to be able to maintain the connection for the following scenarios:

- The terminal moves into the area of a cell which only supports Release 99 dedicated channels. In this case the network can instruct the terminal to perform a handover into the new cell and establish a DCH instead of an E-DCH. As an uplink DCH is limited to 64-128 kbit/s or 384 kbit/s in certain cases, the user might notice that the uplink speed has decreased.
- Due to capacity reasons, an operator can use several 5 MHz carriers per cell. One carrier might be used by the operator to handle voice and video calls and additionally Release 99 dedicated channels for packet transfer while the second carrier is reserved for HSDPA and HSUPA. When setting up a high-speed connection, the network can instruct the terminal to change to a different carrier. If the terminal then moves to a cell in which only a single carrier is used, an inter-frequency handover is necessary to jump back to the basic carrier.
- In the worst case a user might roam outside the coverage area of the UMTS network altogether. If a GSM network is available in this area, the network will then perform a handover into the GSM/GPRS network. This is called an inter-RAT handover.

3.11.5 E-DCH Terminals

New E-DCH capable terminals again require increased processing power and memory capabilities compared to Release 99 or even HSDPA terminals in order to sustain the high data rates offered by the system in both downlink (HSDPA) and uplink (HSUPA) directions. In order to benefit from the evolution of terminal hardware and to be able to offer terminals with low power consumption and thus longer standby times, the standard defines a number of terminal categories that limit the maximum number of spreading codes that can be used for an E-DCH and their maximum length. This limits the maximum speed that can be achieved with the terminal in the uplink direction. Table 3.8 shows a number of typical E-DCH terminal categories and their maximum transmission speeds under ideal transmission conditions. The highest number of simultaneous spreading codes an E-DCH terminal can use is four, with two codes having a spreading factor of two and two codes having a spreading factor of four. The maximum user data rates are slightly lower then the listed transmission speeds as the transport block also includes the frame headers of different protocol layers. Under less ideal conditions, the terminal might not have enough power to transmit using the

Table 3.8 Spreading code sets and maximum resulting speed of different E-DCH categories

Max. E-DPDCH set of the terminal category	Maximum transport block size for 10 ms TTI	Maximum resulting transmission speed
1x SF-4	7.296 bits	729 kbit/s
2x SF-4	14.592 bits	1.459 Mbit/s
2x SF-2	20.000 bits	2.000 Mbit/s
2x SF-2 + 2x SF4	20.000 bits	2.000 Mbit/s

maximum number of codes allowed and might also use a more robust channel coding method which uses smaller transport block sizes, as more bits are used for redundancy purposes. Furthermore, the Node-B can also restrict the maximum power to be used by the terminal as described above in order to distribute the available uplink capacity of the cell among the different active users.

3.12 UMTS and CDMA2000

While UMTS is the dominant 3G technology in Europe it shares the market with a similar system called CDMA2000 in other parts of the world such as the USA. This section compares CDMA2000 and its evolution path to the GSM, GPRS and UMTS evolution path that has been discussed in Chapters 1 to 3.

IS-95A, which is also called CdmaOne, was designed like GSM to be mostly a voice-centric mobile network. Like GSM, it offers voice and circuit-switched data services of speeds up to 14.4 kbit/s. However, IS-95A and all evolutions of that standard are not based on GSM and so both radio and core network infrastructure and protocols are fundamentally different. In particular the radio network is fundamentally different to GSM as it is not based on frequency and time division multiple access. IS-95A was the first system to use the code division multiple access (CDMA) approach for the air interface that was later also used in the UMTS standards where it is referred to as wideband CDMA or W-CDMA for short.

IS-95B is a backward-compatible evolution of the system which offers increased user data rates and packet data transmission of up to 64 kbit/s. Thus it can be roughly compared to a GSM network that offers GPRS services. Just like the earlier version of CdmaOne it uses carriers with a bandwidth of 1.25 MHz which multiple subscribers share by code multiplexing.

The next step in the evolution path is CDMA2000 1xRTT (radio transmission technology) which can roughly be compared to UMTS. While offering theoretical data rates of 307 kbit/s in the downlink direction most deployments limit the maximum speed to about 150 kbit/s. From the overall system point of view there are many similarities between CDMA2000 and UMTS. These include:

- use of CDMA on the air interface;
- use of QPSK for modulation;
- variable length codes for different data rates;
- soft handover;
- continuous uplink data transmission.

As both UMTS and CDMA2000 need to be backward compatible with their respective evolution paths, there are also many differences which include:

- UMTS uses a W-CDMA carrier with a bandwidth of 5 MHz while CDMA2000 uses a multi-carrier approach with bandwidths of multiples of 1.25 MHz. This was done in order to be able to use CDMA2000 in the already available spectrum for IS-95, while UMTS had no such restriction due to the completely new implementation of the air interface and availability of a dedicated frequency band for the new technology.
- UMTS uses a chip rate of 3.84 MChip/s while CDMA2000 uses a chip rate of 1.2288 MChip/s. In order to increase capacity a base station can use several 1.25 MHz

carriers. Up to the latest revision of the standard described in this book (1xEV-DO see below), a subscriber is limited to a single carrier.

- UMTS uses a power control frequency of 1500 Hz compared to CDMA2000 that uses an 800 Hz cycle.
- UMTS uses unsynchronized base stations while in CDMA2000 all base stations are synchronized by using the clock of the global positioning system (GPS).
- As UMTS uses unsynchronized base stations, a three-step synchronization process is used between the terminal and the network as described in Section 3.4.4. CDMA2000 achieves synchronization based on a time-shift process that adapts the clock of the terminal to the network.
- While UMTS has a minimal frame length of 10 milliseconds, CDMA2000 uses 20 millisecond frames for user data and signaling and 5 millisecond frames if only signaling has to be sent.

As has been discussed in Section 3.10, the UMTS evolution towards higher data rates is called high-speed data packet access (HSDPA). A similar technology to increase data rates for CDMA2000 is called 1xEV-DO (evolution – data only) revision 0 which reflects the fact that the system uses one or more 1.25 MHz carriers exclusively for high-speed packet data transmission with data rates similar to those of HSDPA. In a further evolution of the standard, which is called revision A, a boost to uplink performance similar to UMTS HSUPA is introduced. Additional QoS features enabling the use of voice over IP and other real-time applications over the packet-switched network further extends the functionality.

In a separate evolution path from 1xEV-DO, the 1xEV-DV (evolution – data/voice) optimizes the use of the air interface to enable a single carrier to be used for both high-speed data and voice services which is not possible with 1xEV-DO. Revision C is the first evolution of the standard with speeds similar to HSDPA. 1xEV-DV revision D increases uplink speeds similarly to HSUPA. The main difference between the two CDMA2000 evolution paths is the fact that only 1xEV-DV supports circuit-switched voice and packet-switched data on the same carrier. 1xEV-DO compensates for this lack with QoS functionality to enable voice over IP and other real-time applications in the future.

To summarize the different evolutionary steps of CDMA2000, Table 3.9 gives an overview of the different steps and compares them to the evolution path of GSM/UMTS. It should be noted that the comparison is only qualitative as properties such as the maximum packet data rate per user are only roughly equal to the corresponding step of the other technology.

Table 3.9 Approximate comparison between the GSM and CdmaOne evolution path

GSM	IS-95A (CdmaOne)
GSM with (E-)GPRS	IS-95B / CDMA2000 1xRTT
UMTS	CDMA2000 1xRTT
UMTS – HSDPA	CDMA2000 1xEV-DO revision 0
UMTS – HSDPA and HSUPA	CDMA2000 1xEV-DO revision A
UMTS – HSDPA	CDMA2000 1xEV-DV revision C
UMTS – HSDPA and HSUPA	CDMA2000 1xEV-DV revision D

3.13 Questions

1. What are the main differences between the GSM and UMTS radio network?
2. Which advantages does the UMTS radio network have compared to previous technologies for users and network operators?
3. What are the data rates for a packet-switched connection that is offered by a Release 99 UMTS network?
4. What does OVSF mean?
5. Why is a scrambling code used additionally to the spreading code?
6. What does 'cell breathing' mean?
7. What are the differences between the Cell-DCH and the Cell-FACH RRC state?
8. In which RRC states can a terminal be in PMM-connected mode?
9. How is a UMTS soft handover performed and what are the advantages and disadvantages?
10. What is an SRNS relocation?
11. How is the mobility of a user managed in Cell-FACH state?
12. What is the compressed mode used for?
13. What are the basic HSDPA concepts to increase the user data rate?
14. How is a circuit-switched voice connection handled during an ongoing HSDPA session?
15. What are the advantages of the enhanced-DCH (E-DCH) concept?
16. Which options does the Node-B have to schedule the uplink traffic of different E-DCH terminals in a cell?

Answers to these questions can be found on the companion website for this book at http://www.wirelessmoves.com.

References

[1] 3GPP TS 25.331, Radio Resource Control (RRC) Protocol Specification.
[2] 3GPP TS 25.211, Physical Channels and Mapping of Transport Channels onto Physical Channels (FDD).
[3] 3GPP TS 25.931, UTRAN Functions, Examples on Signaling Procedures.
[4] M. Degermark, B. Nordgren and S. Pink, 'RFC 2057-IP Header Compression', Internet RFC Archives, February 1999.
[5] 3GPP TS 25.427, UTRAN Iur and Iub Interface User Plan Protocols for DCH Data Streams.
[6] 3GPP TS 25.413, UTRAN Iu Interface Radio Access Network Application Part (RANAP) Signaling.
[7] 3GPP TS 26.071, AMR Speech Codec: General Description.
[8] M. Chuah, Wei Luo and X. Zhang, 'Impacts of Inactivity Timer Values on UMTS System Capacity', Wireless Communications and Networking Conference, 2002, IEEE, Vol. 2, March 17–21, 2002.
[9] 3GPP TS 25.308, UTRAN High-Speed Downlink Packet Access (HSDPA); Overall Description; Stage 2.
[10] 3GPP TR 25.858, Physical Layer Aspects of UTRAN High-Speed Downlink Packet Access.
[11] 3GPP TS 25.214, Physical Layer Procedures.
[12] 3GPP TR 25.877, High-Speed Downlink Packet Access (HSDPA) Iub/Iur Protocol Aspects.
[13] Ramon Ferrús et al., 'Cross Layer Scheduling Strategy for UMTS Downlink Enhancement', IEEE Radio Communications, June 2005.
[14] Lorenzo Caponi, Francesco Chiti and Romano Fantacci, 'A Dynamic Rate Allocation Technique for Wireless Communication Systems', IEEE International Conference on Communications, Vol. 7, June 20–4, 2004.
[15] 3GPP TS 25.306, UE Radio Access Capabilities Definition.
[16] 3GPP TR 25.896, Feasibility Study for Enhanced Uplink for UTRAN FDD.
[17] 3GPP TS 25.309, FDD Enhanced Uplink: Overall Description, Stage 2.
[18] 3GPP TS 25.213, Spreading and Modulation (FDD).

4

Wireless Local Area Network (WLAN)

In the mid-1990s, the first wireless LAN devices appeared on the market, but did not get a lot of consumer attention. This changed rapidly at the beginning of this decade, when the hardware became affordable, and wireless LAN quickly became the standard technology to interconnect computers wirelessly with each other and the Internet. Chapter 4 takes a closer look at this system, which was standardized by the IEEE (Institute of Electrical and Electronics Engineers) in the 802.11 specification [1]. The first part of this chapter describes the fundamentals of the technology. Apart from wireless Internet access at home and in public hotspots, topics like roaming and wireless bridging are also discussed. Once the system became popular, a number of inherent security flaws were discovered. The chapter therefore also focuses on these issues and shows how wireless LAN can be used securely. Wireless LAN and UMTS are often compared because they have many things in common. However, there are many differences as well. Therefore, the two systems are compared at the end of the chapter to show which applications are best suited for each.

4.1 Wireless LAN Overview

Wireless LAN received its name due to the fact that it is primarily based on existing LAN standards. These standards were initially created by the IEEE for wired interconnection of computers and can be found in the 802.X standards (e.g. 802.3 [2]). Generally, these standards are known as 'Ethernet' standards. The wireless variant, which is generally known as wireless LAN (WLAN), is specified in the 802.11 standard. As shown in Figure 4.1, its main application today is to transport IP packets over layer 3 of the OSI model. Layer 2, the data link layer, has been adapted from the wired world with relatively few changes. To address the wireless nature of the network, a number of management operations have been defined, which are described in Section 4.2. Only layer 1, the physical layer, is a new development, as WLAN uses airwaves instead of cables to transport data frames.

Communication Systems for the Mobile Information Society   Martin Sauter
© 2006 John Wiley & Sons, Ltd

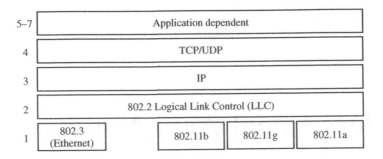

Figure 4.1 The WLAN protocol stack

4.2 Transmission Speeds and Standards

Since the creation of the 802.11 standard, various enhancements have followed. Therefore, a number of different physical layers exist today, abbreviated as 'PHY' in the standard documents. Each PHY has been defined in a different document and a letter has been put at the end of the initial 802.11 document name to identify the different PHYs. See Table 4.1.

The breakthrough for WLAN was the emergence of the 802.11b standard that offers data rates from 1 to 11 Mbit/s. The maximum data rate that can be achieved in a real environment mainly depends on the distance between sender and receiver as well as on the number and kind of obstacles between them such as walls or ceilings – 11 Mbit/s can only be achieved over short distances of a few meters.

In order to ensure connectivity over a larger distance, the number of bits used for redundancy is automatically adapted. This reduces the speed down to 1 Mbit/s under very bad conditions. Many vendors specify a maximum range of their WLAN adapters of up to 300 m. In practice, such a distance is only achieved outdoors where no obstacles absorb signal energy and only at a speed of 1 Mbit/s.

The 802.11b standard uses the 2.4 GHz ISM (industrial, scientific, and medical) band, which can be used in most countries without a license. One of the most important conditions for the license-free use of this frequency band is the limitation of the maximum transmission power to 100 mW. It is also important to know that the ISM band is not technology restricted. Other wireless systems such as Bluetooth also use this frequency range.

Table 4.1 Different PHY standards

Standard	Frequency band	Speed
802.11b [7]	2.4 GHz (2.401–2.483 GHz)	1–11 Mbit/s
802.11g [8]	2.4 GHz (2.401–2.483 GHz)	6–54 Mbit/s
802.11a [9]	5 GHz (5.170–5.250 GHz)	6–54 Mbit/s

The 802.11g standard specifies a much more complicated PHY as compared to the 802.11b standard, in order to achieve data rates of up to 54 Mbit/s. This variant of the standard also uses the 2.4 GHz ISM band and has been designed in a way to be backward compatible to older 802.11b systems. This ensures that 802.11b devices can communicate in new 802.11g networks and vice versa. More about the different PHYs can be found in Section 4.6.

Another frequency range was opened for WLANs in the 5 GHz band in addition to the 2.4 GHz ISM band. This frequency band is used by the 802.11a standard. This standard also specifies data rates of up to 54 Mbit/s. As a new frequency range is used, pure 802.11a devices are not backward compatible to devices that only operate in the 2.4 GHz band. Many vendors therefore offer dual-mode devices that can be used in both the 2.4 and 5 GHz bands. Therefore, care should be taken when buying an 802.11a device, as most public hotspots only operate in the 2.4 GHz band.

Some vendors are also offering products with their own proprietary extensions to increase the transmission speeds. These higher speeds can only be used if sender and receiver are from the same manufacturer.

Additional 802.11 standards, which are shown in the Table 4.2, specify a number of additional optional WLAN capabilities.

Table 4.2   Additional 802.11 standard documents that describe optional functionality

Standard	Content
802.11e [10]	The most important new functionalities of this standard are methods to ensure a certain quality of service (QoS) for a device. Therefore it is possible to ensure a minimum bandwidth and fast media access for real-time applications like voice over IP (VoIP) even during network congestion periods. Furthermore this standard also specifies the direct link protocol (DLP), which enables two WLAN devices to exchange data directly with each other instead of communicating via an access point. DLP can effectively double the maximum data transfer speed between two devices
802.11f [11]	This standard specifies the exchange of information between access points to allow seamless client roaming between cooperating access points. It is used in practice to extend the range of a WLAN network. More about this topic can be found in Section 4.3.1
802.11h [12]	This extension adds power control and dynamic frequency selection for WLAN systems in the 5 GHz band. In Europe only 802.11a devices can be sold that comply with the 802.11h extensions
802.11i [13]	This standard describes new authentication and encryption methods for WLAN. The most important part of 802.11i is 802.1x. More about this topic can be found in Section 4.7

4.3 WLAN Configurations: From Ad-hoc to Wireless Bridging

All devices that use the same transmission channel to exchange data with each other form a basic service set (BSS). The definition of the BSS also includes the geographical area covered by the network. There are a number of different BSS operating modes.

4.3.1 Ad-hoc, BSS, ESS, and Wireless Bridging

In ad-hoc mode, also referred to as independent BSS (IBSS), two or more wireless devices communicate with each other directly. Every station is equal in the system and data is exchanged directly between two devices. The ad-hoc mode therefore works just like a standard wireline Ethernet, where all devices are equal and where data packets are exchanged directly between two devices. As all devices share the same transport medium (the airwaves), the packets are received by all stations that observe the channel. However, all stations except the intended recipient discard the incoming packets because the destination address is not equal to their hardware address. All participants of an ad-hoc network have to configure a number of parameters before they can join the network. The most important parameter is the service set ID (SSID), which serves as the network name. Furthermore, all users have to select the same frequency channel number (some implementations select a channel automatically) and ciphering key. While it is possible to use an ad-hoc network without ciphering, it poses a great security risk and is therefore not advisable. Finally, an individual IP address has to be configured in every device, which the participants of the network have to agree on. Due to the number of different parameters that have to be set manually, WLAN ad-hoc networks are not very common.

One of the main applications of a WLAN network is the access to a local network and the Internet. For this purpose, the infrastructure BSS mode is much more suitable then the previously described ad-hoc mode. In contrast to an ad-hoc network, it uses an access point (AP), which takes a central role in the network as shown in Figure 4.2.

The access point can be used as a gateway between the wireless and the wireline network for all devices of the BSS. Furthermore, devices in an infrastructure BSS do not communicate directly with each other. Instead they always use the access point as a relay. If device A, for example, wants to send a data packet to device B, the packet is first sent to the access point. The access point analyzes the destination address of the packet and then forwards the

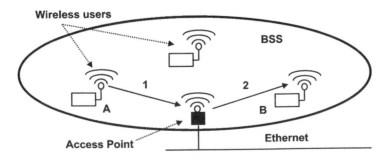

Figure 4.2 Infrastructure BSS

packet to device B. In this way it is possible to reach devices in the wireless and wireline network without knowledge of where the client device is. The second advantage to using the access point as a relay is that two wireless devices can communicate with each other over larger distances with the access point in the middle. In this scenario, shown in Figure 4.2, the transmit power of each device is enough to reach the access point but not the other device because it is too far away. The access point, however, is close enough to both devices and can thus forward the packet. The disadvantage of this method is that a packet that is transmitted between two wireless devices has to be transmitted twice over the air. Thus the available bandwidth is cut in half. Due to this reason, the 802.11e standard introduces the DLP. With DLP, two wireless devices can communicate directly with each other while still being members of an infrastructure BSS. However, this functionality is declared as optional in the standard.

WLAN access points usually fulfill a number of additional tasks. Here are some examples:

- 10/100 Mbit/s ports for wireline Ethernet devices. Thus, the access point also acts as a layer 2 switch.
- At home a WLAN access point is often used as an IP router to the Internet and can be connected via Ethernet to a DSL- or cable modem.
- To configure devices automatically, a DHCP (dynamic host configuration protocol) server [3] is usually also integrated into an access point. The DHCP server returns all necessary configuration information like the IP address for the device, the DNS server IP address, and the IP address of the Internet gateway.

Furthermore, WLAN access points can also include a DSL or cable modem. This is quite convenient as fewer devices have to be connected to each other and only a single power supply is needed to connect the home network to the Internet. A block diagram of such a fully integrated access point is shown in Figure 4.3.

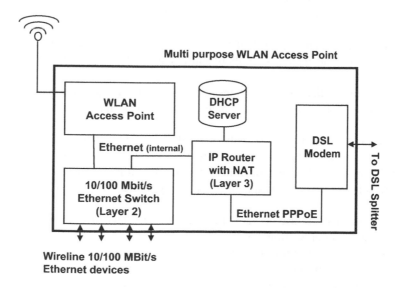

Figure 4.3   Access point, IP router, and DSL modem in a single device

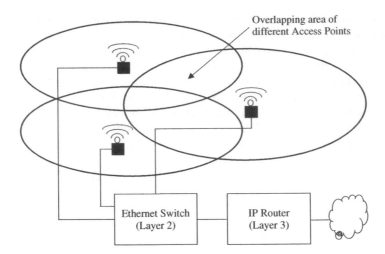

Figure 4.4 ESS with three access points

The transmission power of a WLAN access point is low and can thus only cover a small area. To increase the range of a network, several access points can be used that cooperate with each other. If a mobile user changes his position and the network card detects that a different access point has a better signal quality, it automatically registers with the new access point. Such a configuration is called an extended service set (ESS) and is shown in Figure 4.4. When a device registers with another access point of the ESS, the new access point informs the previous access point of the change. This is usually done via a direct Ethernet connection between the access points of an ESS, and referred to as the 'distribution system'. Afterwards, all packets arriving in the wired distribution system, e.g. from the Internet, will be delivered to the wireless device via the new access point. As the old access point was informed of the location change, it ignores the incoming packets. The change of access points is transparent for the higher layers of the protocol stack on the client device. Therefore, the mobile device can keep its IP address and only a short interruption of the data transfer will occur.

In order to allow a client device to transparently switch over to a new access point of an ESS, the following parameters have to match on all access points:

- All access points of an ESS have to be located in the same IP subnet. This implies that no IP routers can be used between the access points. Ethernet hubs, which switch packets on layer 2, can be used. In practice, this limits the maximum coverage area substantially because IP subnets are only suitable to cover a very limited area like a building or several floors.
- All access points have to use the same BSS service ID, also called an 'SSID'. More about SSIDs can be found in Section 4.3.2.
- The access points have to transmit on different frequencies and should stick to a certain frequency repetition pattern as shown in Figure 4.5.
- Many access points use a proprietary protocol to exchange user information with each other if the client device switches to a new access point. Therefore, all access points

of an ESS should be from the same manufacturer. To allow the use of access points of different manufacturers the IEEE released the 802.11f standard (Recommended Practice for Multi-Vendor Access Point Interoperability) at the beginning of 2003. However, this standard is optional and by no means binding for manufacturers.

- The coverage area of the different access points should overlap somewhat for client devices not to lose coverage in border areas. As different access points send on different frequencies, the overlapping poses no problem.

Another WLAN mode is wireless bridging, sometimes also referred to as a wireless distribution system. In this mode, the access points of an ESS can wirelessly forward packets they have received from client devices between each other. In practice, this mode is used if only one connection to the wired network exists but a single access point is unable to cover the desired area on its own. Usually, a wireless bridging access point also supports simultaneous BSS functionality. Therefore only a single access point is required to offer service at a certain location to users and to backhaul the packets to the access point connected to the Internet.

4.3.2 SSID and Frequency Selection

When an access point is configured for the first time, there are two basic parameters that have to be set.

The first parameter is the basic service set ID (SSID). The SSID is periodically broadcast over the air interface by the access point inside beacon frames, which are further discussed in Section 4.4. Note that the 802.11 standard uses the term 'frame' synonymously for 'packet' and this chapter also makes frequent use of it. The SSID identifies the access point and allows the operation of several access points at the same location for access to different networks. Such a configuration of independent access points should not be confused with an ESS, in which all access points work together and have the same SSID. Usually the SSID is a text string in a human readable form, because during the configuration of the client device the user has to select an SSID if several are found. Many configuration programs on client devices also refer to the SSID as the 'network name'.

The second parameter is the frequency or channel number. It should be set carefully if several access points have to coexist in the same area. The ISM band in the 2.4 GHz range uses frequencies from 2.410 MHz to 2.483 MHz. Depending on national regulations, this range is divided into a maximum of 11 (US) to 13 (Europe) channels of 5 MHz each. As a WLAN channel requires a bandwidth of 25 MHz, different access points at close range should be separated by at least five ISM channels. As can be seen in Figure 4.5, three infrastructure BSS networks can be supported in the same area or a single ESS with overlapping areas of three access points. For infrastructure BSS networks, the overlapping is usually not desired but cannot be prevented if different companies or home users operate their access points close to each other. In order to be able to keep the three access points at least five channels apart from each other, channels 1, 6, and 11 should be used.

In practice, channels 12 and 13 are only allowed for use in Europe. Unfortunately many WLAN card drivers do not ask during software installation in which country the device is going to be used and block these channels by default. If it is unclear during the installation

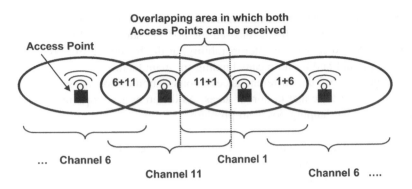

Figure 4.5 Overlapping coverage of access points forming an ESS

of a new access point which devices will be used in the network, channels 12 or 13 should not be selected to enable all client devices to communicate with the access point.

802.11a systems use the spectrum in the 5 GHz range between 5.170–5.250 GHz for data transmission. As a single WLAN channel uses 20 MHz, up to four channels can be used in an overlapping fashion without interference. Unlike access points in the 2.4 GHz range, many access points for this frequency range can only be configured for channels 36, 40, 44, and 48. This makes the selection of a correct channel easier and prevents a partial overlap of independent networks and the resulting interference.

On a client device, the basic configuration for joining a BSS or ESS network is a lot simpler. To join a new network, the device automatically searches for active access points on all possible frequencies and presents the SSIDs it has discovered to the user. The user can then select the desired SSID of the network to join. Selecting a frequency is not necessary, as the client device will always scan all frequencies for the configured SSID during power up. If more than one access point is found with the same SSID during the network search procedure, the client device assumes that they belong to the same ESS. If the user wants to join such a network, the device then selects the access point on the frequency on which the beacon frames are received with the highest signal strength. Further details about this process can be found in the Section 4.4.

It is also possible to leave the SSID field blank on the client device. In this case the device will automatically register with any access point it finds which does not have encryption turned on. Such a configuration is helpful if a device is mainly used in public hotspots of different operators.

Many devices offer to store several network configurations. This is especially useful for mobile devices like PDAs or notebooks, which are often used in different networks.

The user interface for configuring WLAN access is not standardized and thus the implementation depends on the device and the operating system. Some devices are locked to a specific profile until the user manually changes to another profile. The WLAN configuration support of the Windows XP operating system on the other hand behaves quite differently. Here, one of the pre-configured profiles is automatically selected after activation of the WLAN card depending on the SSIDs found during the network search procedure. See Figure 4.6.

In addition to configuring the SSID and frequency channel, activating encryption for the air interface is the third important step while setting up a BSS or ESS. Most access points

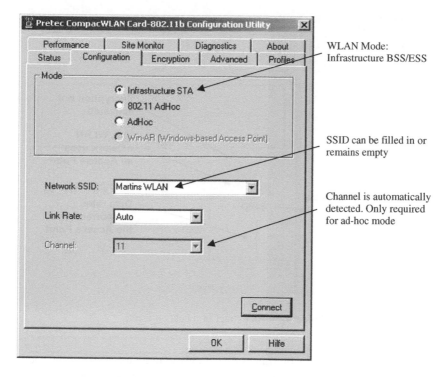

Figure 4.6   Client device configuration for a BSS or ESS

have encryption turned off by default when installed for the first time. This poses a great security risk and the user should therefore turn on encryption immediately during the initial configuration. Encryption is discussed in more detail in Section 4.7.

4.4 Management Operations

In a wired Ethernet it is usually sufficient to connect the client device via cable to the nearest hub or switch to get access to the network. Physically connecting a wireless device to a WLAN network is of course not possible, as there is no cable. Also, a WLAN device has the ability to automatically roam between different access points of an ESS and is able to encrypt data packets on layer 2 of the protocol stack. As all of these WLAN operations have to be coordinated between the access points and the user devices, the 802.11 standard specifies a number of management operations and messages on layer 2, as well as additional information elements in the MAC header of data packets which are not found in a wired Ethernet.

The access point has a central role in a BSS and is usually also used as a bridge to the wired Ethernet. Therefore wireless clients always forward their packets to the access point, which then forwards them to the wireless or wired destination devices. In order to allow wireless clients to detect the presence of an access point, beacon frames are broadcast by the access point periodically. A typical value of the beacon frame interval is 100 milliseconds.

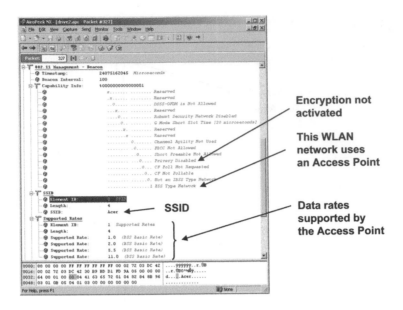

Encryption not activated

This WLAN network uses an Access Point

SSID

Data rates supported by the Access Point

Figure 4.7   An extract of a beacon frame

As can be seen in Figure 4.7, beacon frames do not only contain the SSID of the access point, but also inform the client devices about a number of other functionalities and options in a number of information elements (IEs). One of these information elements is the capability IE. Each bit of this two-byte IE informs a client device about the availability of a certain feature. As can be seen in Figure 4.7, the capability IE informs the client device in the fifth bit for example that ciphering is not activated (privacy disabled). Other IEs in the beacon frame are used for parameters that require more than a single bit. Each type of IE has its own ID which indicates to the client devices how to decode the data part of the information element. IE 0 for example is used to carry the SSID, while IE 1 is used to carry information about the supported data rates. As IEs have different lengths, a length field is included in every IE header. By having an identifier and a length field at the beginning of each IE, a client device is able to skip over optional IEs it does not recognize. Such IEs might be present in beacon frames of new access points that offer functionality that older client devices might not have implemented. This ensures backward compatibility to older devices.

During a network search, a client device has two ways to find available access points. One way is to passively scan all possible frequencies and just wait for the reception of a beacon frame. In order to speed up the search, a device can also send probe request frames to trigger an access point to send its system information in a probe response frame, without waiting for the beacon frame interval to expire. Most client devices make use of both methods to scan the complete frequency range as quickly as possible.

Once a client device has found a suitable access point, it has to perform an authentication procedure. Two authentication options have been defined in the standard.

The first authentication option is called open system authentication. The name is quite misleading as this option performs no authentication at all. The device simply sends an authentication frame with an authentication request to the access point, asking for open system authentication. No further information is given to the access point. If the access point allows this 'authentication' method, it returns a positive status code and the client device is 'authenticated'.

The second authentication option is called shared key authentication. This option uses a shared key to authenticate client devices. During the authentication procedure, the access point challenges the client device with a randomly generated text. The client device then encrypts this text with the shared key and returns the result to the access point. The access point performs the same operation and compares the result with the answer from the client device. The results can only match if both devices have used the same key to encrypt the message. If the access point was able to validate the client's response, it finishes the procedure as shown in Figure 4.8 and the client is authenticated. Note that the use of the same key for all client devices can be a great security risk. This is further discussed in Section 4.7.

Once authenticated successfully, the client device has to perform an association procedure with the access point. The access point answers an association request message by returning an association response message, which once more contains all necessary information about the wireless network, for example the capability IE. Furthermore, the access point assigns an association ID, which is also included in the association response message. It is used later by the client device to enter power-saving mode. Authentication and association with an access point are two separate procedures. This allows a client device to quickly roam between different access points. Once a device is authenticated by all access points it only has to perform an association procedure to roam from one access point to another.

Figure 4.8 shows the message flows of the authentication and association procedures. Acknowledgment frames (see Section 4.5) are not shown for clarity.

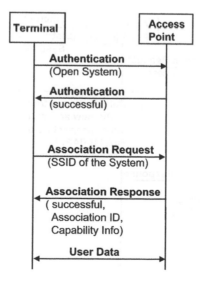

Figure 4.8 Authentication and association of a client device with an access point

Once the association with an access point has been performed successfully, user data packets can be exchanged. As a client device is informed via the capability IE if wired equivalent privacy (WEP) encryption is activated for the network, it automatically starts ciphering all subsequent frames if the corresponding bit is set in the IE. As standard WEP encryption contains a number of severe security flaws, new algorithms and procedures have been standardized which are becoming more available in new products. More about this topic can be found in Section 4.7.

Authentication and encryption are independent from each other. Therefore access points are usually configured to use the open system 'authentication' and only use the shared secret key for encryption of the data packets. Devices that do not know the shared secret key or use an invalid key can therefore authenticate and associate successfully with an access point but cannot exchange user data, as the encrypted packets cannot be decrypted by the receiver. Some access point manufacturers offer the option of specifically activating the shared authentication procedure. However, this does not increase the security of the system in any way. Usually, it just further complicates the initial configuration of a client device because the shared authentication procedure must be manually activated by the user.

If a client device resides in an ESS with several access points (see Figure 4.4) it can change to a different access point which is received with a better signal level at any time. The corresponding reassociation procedure is shown in Figure 4.9. In order to be able to find the access points of an ESS, the client device scans the frequency band for beacon frames of other access points while no data has to be transmitted. As all access points of the same ESS transmit beacon frames containing the same SSID, client devices can easily distinguish between access points belonging to the current ESS and access points of other networks. In order to change to a new access point, the client device changes to the send/transmit frequency of the new access point and sends a reassociation request frame. This frame is similar to the association request frame and only contains an additional IE which contains the

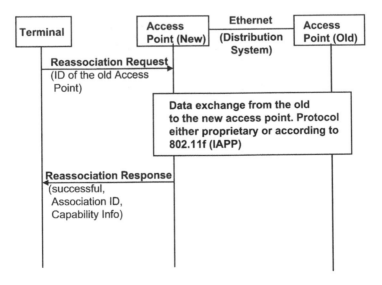

Figure 4.9 Reassociation (acknowledgment frames not shown)

ID of the access point to which the client device was previously connected to. The new access point then informs the previous access point via the wired Ethernet (distribution system) that the user has changed its association. The previous access point then acknowledges the operation and sends any buffered packets for the device to the new access point. Afterwards it deletes the hardware address and association ID from its list of served devices. In the future, all packets arriving for the client device via the wired distribution system will be ignored by the previous access point and are only forwarded to the client device by the new access point. In a last step of the procedure, the new access point sends a reassociation response message to the client device.

At first, only the message exchange between the client device and the access point were standardized for the reassociation procedure. No standard existed for the wired network between the two access points that are part of the procedure. Therefore manufacturers developed their own proprietary messages to fill the gap. This is the reason why today only access points from the same manufacturer should be used to form an ESS in order to ensure a flawless roaming of the client devices. Recently the 802.11f inter access point protocol (IAPP) recommendation was released by the IEEE, which finally standardizes the exchange of messages between access points. As the implementation of the 802.11f standard is optional, it is up to the user to verify the compliancy of an access point to this standard.

The 802.11 standard also offers a power-saving (PS) mode in order to increase the operation time of battery-driven devices. If a device enters power-saving mode, the data transmission speed is decreased somewhat during certain situations. This is only a small disadvantage compared to the substantial reduction in power consumption that can be achieved.

The client device can enter PS mode if its transmission buffer is empty, and no data has been received from the access point for some time. In order to inform the access point that it will enter PS mode, the client device sends an empty frame to the access point with the PS bit set in the MAC header. When the access point receives such a frame, it will buffer all incoming frames for the client device for a certain time. During this time, the client device can power down the receiver. The time between reception of the last frame and activation of the PS mode is controlled by the client device. Many devices use a timeout in the order of 20–25 seconds. Shorter periods are possible as well and might be useful for devices like PDAs, which only send and receive data in very irregular intervals and only for short durations.

If a client device wants to resume the data transfer, it simply activates its transceiver again and sends an empty frame containing a MAC header with the PS bit deactivated. Afterwards, the data transfer can resume immediately. See Figure 4.10.

For most applications used on mobile devices, like web browsing, data will only be delivered in rare cases once the PS mode has been activated. In order not to lose frames, they are buffered on the access point. Thus, a device in PS mode has to periodically activate its transceiver so it can be notified of buffered frames by the access point. This is done via the traffic indication map (TIM) IE, which the access point includes in every beacon frame. Each device has its own bit in the TIM, which indicates if buffered frames are waiting. The client device identifies its bit in the TIM via its association ID (AID), which was assigned by the access point to the client device during the association procedure. Up to 2007 AIDs can be assigned by each access point. Therefore the maximum size of the TIM IE is 2007 bits. In order to keep the beacon frames as small as possible, not all bits of the TIM are sent.

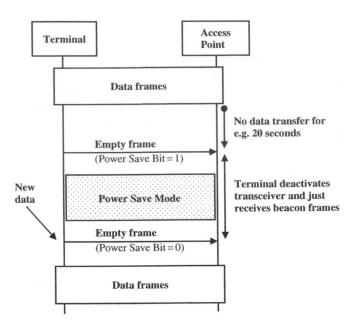

Figure 4.10   Activation and deactivation of the PS mode (acknowledgment frames not shown)

The TIM therefore contains a length and offset indicator. This makes sense as in practice only few devices are in PS mode and therefore only a few bits are required.

As beacon frames are sent in regular intervals (e.g. every 100 ms), the access point and client device agree on a listen interval during the association procedure after which the TIM has to be read. In order to negotiate the listen interval, the client device proposes an interval to the access point. If the access point accepts the proposed interval, it has to buffer any incoming frames for the device for this duration once the device activates the PS mode. It can be observed that a common listen interval is three for example. The value implies that the client device has to check only every third beacon frame and can thus switch off its transceiver for 300 ms at a time. When the client device exits PS mode temporarily to receive a beacon frame, and the TIM bit for the device is not set, the transceiver is again switched off for 300 ms before the procedure is repeated.

If the TIM bit is set, the client device does not go back to PS mode directly. Instead, a PS-poll frame is sent to the access point. The access point will send a single buffered frame to the client device for every PS-poll frame received. To inform the client device of further waiting frames, the 'more' bit in the MAC header of the frame is set. The client device then continues to send PS-poll frames as long as the 'more' bit is set in incoming frames.

Broadcast and multicast frames are buffered by the access point as well if at least one client device is currently in PS mode. Broadcast frames are not saved for every client device individually. Instead the first bit of the TIM (AID = 0) is used as an indicator by client devices in PS mode if broadcast data is buffered. These frames are then automatically sent after a beacon frame which includes a delivery TIM (DTIM) instead of an ordinary TIM.

In order for client devices to be able to activate the receiver at the right time for buffered broadcast frames, a countdown timer inside the TIM announces the transmission of a DTIM.

4.5 The MAC Layer

The medium access control protocol (MAC) on layer 2 has similar tasks in a WLAN as in a fixed-line Ethernet:

- It controls access of the client devices to the air interface.
- A MAC header is put in front of every frame which contains among other parameters the (MAC) address of the sender (source) of the frame and the (MAC) address of the recipient (destination).

4.5.1 Air Interface Access Control

As the air interface is a very unreliable transmission medium, a recipient of a packet is required to send an acknowledgment (ACK) frame to inform the sender of the correct reception of the frame. This is a big difference to a wired Ethernet, where frames are not acknowledged. In all previous figures in this chapter, ACK frames were not shown for easier interpretation of the content. Figure 4.11 shows for the first time how frames are exchanged between a client device and an access point including the ACK frames. Each frame, regardless of whether it contains user data or management information (authentication, association, etc.), has to be acknowledged with an ACK frame. The same or a different client device is only allowed to send the next frame once the ACK frame has been received. If no ACK frame is received within a certain time, the sender assumes that the frame was lost and thus resends the frame.

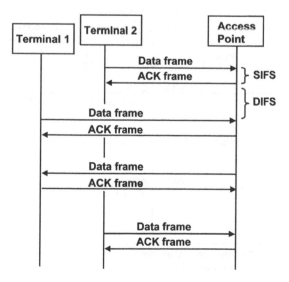

Figure 4.11 Acknowledgment for every frame and required interframe space periods

To ensure that the ACK frame can be sent before another device attempts to send a new data frame, the ACK frame is sent almost immediately after the data frame has been received. There is only a short delay between the two frames, the short interframe space (SIFS). All other devices have to delay their transmission by at least a DCF interframe space (distributed coordination function interframe space, or DIFS for short).

Optionally devices can also reserve the air interface prior to the transmission of a data frame. This might be useful in situations where devices can reach the access point but are too far away from each other to receive each other's frames. Under these circumstances it can happen that two stations might attempt to send a frame to the access point at the same time. As the two frames will interfere with each other, the access point will not be able to receive either of the frames correctly. This scenario is also known as the 'hidden station problem'. To prevent such an overlap, a device can reserve the air interface as shown in Figure 4.12 by sending a short RTS (ready to send) frame to the access point. The access point then answers with a CTS (clear to send) frame and the air interface is reserved. While the RTS frame might not be seen by all client devices in the network due to the large distance between them, the CTS frame can be seen by all devices because the access point is the central point of the network. Both RTS and CTS frames contain a so-called network allocation vector (NAV) to inform other devices for which period of time the air interface is reserved. If a device uses an RTS/CTS sequence before sending, a frame can be configured in the driver settings dialog box of the network card. However, RTS/CTS sequences slow down the throughput of a device. Therefore this mechanism should only be used if a very high network load is expected and the client devices are dispersed over a wider area.

As in a wired network, there is no central instance that controls which device is allowed to send a frame at a certain time. Every device has to decide on its own when it can send a frame. To minimize the chance of a collision with frames of other devices, a

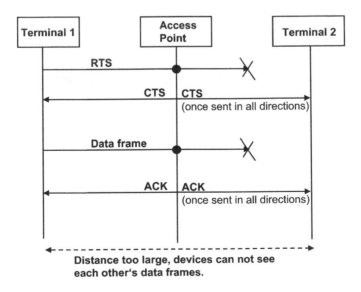

Figure 4.12 Reservation of the air interface via RTS/CTS frames

coordination function is necessary. In WLAN networks, the distributed coordination function (DCF) is used for this purpose. Distributing coordination to access the air interface is a completely different approach from that taken by all other systems described in this book. The other systems use a central logic that decides which user device is allowed to send at a certain time and for how long. The advantage of the DCF, however, is the easy implementation in all devices. The biggest disadvantage is the fact that no bandwidth can be reserved or guaranteed. This is mainly a problem for real-time applications like voice or video telephony if the network is already highly loaded with other traffic. As voice and video telephony over IP and over WLAN becomes more and more popular, the IEEE has released the 802.11e standard for devices and applications that require a constant bandwidth and a deterministic medium access time. With this enhancement, devices can request a certain quality of service from the access point to get precedence over transmissions from other devices. The enhancements also include a method to assign the air interface to a device for a specific time and thus guarantee a certain bandwidth and a maximum medium access time. 802.11e is backward compatible to today's 802.11b, g, and a standards. Older devices that do not support the new standard can still be used in such a network without degrading the new quality of service mechanism offered by the 802.11e standard.

Back to the standard 802.11b DCF medium access scheme: DCF uses carrier sense multiple access/collision avoidance (CSMA/CA) to detect if another device is currently transmitting a frame. This method is quite similar to CSMA/CD (CSMA/collision detect) which is used in fixed-line Ethernet but offers a number of additional functionalities to avoid collisions.

If a device wants to send a data packet and no activity is detected on the air interface, the packet can be sent without delay. If another device is already sending a data packet, the device has to wait until the data transfer has finished. Afterwards, the device has to observe another delay time, the DIFS period, which has been described above. Afterwards, the device yet again defers sending its packet for additional backoff time, which is generated by a random number generator. Therefore it becomes very unlikely that several devices attempt to send data waiting in their output queue at the same time. The device with the smallest backup time will send its data first. All other devices will see the transmission, stop their backup timer and repeat the procedure once the transmission is over. If despite of this procedure two devices still attempt to send packets at the same time the transmissions will interfere with each other and thus no acknowledgment frame is sent. Both stations then have to retransmit their packet. If a collision occurs, the maximum possible backup time from which the random generator can choose is increased in the affected devices. This ensures that even in a high-load situation the number of collisions remains small.

The backoff time is divided into slots of 20 microseconds. For the first transmission attempts, the random generator will select one of 31 possible slots. If the transmission fails, the window size is increased to 63 slots, afterwards to 127 slots and so on. The maximum window size is 1023 slots, which equals 2 milliseconds.

In addition to the detection of an ongoing transmission and the use of a backoff time, each packet header contains an NAV field to inform the other devices of the time required to send the current frame and the following ACK frame. This additional feature is especially useful if the air interface is reserved via RTS and CTS frames as shown in Figure 4.12. Here, the first RTS frame contains the duration required to send the subsequent CTS frame, the actual data frame, and the final ACK frame. The following CTS frame of the other device contains a slightly smaller NAV, which only contains the transmission duration for the subsequent data frame and the final ACK frame.

4.5.2 The MAC Header

The most important function of the MAC header is to address the devices in the local network. This is done by using 48-bit MAC addresses for the sender (source) and receiver (destination). The WLAN MAC addresses are identical to the MAC addresses that are used in a wired Ethernet. In a WLAN basic service set (BSS), however, a frame is not sent directly from sender to receiver but is always sent to the access point first. Due to this, three MAC addresses are part of the MAC header as shown in Figure 4.13. The third MAC address is the access point address. When the access point receives a frame, it uses the destination address

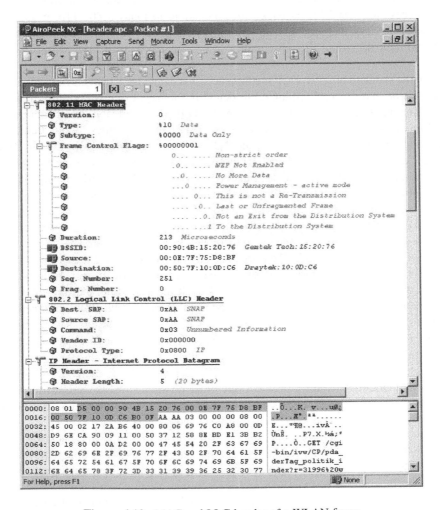

Figure 4.13 MAC and LLC header of a WLAN frame

to decide if the receiver is a fixed or a wireless client and forwards the frame accordingly. Therefore a client device does not need to know if the destination device is a wireless or fixed Ethernet device.

Other important fields of the MAC header are the frame type and subtype. The frame type field informs the receiver if the current frame is a user data frame, a management frame (e.g. association request), or a control frame (e.g. ACK). Depending on the type of frame, the subtype field contains further information. For management frames, it indicates which management operation is contained in the frame (e.g. authentication, association, beacon frame, etc.).

The frame control flags are used to exchange additional management information between two devices. They are used for example to indicate to the destination if the user data is encrypted (WEP-enabled bit), if the device is about to change into PS mode (power management bit), or if the frame is intended for an access point ('to distribution system' bit).

If the frame contains user data, the logical link control header (LLC header, layer 2) follows the MAC header. The most important job of the LLC header is to identify which protocol is used on layer 3.

4.6 The Physical Layer

On layer 1, the physical layer, there are different modulation standards as shown in Section 4.2, which are defined in the IEEE 802.11b, g and a standards.

4.6.1 IEEE 802.11b – 11 Mbit/s

The breakthrough of WLAN in the consumer market was triggered by the introduction of devices compliant to the 802.11b standard with a maximum speed of up to 11 Mbit/s. More recent physical layers described in the 802.11g or 802.11a standards can achieve even higher speeds with the same bandwidth requirement of 22 MHz. More about the different physical layers can be found later on in this chapter. The following list shows some basic WLAN parameters and compares some of them to similar parameters of other systems:

- WLAN maximum transmission power is limited to 0.1 Watt. Mobile phone power on the other hand is limited to 1–2 Watt. GSM and UMTS base stations have a typical power output of 40–60 Watt per sector and frequency.
- Each channel has a bandwidth of 22 MHz. Up to three access points can be used at close range in the ISM band without interfering with each other. GSM uses 0.2 MHz (200 kHz) per channel, UMTS uses 5 MHz.
- Framesize: 4–4095 bytes. However, IP frames do not usually exceed 1500 bytes. This value is especially interesting to compare to other technologies: a GPRS packet for example, as shown in Section 2.3.3, consists of four bursts of 114 bits each and thus can only contain 456 bits. If coding scheme 2 for error detection and correction is used, only 240 bits or 30 bytes remain for the actual packet. Therefore an IP packet can be transmitted over a single WLAN frame, but it has to be split into several packets if it is transmitted over the air interface of a GPRS network.
- Transmission time of a large packet: the transmission time depends on the size of the packet and the transmission speed. If a large packet with a payload of 1500 bytes is

transmitted with a speed of 1 Mbit/s, the transmission takes about 12 milliseconds. If reception conditions are good and the packet is sent with a transmission speed of 11 Mbit/s, the same transmission takes only 1.1 milliseconds. Note, that the SIFS and the time it takes to send a short ACK frame as confirmation has to be added to these values to calculate the precise transmission time.

- Time between a data frame and an ACK frame (SIFS): 10 microseconds or 0.01 milliseconds.
- If a transmission error occurs, a back-off procedure is performed as described in previous section. A back-off slot (of which 63 exist for the first retry) has a length of 20 microseconds or 0.02 milliseconds.
- At the beginning of the frame, a preamble is sent which notifies all other devices that the transmission of a frame is about to start. The preamble is necessary in order to synchronize all listeners to the start of the frame. The preamble has a length of 144 microseconds or 0.144 milliseconds.

The preamble mentioned in the list above is part of the physical layer convergence procedure (PLCP) header, which is sent at the start of every frame. The PLCP header also contains information about the data rate used for the subsequent MAC frame. With the 802.11b standard, the MAC frame can be sent with 1, 2, 5.5, and 11 Mbit/s. This flexibility is necessary, as devices experiencing bad radio conditions can only send and transmit with a lower speed to compensate unfavorable radio conditions with a higher redundancy. In practice, the sender decides on its own which coding to use for a frame. Some devices also offer the possibility for the user to manually lock the speed to a fixed value (e.g. 5.5 Mbit/s). This helps in situations when the automatic speed selection algorithm cannot find a good value on its own. Access points by some manufacturers note the speed at which individual stations send their frames and then use the same speed for subsequent packets to the clients. Beacon frames are usually sent at a speed of 1 or 2 Mbit/s. This allows even distant devices to detect the presence of an access point. However, this behavior is not mandatory and some access points send their beacon frames at a speed of 11 Mbit/s. This increases the overall speed of the network slightly, but has some disadvantages for distant devices.

For the coding of the actual user data in a frame, the direct sequence spread spectrum (DSSS) method is used for transmission speeds of 1 and 2 Mbit/s. Instead of transferring the bit itself, the DSSS algorithm converts the bit into 11 chips which are then transmitted over the air. Instead of sending a bit with the value of '1' the chip sequence '0,1,0,0,1,0,0,0,1,1,1' is transmitted. For a bit with the value of '0', the sequence is '1,0,1,1,0,1,1,1,0,0,0'. These sequences are also known as Barker code. As 11 values are transmitted instead of only one, redundancy increases substantially. Thus, a bit can still be received correctly even if some of the chips cannot be decoded correctly at the receiver site.

UMTS also makes use of this technique, also known as spreading, in order to increase redundancy. However, there is an important difference. In a WLAN network, only a single station sends at a time (time division multiple access). UMTS additionally uses spreading to allow several devices to send at the same time (code division multiple access). This is possible in UMTS, as shown in Chapter 3, as orthogonal codes are used instead of fixed Barker sequences.

Once a bit has been converted into chips, the Barker chip sequence is sent over the air using differential binary phase shift keying (DBPSK) with a transmission speed of 11 MChips/s.

The resulting bit rate is 1 Mbit/s. To transmit chips, DBPSK changes the phase of the signal for a '1' chip 180 degrees. For a '0' chip, no phase of the carrier frequency remains unchanged.

In order to achieve a bit rate of 2 Mbit/s, DBPSK is replaced with differential quadrature phase shift keying (DQPSK) modulation. Instead of one chip per transmission step, two chips are transmitted. The four (quadrature) possible values (00, 01, 10, or 11) of the two chips are encoded into 90 degree phase shifts of the carrier frequency per transmission step.

To further increase the data transmission speed without increasing the necessary bandwidth, the 802.11b standard introduces complementary code keying (CCK), also known as high rate DSSS (HR/DSSS). Instead of coding a single bit into an 11-chip Barker sequence, CCK encodes the bits as follows.

For a data rate of 11 Mbit/s, all bits of a frame are arranged into blocks of eight bits as shown in Figure 4.14. The first two bits of a block are then transmitted using DQPSK and are encoded in phase shifts of 90 degrees.

The remaining six bits are used to generate an eight-chip symbol. As six bits are coded in an eight-bit symbol, the process adds some redundancy. The symbol is then split into four parts of two chips each, which are then modulated onto the carrier frequency as phase changes.

As the chip rate remains the same as for 1 and 2 Mbit/s transmissions, CCK raises transmission speeds to 11 Mbit/s. A disadvantage compared to lower transmission speeds, however, is the fact that there is less redundancy in the resulting chip stream.

Different devices can send at different speeds depending on their reception conditions. Therefore it has to be ensured that even devices that cannot receive high data rate frames correctly are not causing collisions because of their inability to detect an ongoing transmission of a high data rate frame. Therefore it has to be ensured that at least the beginning of the frame can be received correctly by all devices. Thus the PLCP header of a frame is always sent at a speed of 1 Mbit/s regardless of the speed and coding scheme of the rest of the frame. To inform other stations about the duration of the transmission, the PLCP header also contains information about the total duration of the transmission.

If the actual speed of an 11 Mbit/s WLAN network is compared to a 10 Mbit/s fixed Ethernet, it becomes apparent that is not quite as fast as its fixed-line counterpart. A 10 Mbit/s

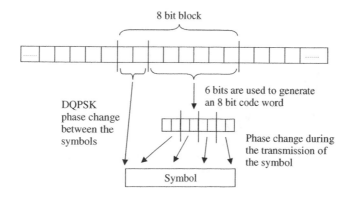

Figure 4.14 Complementary code keying for 11 Mbit/s transmissions

fixed-line Ethernet can reach a maximum speed of about 700–800 kbytes/s under ideal conditions. In an 11 Mbit/s WLAN, the maximum speed is about 300 kbytes/s between two wireless devices. This is due to the following properties that have already been described in this chapter:

- The PLCP header of a WLAN frame is always transmitted at 1 Mbit/s.
- Each frame has to be acknowledged with an ACK frame.
- In a wired network, a frame is sent directly from the source to the destination device. In a WLAN BSS, a frame is sent from the source to the access point, which then forwards the frame to the destination. The frame therefore traverses the air interface twice. In practice, this cuts the maximum transmission speed in half.

4.6.2 IEEE 802.11g with up to 54 Mbit/s

In order to further increase data transmission speeds, the 802.11g standard introduces a new modulation scheme called orthogonal frequency division multiplexing (OFDM). This modulation scheme enables speeds up to 54 Mbit/s while using almost the same bandwidth as the 802.11b standard.

The OFDM modulation scheme is fundamentally different from the modulation schemes used in the 802.11b standard. As shown in Figure 4.15, OFDM divides the bandwidth of a single channel of about 20 MHz into 52 subchannels.

The subchannels are 'orthogonal', as the amplitudes of the neighboring subchannels are exactly zero at the middle frequency of a subchannel. Therefore they do not influence the amplitude of neighboring subchannels. OFDM does not transmit data by changing the phase of the carrier but by changing the amplitudes of the subchannels. Depending on the reception quality, a varying number of amplitude levels are used to encode a varying number of bits.

In order to demodulate the signal, the receiver performs an FFT (fast Fourier transformation) analysis for each transmission step. This method calculates the signal energy (amplitude) over the frequency band. The simplified result of an FFT analysis is shown in Figure 4.15. The x-axis represents the frequency band instead of the time as in most other graphs. The amplitude of each subchannel is shown on the y-axis.

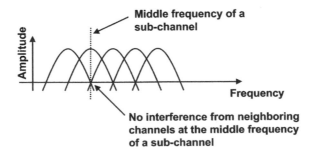

Figure 4.15 Simplified representation of OFDM subchannels

Table 4.3 802.11g data rates

Speed (Mbit/s)	Modulation and coding	Coded bits per channel	Coded bits in 48 channels	Bits per step
6	BPSK, R = 1/2	1	48	24
9	BPSK, R = 3/4	1	48	36
12	QPSK, R = 1/2	2	96	48
18	QPSK, R = 3/4	2	96	72
24	16-QAM, R = 1/2	4	192	96
36	16-QAM, R = 3/4	4	192	144
48	64-QAM, R = 2/3	6	288	192
54	64-QAM, R = 3/4	6	288	216

Table 4.3 gives an overview of the data rates offered by the 802.11g standard. In practice, an algorithm dynamically selects the best settings depending on reception conditions.

Under ideal transmission conditions, the 64-quadrature amplitude modulation (64-QAM) can be used. Together with a 3/4 convolutional coder (three data bits are coded in four output bits) and a symbol speed of 250000 symbols/s, a maximum speed of 54 Mbit/s is reached (216 bits per step × 250000 symbols/s = 54 Mbit/s). Note: a similar convolutional coder for increasing redundancy is also used for GSM and UMTS (see Section 1.7.5 and Figure 1.35).

802.11g client devices and access points are backward compatible to slower 802.11b devices. This means, that 802.11g access points also support 802.11b client devices that can only communicate with up to 11 Mbit/s. 802.11g client devices can also communicate with older 802.11b access points. However, the maximum data rate is then of course limited to 11 Mbit/s. As slower 802.11b devices are not able to decode OFDM modulated frames, it is necessary for 802.11g devices in mixed configurations to use the RTC/CTS mechanism for the reservation of the air interface prior to sending a frame. This ensures that 802.11b and g devices can be used simultaneously in a BSS. Furthermore, the PLCP header of a frame is sent at 1 Mbit/s in order for all devices to be able to receive the header correctly. While these procedures ensure interoperability, performance is reduced by about 20% due to the extra overhead of the RTS and CTS packets which can only be sent at a maximum speed of 11 Mbit/s. Due to these disadvantages, a 'G-only' mode can be activated in some access points to avoid this extra overhead. This is especially useful for home networks in which only 802.11g devices are used.

Under ideal conditions, a maximum transfer rate of about 2500 kbytes per second can be observed in an 802.11g BSS. If two wireless devices communicate with each other, the maximum speed drops to about 1200 kbyte/s, as all frames are first sent to the access point which then forwards the frames to the wireless destination device. As mentioned before, the 802.11e standard aims to overcome this problem by standardizing direct client-to-client communication in a BSS, provided the devices are sufficiently close to each other. Compared to the transfer speeds of the 802.11b standard, the 802.11g data rates are a dramatic improvement on older networks with a throughput of 600 kbyte/s, or 300 kbytes/s between two wireless client devices. There is still a big gap to 100 Mbit/s wired Ethernet, in which maximum transfer speeds of over 7000 kbyte/s can be achieved.

4.6.3 IEEE 802.11a with up to 54 Mbit/s

The 802.11a standard is almost identical to the 802.11g standard. The main difference is the use of channels in the 5 GHz band, which makes it incompatible to 802.11b and g networks. Due to the fact that a different frequency band is used, 802.11a devices do not have to be backward compatible. Therefore, PLCP headers can be sent with 6 Mbit/s instead of 1 Mbit/s. 802.11a networks are thus faster than mixed 802.11g/b networks and also have a slight advantage over 802.11g networks due to transmitting the PLCP header at a higher speed.

4.7 WLAN Security

WLAN security is a widely discussed topic because the default security settings and methods do not reliably protect the user. In most cases, access points are sold with security turned off by default. If encryption is not specifically configured by the owner of the network, any wireless device can access the network without prior authorization. This configuration is used in most public hot spots to allow users to easily roam into the network. As the frames are not encrypted, however, it is easy to eavesdrop on the activities of the users. Even more critical is the use of such an open configuration for private home networks that use the wireless network to access the Internet. If encryption is not specifically turned on, neighbors can use the Internet connection without the knowledge of the owner of the Internet connection. Furthermore, it is possible to spy on the transmitted frames to collect passwords. As such an open access point also allows an eavesdropper to have access to any PC which is also connected to the wireless network, it enables him to exploit weaknesses of the operating systems such as being able to read, modify, or destroy files on the computers. A test drive revealed the reality of this problem: 12 access points were found within a few minutes, of which five were used without proper encryption setting.

4.7.1 Wired Equivalent Privacy (WEP)

To protect WLAN networks from unauthorized use and eavesdropping, WEP encryption is part of the 802.11b, g, and a standards. Similar to GSM and UMTS, this encryption method is based on a stream-ciphering algorithm (see Figure 1.37) which encrypts a data stream with a ciphering sequence. The ciphering sequence is calculated for each frame by using a key and an initial vector (IV) as shown in Figure 4.16. The initial vector changes for every frame to prevent easy reconstruction of the secret key by an attacker. In contrast to GSM or UMTS, however, WEP uses the same key for all users. While a single key is easy to manage, it creates a big problem especially if a WLAN is used by a company. As the same key has to be manually configured by all users in their devices, it is not possible to keep the key secret. In GSM or UMTS, the individual private key of each user is securely stored on the SIM card.

An even more serious problem is that the first bytes of an encrypted frame always contain the same information for the LLC header. In combination with certain IVs, which are transmitted as clear text, it is possible for an attacker to calculate the key. About 5–6 million frames are necessary to calculate the key with this approach. The length of the ciphering

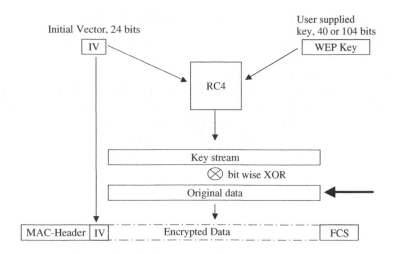

Figure 4.16 WEP encryption

key only plays a secondary role. Tools that automate this process are freely available on the Internet. The attacker therefore simply has to bring his eavesdropping device into the range of a WLAN and wait until enough frames have been collected. The number of required frames sounds very high at first. However, the contrary is the case. If we assume for a moment that each of the 5 million frames contains about 300 bytes of user data, the key can be calculated by collecting 5,000,000 frames $\times$ 0.3 kbyte = 1.5 Gbyte. Depending on the network load, the key can thus be generated in a timeframe ranging from several weeks to only a few hours. It can be assumed that the larger time frame usually applies for home networks due to their low traffic rates. In company networks, however, many devices usually communicate with a server and thus create a high traffic volume. Therefore, the WEP encryption will only protect the network for a short time. Companies should therefore take additional security measures or use other encryption methods to secure their networks, as described in the next paragraph.

To increase the security of a WLAN network, many access points offer a number of additional security features. By activating the 'Hide SSID' option, an access point leaves the SSID fields of beacon frames empty. In this way, the access point is only visible to users that know the SSID of the network and are able to manually configure the SSID in their device. Another security feature is the MAC address filter, which prevents devices from connecting to the network which have not been previously allowed to do so by the administrator. These features, however, do not prevent an ambitious hacker with the tools described above to collect and automatically analyze frames to generate the ciphering key. The hacker can also easily retrieve the SSID that is no longer broadcast in the beacon frames during the association procedure. Statically defined MAC hardware addresses in the access point are also an easy measure to circumvent for a hacker, as a network trace reveals the MAC addresses of devices that are allowed to communicate with the network. This information can then be used to manually change the MAC address on the hacker's device to match the MAC address of a device that was previously used in the network.

4.7.2 Wireless Protected Access (WPA), WPA2, and 802.11i

Because of the security problems presented above, the IEEE 802.11i working group created the 802.1x standard, which offers a solution to all security problems that have been previously found. As the ratification of the 802.11i was considerably delayed, the industry created the wireless protected access (WPA) standard. See Figure 4.17. WPA contains all the important features of 802.11i and has been specified in a way that allows vendors to implement WPA on hardware that was originally designed for WEP encryption only. With WPA, it is possible for the first time since the creation of the WLAN standards to use the technology in a secure way both at home and in companies. The weaknesses of the previous approaches are solved by WPA using the following measures.

While current standards leave authentication of client devices optional and usually disabled, it has been defined as mandatory in WPA. Unlike in previous approaches, every device may now have its own key. This fixes a big security hole especially for companies. The WPA standard offers three possibilities to store the keys and authenticate a client device: the simplest way to store the key, which mainly addresses home network users and small companies, is to directly store them in the access point. This kind of key management is referred to as pre-shared key (PSK) management. In practice, it can be observed that despite the possibility of using a unique key for each device, access point vendors have tailored their software in a way that requires all clients to use the same key.

Bigger companies on the other hand usually store their user credentials on a central server. In order to support this scheme, WPA implements the extensible authentication protocol – transport layer security (EAP-TLS) mechanism. This mechanism uses a certificate authority (CA), like a Windows 2000 server, to assist in the authentication of a user. The access point only acts as a broker between the user devices and the CA. The CA not only stores the keys, but also implements the authentication algorithm. Alternatively, WPA offers a procedure called PEAP (protected EAP) to include a RADIUS (remote authentication dial-in server) authentication server that stores the user keys and is used as a central instance for the authentication process. While being more complex, these methods offer the possibility to store all keys at a central place. This is especially useful for companies, as they usually use more then one access point to cover their office space. Without a central administration, user

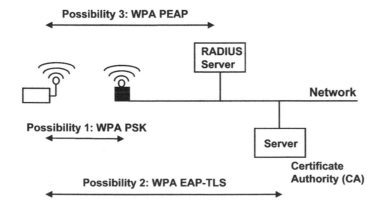

Figure 4.17   Three authentication methods defined by WPA

keys would have to be stored in each access point, which would increase the administrative overhead.

WPA also replaces the insecure WEP encryption with the new temporal key integrity protocol (TKIP) ciphering algorithm. TKIP effectively prevents attackers from generating the ciphering key by launching a brute-force attack on the network. The encryption is not as strong as required in the 802.11i standard, but can be implemented on current WLAN chipsets designed for WEP.

In theory, access points can support WPA and WEP devices at the same time. In practice, only a few access points available on the market support this. As the security of the network is only as good as the weakest link, the use of 'WPA only' makes a lot of sense for company networks. For private networks, the concurrent use of WPA an WEP could be desirable if the access point and all PCs of the network are already WPA compatible, while other devices like PDAs only support WEP. PDAs usually transmit only small amounts of data and thus do not pose a great security risk.

After final publication of the 802.11i standard, the WiFi Alliance, which promotes the IEEE WLAN standards (http://www.wifi.org), adapted its WPA certification program accordingly. WPA2 is an implementation of the 802.11i standard and is backward compatible to WPA. This means that a WPA2 certified access point is also able to support WPA devices. For further backward compatibility, a WPA2 access point can also support legacy WEP clients if WPA/WPA2 is turned off. In addition to WPA's TKIP encryption algorithm, WPA2 now supports the stronger AES (advanced encryption standard) encryption. WPA2 comes in two flavors: if a device is personal mode certified, a client is authenticated by an access point by using a pre-provisioned password as shown in Figure 4.17 for WPA. For use in companies with potentially several access points that serve many subscribers in a larger area, an access point should be WPA2 enterprise mode certified. In addition to pre-provisioned keys, such devices can also use the 802.1x authentication framework to communicate with an external authentication server as already described for WPA.

4.8 Comparison of WLAN and UMTS

WLAN and UMTS can both be used for a number of applications despite their completely different architectures. The media often uses this fact to prove the superiority of one system over the other or even to doubt that one of the two systems will survive in the marketplace. Such reports easily forget the number of applications that are suitable for only one of the systems.

The press usually likes to point to the 'huge' speed difference between WLAN and UMTS. The maximum transfer rates of 11–54 Mbit/s of WLAN hotspots compared to the 384 kbit/s of UMTS Release 99 or 800 kbit/s to 3 Mbit/s of the initial HSDPA implementations seem to be quite different at first. This is certainly true if a WLAN is used for a wireless connection at the office. In this scenario, the network is usually used to exchange data between the PCs and a server in a local network. Nobody would ever consider doing this via the UMTS network. Clearly, WLAN as a local area network (LAN) is the technology of choice for this application.

WLAN hotspots in public buildings are another application for WLAN technology. Here, the speed is usually not limited by the maximum speed of the WLAN link itself, but by the speed of the backhaul link to the Internet. For most hotspots, DSL is used as backhaul, which limits the downlink speed to about 1–8 Mbit/s. This bandwidth has to be shared by all users of the hotspot. In practice, the uplink bandwidth is usually even smaller and in most cases limited to 256 kbit/s to 1 Mbit/s. Even hotspots with a fast backhaul connection rarely achieve more than 2 Mbit/s on the uplink. This means that the maximum transmission speed of the WLAN can never be fully used in such a scenario. Using UMTS on the other hand, the speeds described above of 384 kbit/s up to 3 Mbit/s with HSDPA are available per user. See Figure 4.18. A well-equipped UMTS Node-B with three sectors can have a transmission capacity of about 12 Mbit/s. Seen from this point of view, the picture reverses and a UMTS Node-B can handle more users than a WLAN hotspot with an adequate speed for each user. To complete the picture, it has to be noted that a Node-B usually covers a much larger geographical area than a WLAN hotspot. Additionally, it has to be taken into account that a Node-B is also responsible for voice traffic, which also reduces the amount of bandwidth available for data. Combining these facts, it is obvious that the simple statement that WLAN is faster then UMTS is only true for the home or office network scenario.

At the time of writing, WLAN is available at many public places such as hotels, airport lounges, and train stations. It is expected that the number of such hotspots will further increase in the future. Due to the short range of WLAN, it cannot be guaranteed, however, that a hotspot is always available when a user wants to go online. Even though the number of hotspots will grow, only certain areas in hotels are covered by WLAN. A traveler therefore has to plan ahead carefully or simply leave it to luck if WLAN coverage is available at his destination or not. UMTS networks on the other hand cover large areas. They mostly cover larger cities at the time of writing but are about to spread into smaller cities as well and will also reach rural areas eventually. A fallback to GSM and GPRS ensures that access to the

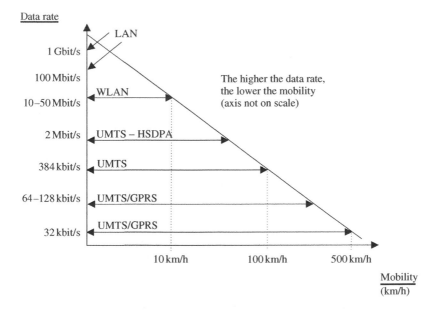

Figure 4.18   Data rates and their dependence on mobility

Internet is also available where no UTMS coverage yet exists and also ensures connectivity when traveling abroad (international roaming).

As the UMTS core network is an evolution of the already existing GSM and GPRS networks, a functioning world-wide billing solution already exists. WLAN on the other hand does not have a standardized billing solution. This is due to the fact that for many scenarios like for home and office use, for which the WLAN standard was initially conceived, no billing was necessary. For commercial hotspots, like in hotels, however, billing is an essential task. Due to missing standards and the vast number of hotspot operators, a number of different billing methods are appearing on the market. These range from scratch cards that can be bought at the hotel's reception desk, online credit card payment, and billing via the GSM or UMTS. The later billing method can only be used if the WLAN hotspot is operated by the mobile operator of the user. In most cases, a user is therefore not able to use the hotspot right away but has to deal with billing first.

An open issue for public use of WLAN is the technical realization of lawful interception by the authorities. This contrasts other telecommunication networks including GSM, GPRS, and UMTS, for which most countries have passed laws and standardized methods to allow access by police and other organizations to the data that a user transfers. This process has not yet started for WLAN hotspots and is also not easily achievable due to the current user authentication architecture. With the increasing success of WLANs it is likely that laws will be put into place for this technology as well. This will force many WLAN hotspot operators to redesign their current user authentication and data routing functionality.

WLAN has been designed for small coverage areas. This area can be somewhat increased by using several access points to form an ESS. As all access points have to be in the same IP subnet (see Section 4.4 and Figure 4.9), the maximum coverage area is still limited to the size of a single building. For most WLAN applications, this limitation is acceptable, especially because automatic access point changes are possible. UMTS on the other hand has been designed for nationwide coverage. Furthermore, the standard has been designed (see Chapter 3) for seamless handovers between cells to maintain connections over long periods and distances as well as at high speeds of up to 500 km/h. Only these methods enable users to make calls while being on the move or to connect their PDAs or notebooks to the Internet while traveling in trains or cars.

The size of cells also differs greatly between WLAN and UMTS. WLAN is limited to a few hundred meters due to its maximum transmission power of 0.1 Watt. Inside buildings, the range is further reduced due to obstacles like walls. UTMS cells in practice can stretch for several kilometers but can also be used to cover only certain buildings or floors (pico-cells), for example shopping centers, etc.

Strong security and encryption were only added to the WLAN standards once the system was already popular. While WPA and WPA2 (802.1x) offer good security and privacy for private and company networks, security is still a problem for public hotspots. Especially in this market, WPA will most likely not be introduced, as keys would have to be manually configured by the user.

As all users of a hotspot get an IP address in the same subnet, a user should ensure that his notebook is protected against hacker attacks from the same subnet. An adequately configured firewall and an up-to-date virus scanner on a client device is an absolute must. Some access points offer to protect users by preventing direct communication between devices of the hotspot. The 'client isolation' feature is based on layer 2 MAC filtering. In practice, however, there is no guarantee that such a feature has been implemented or activated in an access point. UMTS devices can also be accessed by other devices in the network. Different users in the same area, however, do not usually belong to the same subnet. A UMTS user has no means of finding out which IP addresses have been given to devices in the local area thus preventing him from launching a specific attack. As security is part of the overall design of UMTS, a user does not have to take care if and how the connection to the network is encrypted as the system automatically encrypts the link to the user. The user also does not have to worry about key management, as the key for authentication and ciphering is stored on the SIM card.

Telephony is another important application. The circuit-switched part of the UMTS network has been specifically designed for voice and video telephony. These two services are not covered by WLAN hotspots today. However, a clear trend can be seen towards voice (and video) over IP (VoIP). UMTS addresses this with its IMS architecture (see Chapter 3). Wireless hotspots benefit from this trend as well. Various VoIP software clients, together with a notebook, enable the user to make calls via WLAN at home, in the office, or at a public hotspot. Recently, devices like the Nokia Communicator have introduced WLAN connectivity in addition to GSM and UMTS access. To ensure a good quality of service for telephony applications in heavily loaded hotspots, an extension to the DCF of access points is required (see Section 4.5) to ensure a constant bandwidth and latency for the call. A solution for this problem has already been standardized in the 802.11e specification, but it will still take a number of years before these features are available in public hotspots and client devices. It also should be noted that the majority of public hotspots are connected to the Internet via DSL lines with limited uplink bandwidths of only a few hundred kilobits per second. This limits the number of simultaneous voice calls to two or three. Due to these reasons, telephony over public WLAN hotspots will only complement the current voice-call capabilities of GSM and UMTS networks. To standardize VoIP using public hotspots, the 3GPP community has worked on an extension of the UMTS standard in the technical specifications TS 22.234 [4], 23.234 [5] and 24.234 [6]. These Release 6 standards describe how the UMTS IP multimedia subsystem (IMS) can be extended to public WLANs.

In summary, WLAN is a hotspot technology that offers fast Internet access to users in a small area for a limited amount of time. Due to the simplicity of the technology compared to UMTS, as well as the use of license-free bands, costs for installation and operation of WLAN hot spots are lower than for a UMTS cell. Together with a fast backhaul connection to the Internet, WLAN can offer fast data transmission capabilities for private, office, and public use. In practice, WLAN is the standard connection technology for notebooks and PDAs today. WLAN reaches its technical limits in cars or trains and due to its maximum coverage area, which is typically the size of a building. Due to these limitations, the term 'nomadic Internet' is sometimes used for WLAN Internet access. Users typically move into the coverage area of a cell for some time during which they will be mostly stationary, before leaving the area again.

UMTS, on the other hand, addresses the needs of mobile users that need to communicate while being on the move. With its fast data transfer rates, UMTS is also ideally suited for accessing the Internet if no WLAN hotspot is available that can be used at a lower price. The complex technology, compared to WLAN, is necessary to support the mobility of users and for applications like telephony at any place any time. This makes UMTS more expensive than WLAN. The huge frequency licensing fees that mobile operators have paid in many countries are also adding a significant amount to the total cost. The main applications for UMTS are therefore mobile voice and video telephony, Internet access if no WLAN hotspot is available, as well as WAP, MMS, video streaming, and instant messaging. Thus, UMTS is considered as the 'mobile Internet', as the technology enables users to communicate at any place, any time, even in cars and in trains.

4.9 Questions

1. What are the differences between the 'ad-hoc' and 'BSS' modes of a WLAN?
2. Which additional functionalities can often be found in WLAN access points?
3. What is an extended service set (ESS)?
4. What is an SSID and in which frames is it used?
5. What kinds of power-saving mechanisms exist in the WLAN standard?
6. Why are acknowledgment frames used in a WLAN?
7. Why do 802.11g networks use the RTS/CTS mechanism?
8. Why are three MAC addresses required in BSS frames?
9. How can a receiving device detect at what speed the payload part of a frame was sent?
10. What is the maximum transfer rate that can be reached in a data transfer between two 802.11g devices in a BSS?
11. Which disadvantages does the DCF method have for telephony and video streaming applications?
12. Which security holes exist in the wired equivalent privacy (WEP) procedures and how are they solved by WPA and WPA2 (802.1x)?

Answers to these questions can be found on the companion website for this book at http://www.wirelessmoves.com.

References

[1] IEEE, 'Part 11: Wireless LAN Medium Access Control (MAC) and Physical Layer (PHY) Specifications', ANSI/IEEE Std 802.11, 1999 Edition (R2003).
[2] IEEE, 'Part 3: Carrier sense multiple access with collision detection (CSMA/CD) access method and physical layer specifications', ANSI/IEEE Std 802.3, March 2002 Edition.
[3] R. Droms, 'RFC 2131 – Dynamic Host Configuration Protocol', RFC 2131, March 1997.
[4] 3GPP, 'Wireless Local Area Network (WLAN) Interworking', TS 22.234, V6.2.0, September 2004.
[5] 3GPP, '3GPP System to Wireless Local Area Network (WLAN) Interworking: System Description', TS 23.234, V6.3.0, December 2004.
[6] 3GPP, '3GPP System to Wireless Local Area Network (WLAN) Interworking: User Equipment (UE) to Network Protocols; Stage 3', V6.1.1, January 2005.
[7] IEEE, 'Part 11: Wireless LAN Medium Access Control (MAC) and Physical Layer (PHY) Specifications: High-Speed Physical Layer Extensions in the 2.4 GHz Band', ANSI/IEEE Std 802.11b, 1999 Edition (R2003).
[8] IEEE, 'Part 11: Wireless LAN Medium Access Control (MAC) and Physical Layer (PHY) Specifications – Amendment 4: Further Higher Data Rate Extensions in the 2.4 GHz Band', ANSI/IEEE Std 802.11g, 2003.
[9] IEEE, 'Part 11: Wireless LAN Medium Access Control (MAC) and Physical Layer (PHY) Specifications – High-Speed Physical Layer Extensions in the 5 GHz Band', ANSI/IEEE Std 802.11a, 1999.
[10] IEEE, 'Part 11: Wireless LAN Medium Access Control (MAC) and Physical Layer (PHY) Specifications – Amendment: Medium Access Control (MAC) Quality of Service Enhancements', IEEE Std P802.11e/D13, January 2005.

[11] IEEE, 'IEEE Trial-Use Recommended Practice for Multi-Vendor Access Point Interoperability via an Inter-Access Point Protocol Across Distribution Systems Supporting IEEE 802.11 Operation', IEEE Std 802.11F, 2003.

[12] IEEE, 'Part 11: Wireless LAN Medium Access Control (MAC) and Physical Layer (PHY) Specifications – Amendment 5: Spectrum and Transmit Power Management Extensions in the 5 GHz Band in Europe', IEEE Std 820.11h, 2003.

[13] IEEE, 'Part 11: Wireless LAN Medium Access Control (MAC) and Physical Layer (PHY) Specifications – Amendment 6: Medium Access Control (MAC) Security Enhancements', IEEE Std 802.11i, 2004.

5

802.16 and WiMAX

In recent years, advances in signal-processing technologies and increased processor speeds have allowed wireless networks to evolve into broadband Internet access technologies. The GSM system was first enhanced by the UMTS radio access network and later with the high speed downlink packet access (HSDPA) standard, which allowed for wireless Internet access at speeds of several megabits per second. CDMA systems have undergone a similar evolution. Several large companies, like Intel for example, which thus far have had no major market share in equipment sales for wireless networks have reacted in support of a new system standardization effort by the Institute of Electrical and Electronics Engineers (IEEE) to create an alternative wireless broadband network. This effort culminated in the ratification of the 802.16-2004 standard [1]. In the press, the 802.16 standard is often referred to as WiMAX (worldwide interoperability for microwave access), though this is not technically accurate as will be explained below.

The capability of WiMAX to deliver high-speed Internet access and telephone services to subscribers enables new operators to compete in a number of different markets. In urban areas already covered by DSL and high-speed wireless Internet access, WiMAX allows new entrants in the telecommunication sector to compete with established fixed-line and wireless operators. The increased competition can result in cheaper broadband Internet access and telephony services for subscribers. In rural areas with limited access to DSL or cable Internet, WiMAX networks can offer cost-effective Internet access and may also encourage HSDPA or 1xEvDO operators to extend their networks into these areas. Developing countries with limited infrastructure connecting subscribers to a central office are another potential market for WiMAX. By connecting them wirelessly, WiMAX allows these markets to bypass fixed-line Internet access technologies. This has already happened for mass-market telephony services with the introduction of wireless GSM networks, which offer phone and messaging services to millions of people in the developing world. Previously, this market was underserved for reasons such as missing infrastructure and lack of competition, which kept prices at unaffordable levels. The introduction of WiMAX also drives the evolution of other high-speed wireless access technologies, as standards bodies like 3GPP or 3GPP2 have to enhance their systems to stay competitive.

This chapter aims to give a technical overview of the 802.16 standard and compares the capabilities and design of the system to other technologies like HSDPA and wireless

Communication Systems for the Mobile Information Society   Martin Sauter
© 2006 John Wiley & Sons, Ltd

LAN (802.11). In this way, the differences and similarities between these systems become apparent allowing us to put the marketing promises into perspective with the real capabilities of the technology.

5.1 Overview

802.16 is part of the 802 local and metropolitan area standards series of the IEEE. Other important network technologies in this series include the 802.3 fixed-line 'Ethernet' standard and the 802.11 wireless LAN standard. While the fixed-line and wireless local area network standards share concepts concerning how the network is managed and how packets are transferred between the devices, 802.16 as a metropolitan area network standard has taken a fundamentally different approach. There are important differences on layer 1 (physical layer, PHY) and layer 2 (data link layer, MAC) of 802.16 compared to 802.11 wireless LAN. The most important ones are:

- An 802.16 network can be operated in several modes. In the point-to-point mode, 802.16 is used to build a bridge between two locations. A second mode, the point-to-multipoint mode, is used to offer Internet access and telephony services to private customers and businesses. As this is the main application for the technology in the years to come, this chapter focuses mainly on this mode.
- In 802.16 point-to-multipoint mode, access to the network by client devices, also referred to as subscriber stations, is managed from a central authority. In 802.11 (WLAN) in comparison, clients can access the network whenever they detect that the air interface is not being used.
- Subscriber stations do not receive individual frames. In the downlink direction (network to subscriber station), data is embedded in much larger frames. During transmission of the frame, the network can dynamically adjust modulation and coding for parts of the frame to serve subscriber stations closer to the base station with higher data rates than those available to subscriber stations with less favorable reception conditions. In the uplink direction, the same concept is used and subscriber stations are assigned individual parts of a frame in which they are allowed to send their data.
- Most 802.11 WLAN networks today do not offer quality of service (QoS) mechanisms for subscriber stations or single applications like voice over IP, which are very sensitive to variations of bandwidth or delay. Most of the time, the available bandwidth of the network and the low number of users per access point compensate for this. The 802.16 standard on the other hand defines in detail how to ensure QoS, as metropolitan networks are usually engineered for high loads and many subscribers per cell.

As in any standardized technology, companies interested in the technology and its success have set up an organization to promote the adoption of the technology in the market and to ensure that devices of different manufacturers are compatible with each other. Interoperability is often hard to achieve, as most standards offer many implementation options and leave things open to interpretation. 802.16 is no exception. The WiMAX forum (http://www.wimaxforum.org) is the organization that aims to ensure interoperability between 802.16 devices of different manufacturers. Apart from promoting the technology, it has defined a number of profiles to ensure interoperability and has launched the WiMAX certification program [2]. Vendors interested in ensuring interoperability with products of other vendors can certify their equipment in WiMAX test labs. Once certified, they can officially claim to be WiMAX compliant, which is a basic requirement of most network vendors. The WiMAX forum for 802.16 therefore fulfills the same tasks as the Wi-Fi alliance (http://www.wi-fi.org) does for 802.11 wireless LAN. Due to this relationship, the remainder of this chapter uses the terms 802.16 and WiMAX interchangeably.

The 802.16 standard uses the protocol layer model shown in Figure 5.1. This chapter will look at the individual layers as follows: first, the physical layer is discussed with the different options the standard offers for different usage scenarios. Then, the physical layer frame structure for point-to-multipoint scenarios is discussed, as this operating mode will be used by operators to offer high-speed Internet access and telephony to consumers and businesses. By comparing the frame structure to the WLAN architecture described in the previous chapter, it will become apparent how the 802.16 standard deals with the additional requirements of a metropolitan area network (MAN).

Due to the many tasks fulfilled by the MAC layer, it has been split into three different sublayers. The privacy sublayer, which is located above the physical layer, deals with the encryption of user data which can be activated after a subscriber has been successfully authenticated by the network. This procedure is described at the end of the chapter.

The MAC common part sublayer deals with the connection establishment of subscribers to the network, and manages individual connections for their lifetime. Furthermore, this layer is responsible for packing user data received from higher layers into packets that fit into the physical layer frame structure.

Finally, the MAC convergence sublayer offers higher layer protocols a standardized interface to deliver user data to layer 2. The 802.16 standard defines interfaces for three different higher layer technologies. The ATM convergence sublayer is responsible for handling the

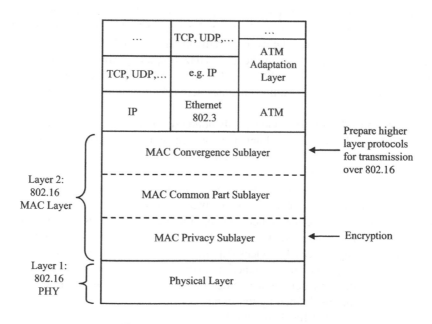

Figure 5.1 The 802.16 protocol stack

exchange of ATM (asynchronous transfer mode) packets with higher layers. This is mainly used to transparently transmit ATM connections via an 802.16 link. The applications for sending ATM frames are point-to-point connections for backhauling large amounts of data, like connecting a UMTS base station to the network. ATM will not be used for communication with the user. Therefore, this part of the standard is not discussed in further detail in this chapter, as the chapter concentrates on point-to-multipoint applications for delivering Internet access and telephony services to end users. For this purpose, the MAC convergence sublayer offers an interface to directly exchange IP packets with higher layers. This makes sense as the Internet protocol is the dominant layer 3 protocol today. Alternatively, higher layer frames can be encapsulated into 802.3 Ethernet frames, as shown in Figure 5.1, before being forwarded to the MAC convergence sublayer. This allows any layer 3 protocol to be transported over an 802.16 protocol, as the header of an 802.3 Ethernet frame contains an information element which informs the receiver of the protocol (e.g. IP) used on the layer above.

5.2 Standards, Evolution, and Profiles

WiMAX comprises a number of standards documents. The 802.16 standard in general addresses the physical layer (layer 1) and the data link layer (layer 2) of the network. In its initial version, 802.16a, the standard only supported line-of-sight connections between devices in the frequency range between 10 and 66 GHz. If WiMAX is operated in point-to-multipoint mode for Internet access, most subscriber stations in cities and even rural areas will not have a free line of sight (LOS) to a WiMAX base station (BS) due to obstructing buildings or landscape. WiMAX was thus extended in the 802.16d standard for non-line of sight (NLOS) operation for the frequency range between 2 and 11 GHz. A single base station only uses a fraction of the frequency ranges given above. The system is very flexible and typical bandwidths per base station are between 3.5 and 25 MHz. The bandwidth allocated to a BS mainly depends on regulatory requirements and available spectrum, as there are many other wireless systems used in the 2–11 GHz frequency range, like UMTS, 802.11 wireless LAN and Bluetooth. In 2004, 802.16a and 802.16d were combined to form the IEEE 802.16-2004 standard, which thus includes network operation in both LOS and NLOS environments.

The first version of the 802.16 standard only addresses non-moving or low mobility users. Subscriber stations either use internal antennas or roof-mounted external antennas if further away from the base station. The 802.16e standard adds mobility to the WiMAX system and allows terminals to roam from base station to base station. The intent of this extension is to compete with other wireless technologies like UMTS, CDMA and WLAN for moving subscribers using devices like notebooks while away from home or the office.

As a first step to foster alternative network topologies, 802.16f adds improved multi-hop functionality for meshed network architectures. It describes how stations can forward packets to other stations so they can reach devices that are outside the radio coverage of a sender.

As shown in Table 5.1, the 802.16 standard covers a wide range of different applications and scenarios. The standard defines a number of profiles that describe how the different physical layers and options defined by the standard are to be used.

The two profiles intended for delivering Internet access to private subscribers and businesses with stationary devices are the wirelessMAN-OFDM (wireless metropolitan area

Table 5.1 802.16 standards documents

Standards document	Functionality
802.16a	Initial standards document, 10-66 GHz LOS operation only
802.16d	NLOS operation at 2–11 GHz
802.16e	Adds mobility to 802.16
802.16f	Introduces multi-hop functionality
802.16-2004	Umbrella document which combines the different subdocuments

network – orthogonal frequency division multiplex) and wirelessMAN-HUMAN (high-speed unlicensed metropolitan area network) profiles. They describe how 802.16 can be used for point-to-multipoint NLOS applications in frequency bands below 11 GHz. The first profile is intended for use in licensed bands where the operator pays for the right to use a certain frequency range. The second profile is intended for license free bands such as the ISM (industrial, scientific, and medical) band, which is also used by various other technologies such as WLAN and Bluetooth. Both profiles use orthogonal frequency division multiplexing (OFDM) for data transmission. This modulation technique is also used in the 802.11g WLAN standard (see Chapter 4), and uses several carriers to transmit data.

The 802.16e extension of the standard uses the wirelessMAN-OFDMA profile to address the requirements of mobile subscribers. Many enhancements and additions have been made to the original profile and radio network and core network designs have been specified by the WiMAX forum network group.

For other applications the standard defines the following profiles, which will not be covered in further detail in this chapter:

- WirelessMAN-SC: use of a single carrier frequency for point-to-point operation on licensed bands between 10 and 66 GHz. Mainly intended for high-capacity wireless backhaul connections.
- WirelessMAN-SCa: use of a single carrier frequency for operation in licensed bands below 11 GHz.

5.3 WiMAX PHYs for Point-to-Multipoint FDD or TDD Operation

To communicate with stationary subscribers in a point-to-multipoint network, the 802.16 standard describes two basic options in the mirelessMAN-OFDM/HUMAN profiles.

For license exempt bands, time division duplex (TDD) is used. This means that the uplink and downlink direction between the base station and a subscriber use the same frequency band. Uplink and downlink are time multiplexed in a similar way as described in Chapter 4 for WLAN systems. The advantage of using a single frequency band for both directions is a flexible partitioning of the available bandwidth for the uplink and downlink directions. For applications like web surfing, the amount of data sent from the network to the subscriber is much higher than in the other direction. For such applications, more transmission time is assigned in the downlink direction than in the uplink direction. Disadvantages of TDD are that devices cannot send and receive simultaneously and that a device has to switch between transmit and receive state. As some time is required to switch between transmitting and

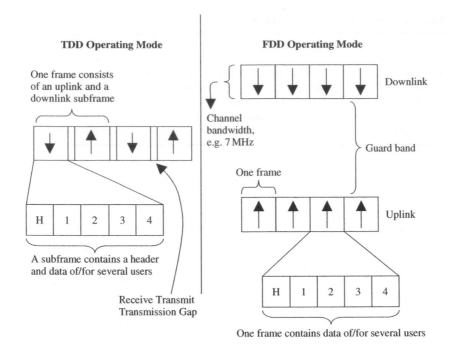

Figure 5.2   802.16 operation modes: TDD and FDD operation

receiving, some bandwidth is wasted during the required gap between the times allocated for sending and times allocated for receiving.

Depending on national regulations, operators can also use licensed spectrum for their network. This will be the rule rather then the exception, as the operation in license-free bands is only allowed with minimal transmit power, usually well below 1 W. This power level is usually not sufficient to cover large areas with a single base station, which is required for economic operation of a network. In licensed bands, operators can choose between the TDD mode described above and frequency division duplex (FDD) (see Figure 5.2). Here, the uplink and downlink data flows use two frequency bands which are separated by a guard band as in GSM, UMTS or CDMA. Full duplex devices can send and receive data at the same time as in UMTS or CDMA. Subscriber stations, which are only half-duplex capable, are only able to send or receive at a time. The 802.16 standard accommodates both types of devices. Hence, subscriber stations have to announce their duplex capabilities during the network entry procedure described further below.

5.3.1 Adaptive OFDM Modulation and Coding

The wirelessMAN-OFDM transmission convergence sublayer, which is part of the physical layer, uses OFDM in both FDD and TDD mode in a similar way as wireless LAN, which was described in Section 4.6.2. For 802.16, data is modulated onto 256 carriers, independent of the overall bandwidth of the channel. Data bits are transmitted not one after another but in parallel over many carriers. All bits transmitted during one transmission step over all carriers are referred to as a symbol. Instead of bit rate, the symbol rate is used as a measurement unit for the speed on the physical layer. For point-to-multipoint operation, the standard defines physical profiles with bandwidths of 1.75, 3, 3.5, 5.5, 7, and 10 MHz. The higher the bandwidth of the channel, the faster the data is transmitted over the air. As the number of OFDM carriers is the same for all bandwidths, the number of symbols per second, i.e. the time it takes to transfer a symbol, varies. In a 10 MHz channel, symbols are transmitted much more quickly than in a 1.75 MHz channel, as the subcarriers are spaced further apart and can thus change their states more quickly without interfering on neighboring subcarriers.

For 1.75 MHz channels, the symbol transmit time has been defined at 128 microseconds, excluding the time required to compensate for the delay spread. For a 3.5 MHz channel, the symbol transmit time is 64 microseconds, a 7 MHz channel requires 32 microseconds per symbol, and a 10 MHz channel requires a symbol transmission time of 22.408 microseconds.

Out of the 256 subcarriers, 193 are used to transfer user data, and 55 subcarriers are set aside for guard bands at the edges of the used frequency band. A further eight subcarriers are used for pilot information, which is used by the receiver for channel approximation and filter parameter calculation to counter signal distortions.

For each transmission step, several bits are coded on each subcarrier. Under ideal transmission conditions, for example when clear line of sight exists between sender and receiver over very short distances, 64-QAM (quadrature amplitude modulation) is used, which codes six bits on a single subcarrier. Under harsher conditions, less demanding modulation schemes like 16-QAM, QPSK and BPSK are used, which code fewer bits on a subcarrier per transmission step. Table 5.2 lists the different modulation schemes, the signal-to-noise ratio required for each, and the number of bits coded on a single subcarrier per transmission step. The signal-to-noise ratio is a figure that describes how much higher the signal energy has to be compared to the noise level in the frequency band.

The modulation schemes used by 802.16 are also used by the 802.11g and 802.11a standards for wireless LAN. Instead of 256 subcarriers, however, WLAN only uses 52 subcarriers, and a fixed bandwidth of 22 MHz instead of 1.75 to 10 MHz. UMTS and HSDPA also make use of QPSK and 16-QAM modulation (HSDPA only). 64-QAM was not specified for HSDPA, as 3GPP considered it very unlikely that this higher order modulation scheme would deliver good performance in rural or urban environments. It is important to note that UMTS and HSDPA use a wideband-CDMA carrier of 5 MHz with only a single carrier frequency (see Chapter 3) in contrast to the 256 subcarrier transmission technique used by 802.16 and varying bandwidths of 1.75 to 10 MHz.

Table 5.2 802.16 modulation schemes

Modulation scheme	Required signal-to-noise ratio	Description
64-QAM	22 dB	6 bits per step, only for LOS and very short distances
16-QAM	16 dB	4 bits per step
QPSK	9 dB	2 bits per step
BPSK	6 dB	1 bit per step, very robust, for harsh environments

The 802.16 standard also keeps the intersymbol guard time very flexible in the range of 3–25% of the total time required to transfer a symbol over the air. During the guard time at the beginning of the transmission interval of a symbol, a valid signal is not ensured as it could be distorted by multipath fading. As different radio environments have different multipath fading behaviors, this flexibility is useful in environments were only low multipath fading occurs and smaller guard times can be used. Thus, more signal energy is available at the receiver side to reconstruct the original signal. This in turn reduces the number of transmission errors, and data can be transmitted faster by using higher order modulation schemes.

Compared to the overall data rate, the actual symbol transfer speed is rather low, as 193 carriers are used for the data transmission. This means that the intersymbol guard time can be relatively small. If fading still occurs after the guard time, it only affects a small part of the overall frequency band. Therefore, only a few OFDM carriers will be affected which can be more easily detected and corrected in comparison to a wideband signal that uses only a single carrier frequency [4].

Another important parameter is the coding rate of the user data stream. The coding rate is the ratio between the number of user data bits and the number of error correction and detection bits sent over the air interface. The PHY transmission convergence sublayer uses Reed–Solomon forward error correction (FEC) schemes similar to those described for GSM, UMTS and HSDPA in Chapters 1 to 3. The lowest coding rate is 3/4. Here, three user data bits are encoded in four bits, which are then sent over the air interface. This coding rate can only be used for exceptionally good signal conditions. For less favorable conditions, which are the norm rather than the exception, coding rates of 2/3 or 1/2 are used. 1/2 coding basically cuts the data rate in half.

In a typical WiMAX cell, users are dispersed and signal conditions vary by a great degree. Therefore, an 802.16 base station needs to adapt modulation and coding schemes per user as will be shown in more detail below. This ensures the best use of the air interface by allowing higher order modulation schemes and few FEC bits to be applied to subscribers close to the base station, while a more conservative combination can be used for distant users and less favorable conditions. Either the network or the user can change the modulation and coding schemes to adapt to changing signal conditions after the initial network access procedure, which is always performed with a conservative modulation and a coding rate of 1/2. Further information on this topic can be found in Section 5.6.2.

As in other systems described in this book, the 802.16 standard makes use of interleaving to disperse consecutive bits over time to disperse faulty bits generated by temporary interference. This improves the capabilities of FEC algorithms which are capable of restoring many faulty dispersed bits but do not work very well for several consecutive erroneous bits. Furthermore, bit randomization is used to minimize the possibility of long sequences of one's or zero's which are difficult to decode and complicate clock synchronization on the receiver side.

In many cases, base stations have a higher transmit power than subscriber stations. This means that the range or transmission speed of a base station is potentially much higher than that of a subscriber station. To compensate for this disparity, the 802.16 standard supports subscriber station sub-channelization. Instead of using all 193 carriers, the base station can assign a set of $n \times 12$ carriers to the subscriber station in the uplink direction. Using fewer carrier frequencies either reduces power consumption of the subscriber station or helps to concentrate the available transmit power on fewer carrier frequencies, which extends the range of the signal. In both cases, the maximum data rate is reduced. Using only 12 carriers increases the link budget by 12 dB. Sub-channelization is implementation dependent and the subscriber station has to inform the base station during the first connection establishment if this functionality is supported.

While other systems like UMTS or HSDPA rely on acknowledged data transfers on lower layers of the protocol stack, automatic retransmission requests (ARQs) of faulty blocks are only optional in WiMAX. The profiles for point-to-multipoint connections specifically define ARQs as an implementation option only. While HSDPA accepts block error rates of 10% due to its very efficient ARQ scheme in exchange for a higher modulation and lower coding scheme, the 802.16 standard has chosen a different route. This means that the system has to ensure a proper modulation scheme and coding setting for all transmission conditions of a subscriber station in order to minimize TCP retransmissions (layer 4), which have a severe impact on the throughput and jitter behavior of the connection. As the 802.16-2004 standard is only intended for stationary use, error-free transmissions might be easier to achieve then with HSDPA, whose mobile subscribers experience far more variability in signal conditions.

In order to reduce both power consumption of subscriber stations and interference, 802.16 networks can instruct subscriber stations to increase or decrease their power output. This is possible because the base station can measure the quality of the uplink signal of each subscriber station. This functionality is also part of MAC layer signaling and is thus performed relatively slowly compared to the fast power adaptations required for CDMA systems described in Chapter 3.

5.3.2 Physical Layer Speed Calculations

Many marketing articles today claim that transmission speeds of 70 Mbit/s or more can be achieved with 802.16 systems. As the following calculation shows, this value can theoretically be reached when using a 20 MHz carrier and 64 QAM modulation with a coding rate of 3/4 (three user data bits are coded in four transmitted bits):

Symbol rate = 1/Symbol transmit time = 1/11 microseconds = 90,909 symbols/s

Raw bit rate = Symbol rate × Number of carriers × Bits per carrier

$$= 90,909 \times 193 \times 6 = 105.27\,\text{Mbit/s}$$

Bit rate after coding = Raw bit rate × Coding rate = 105.27 Mbit/s × (3/4) = 78 Mbit/s

The values used for this calculation are unlikely to be used for point-to-multipoint connections, i.e. for connecting many users via a single base station to the Internet. The highest bandwidth profile specified in the 802.16 standard for the wirelessMAN-OFDM profile is 10 MHz, only half the value used for the calculation above. Furthermore, it is questionable if operators will be able to obtain sufficient bandwidth from the national regulator to operate a single cell with a bandwidth of 10 MHz, as neighboring cells must use a different frequency band in order to avoid interference. Thus, a WiMAX operator has to obtain a license for a much broader frequency band to operate a larger network. In addition, the 64-QAM modulation and coding rate of 3/4 of the example above are not realistic for real environments. For a realistic scenario, the next calculation uses the following parameters: channel bandwidth of 7 MHz, 16-QAM modulation, coding rate of 2/3:

Symbol rate = 1/Symbol transmit time = 1/32 microseconds = 31,250 symbols/s

Raw bit rate = Symbol rate × Number of carriers × Bits per carrier

$$= 31{,}250 \times 193 \times 4 = 24.12 \, \text{Mbit/s}$$

Bit rate after coding = Raw bit rate × Coding rate = 24.12 Mbit/s × (2/3) = 16 Mbit/s

Note that the bit rates after coding of the two examples still include the overhead of higher layers and have been calculated without taking symbol guard times into account.

The two calculations show how much advertised data rates can vary depending on how the system parameters are chosen. The 16 Mbit/s of a real WiMAX cell as calculated in the second example are comparable to the achievable data rates of HSDPA and 1xEV-DO per cell (see Chapter 3), taking into account the slightly higher bandwidth of 7 MHz required compared to 5 MHz for HSDPA. To increase the total bandwidth per base station, all technologies can use sectorization (SDMA), multiple transmission bands (FDMA) and separate uplink and downlink frequencies (mandatory in UMTS FDD and HSDPA) as already described in Chapter 1. Field trials of 802.16 equipment as those described in [5] have resulted in achievable data rates similar to those calculated in the second example above. As the results of these trials were somewhat lower, they might have also taken subscriber stations into account which were not able to use 16-QAM coding due to their distance from the base station.

5.3.3 Cell Sizes

Apart from high speeds for individual users and a high overall capacity of a cell, cell size is another important factor that decides if an 802.16 network can be operated economically. Ideally, a single cell should be as large as possible and should have a very high capacity in order to serve many users simultaneously. However, these goals are mutually exclusive. The larger the area covered by a cell, the more difficult it is to serve remote subscribers. As a consequence, distant subscribers have to be served with a lower modulation and higher coding scheme, which reduces the overall capacity of a cell. A cell serving only users in close proximity can have a much higher capacity, as less time has to be spent sending data packets with lower modulation schemes, which requires more time then sending data packets of the same size with 16- or 64-QAM modulation. In urban and suburban areas, cell sizes will be small because the number of users per square kilometer is high. In rural areas on the other hand, cell sizes need to be much larger in order to cover enough subscribers to make the operation of the network economically feasible. However, the capacity of the cell is reduced as the percentage of subscribers, which are quite distant from the cell, is higher than for the rural scenario. Also, the achievable data rates per user will be lower, especially for more distant subscribers. See Figure 5.3.

WiMAX is a wireless technology, but will mostly be used with stationary terminals until the introduction of the 802.16e extension of the standard, which adds mobility for terminals. Reception conditions can be substantially increased by installing an outdoor directional antenna on the roof of a building, pointing towards the base station. Users with no other

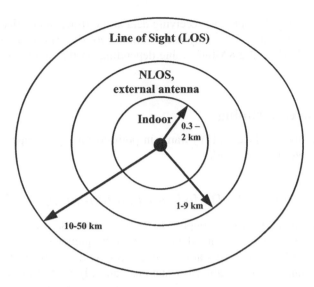

Figure 5.3 Cell sizes depending on type of subscriber station, antenna, site conditions and transmit power

means of getting high-speed Internet service probably accept such a one-time activity, which is similar to installing a satellite dish for television reception. Outdoor antennas can greatly increase the available data rate for a user if cabling is short enough and if a quality cable with a low loss factor is used in order the preserve the gain achieved by using an external antenna. The overall cell capacity also benefits from external antennas as higher order modulation and better coding schemes can be used for the subscriber. Thus, data for this subscriber takes less time to be transmitted over the air interface and the overall capacity of the cell increases.

A number of studies like those performed by the WiMAX forum [4] have analyzed the achievable coverage area of a single base station. The studies have shown that a base station can provide service to indoor equipment with an internal antenna within a radius of 300 meters to 2 kilometers as shown in Figure 5.3. The range mostly depends on the available transmission power of the base station, receive sensitivity, and frequency band used. These values are similar to what can be achieved with a UMTS/HSDPA base station, where most if not all devices will be used indoors with very small antennas.

The study also concluded that an externally mounted directional antenna can extend the range of a cell to up to 9 kilometers. It is assumed that the antenna has no direct line of sight to the base station.

If the antenna can be mounted high enough to have direct line of sight to the base station and the Fresnel zone is undisturbed, a cell could have a range of 10–50 kilometers. This value is purely for academic interest as few distant locations will have a direct line of sight and be high enough for an undisturbed Fresnel zone. The study was conducted for a cell transmitting in the 3.5 GHz band and using a 5 MHz carrier. Maximum downlink throughput close to the center of the cell was calculated to be around 11 Mbit/s in case the subscriber is the only receiver of data for a certain time. At the cell edge, 2.8 Mbit/s are expected, again with the subscriber being the only one receiving data at the time. In practice, throughput per user will be lower as a cell serves many users simultaneously. The total cell capacity will be between the 11 Mbit/s and 2.8 Mbit/s value depending on the number of users and their distribution in the cell.

5.4 Physical Layer Framing

The structure of a physical layer (PHY) frame in point-to-multipoint operation depends on the duplex mode used in the network.

5.4.1 Frame Structure in FDD Mode for Point-to-Multipoint Networks

In licensed bands, operators usually deploy FDD base stations, where data in the uplink and downlink directions are transmitted on different frequencies as shown on the right side of Figure 5.2. While the base station can always send and receive data on the two frequency bands simultaneously, subscriber stations can only be full or half-duplex. While full-duplex devices are slightly more expensive due to independent transmission and reception chains, they are able to support the highest possible transmission rate in both directions simultaneously. Half-duplex devices on the other hand cannot benefit directly from FDD, as they have to stop sending data in order to be able to receive new data from the network. This is problematic if a device has a lot of data to send and receive. In this case, the theoretical bandwidth of a cell that only serves a single subscriber is cut in half, as 50% of the downlink time and 50% of the uplink time cannot be used by the subscriber station. Most applications, like web browsing, are asymmetric and subscribers usually receive more data than they transmit. In these cases, half-duplex devices are not at such a big disadvantage as they mostly receive data and only rarely switch into transfer mode.

In most scenarios more than one subscriber is served by a cell. Thus, a network still benefits from using FDD, even if all subscriber stations are half-duplex only. A base station can ask some devices to receive data when other devices are in the process of sending data and are thus unable to receive data anyway. The network has to be aware which devices are full-duplex capable and which are not in order to schedule data transfers correctly for half-duplex devices.

Figure 5.4 shows how data is transmitted in the downlink direction. On the highest layer of abstraction, chunks of data are packed into frames, which are then transmitted over the air interface. Frames have a fixed size between 2.5 and 20 milliseconds and the selection is usually static. If the frame size is changed by the network, subscriber stations have to resynchronize. While a single frame in other wireless systems contains data to or from a single user, frames are organized in a different way in an 802.16 system. Here, a frame contains data packets for several users. This is organized as follows: At the beginning of a frame, a preamble with known content is sent to allow all devices to synchronize to the beginning of a frame. Next, the FCH (frame control header) informs subscriber stations of the modulation and coding scheme used for the first downlink burst of a frame. The FCH is modulated using BPSK, and a coding rate of 1/2 is used to ensure that even the most distant devices with the worst reception conditions can properly decode this information. All devices are required to decode the first burst following the FCH, as it may contain management information and may inform subscriber stations if and in which burst of the frame they can

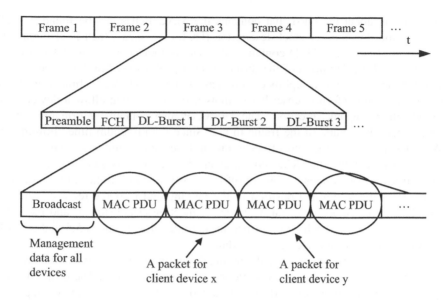

Figure 5.4 FDD downlink frame structure

find their individual user data. The rest of the burst then contains the individual MAC packet data units (PDUs), i.e. the data that is sent to individual subscriber stations. As previously mentioned, different subscriber stations require different modulation and coding schemes in order to receive their data properly. Therefore, a frame can contain several downlink bursts, each modulated with a different modulation scheme in ascending order. The data of subscriber stations experiencing the worst reception conditions are sent in the first burst of a frame, while the data of subscriber stations with good reception conditions is sent in further bursts with a higher modulation scheme. To keep the modulation and coding scheme of the first burst flexible, the FCH contains information about modulation and coding for the first burst. This allows the use of a higher order modulation scheme for the first burst as well in case all subscriber stations are able to receive bursts with a higher modulation scheme than BPSK.

The actual user data packets, i.e. the MAC PDUs, of individual users are marked with a circle in Figure 5.4 to show that several MAC PDUs are contained in a single frame, which is very different to 802.11 WLAN frame encapsulation (see Chapter 4).

Within the management information broadcast to all subscriber stations at the beginning of the first burst, messages informing devices when to expect data and when to send data to the network are most important. For the downlink direction, this is done by the DL-MAP (downlink map) message. The DL-MAP contains a list of all devices to which data will be sent in the current and possibly subsequent frames that do not contain a DL-MAP. Each entry in the list starts with the 16-bit connection id (CID), which identifies a subscriber station and which is later part of the MAC PDU header. Even though a subscriber station has a 48-bit MAC address which is defined in the same way as for fixed-line Ethernet and 802.11 WLAN devices, the MAC address is only used by the subscriber station during connection establishment. Once a device has joined the network, a shorter 16-bit CID is assigned. If a subscriber station detects its CID in the DL-MAP, it analyzes the remainder of the entry.

Here, information about the burst that contains the MAC PDUs can be found as well as a reference to the downlink channel description (DCD) message which is also part of the beginning of the frame. The DCD contains information about the length of the frame, the frame number, and the definition of the different burst profiles used in the frame.

Similar messages exist for the uplink direction as for the downlink direction. The UL-MAP (uplink map) message informs subscriber stations about grants that allow a device to send MAC PDUs in the uplink direction. The UL-MAP also contains information for each subscriber about which burst of the frame to use. Since the minimum time allowed for the UL-MAP allocations to come into effect is one millisecond, uplink resource assignments can be used very quickly. The UCD (uplink channel descriptor) is similar to the DCD for the uplink direction and defines the burst profiles to use in uplink frames. Furthermore, the message contains the length and position of the ranging and resource request windows of the uplink frame, which are used during initial connection establishment and requests for uplink opportunities. These will be described in more detail in Sections 5.5 and 5.6, which deal with QoS and MAC management procedures.

Figure 5.5 shows how data is sent in the uplink direction. Again, a frame structure is used and many subscriber stations can use a single frame to send their data. The instruction about which part of the frame to use to send their data, and which modulation and coding scheme to use was sent to them in the DL-MAP message in one of the previous downlink frames. The figure also shows the contention and UL resource request slots at the beginning of the frame which subscriber stations use for initial ranging and to send their uplink resource requests to the network.

The standard describes two ways for a subscriber station to request resources: The base station can address individual subscriber stations and ask them to report to the network if they require bandwidth in the subsequent uplink frames. The subscriber station then sends a resource request in a dedicated resource request slot. If no resources are required, a resource

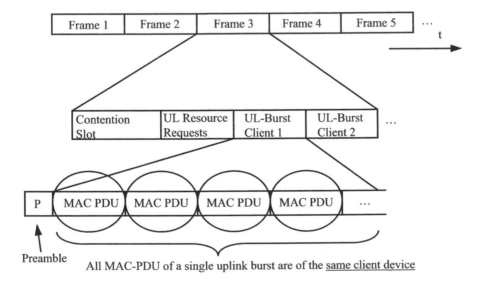

Figure 5.5 FDD uplink frame structure

request for zero bytes has to be sent. The base station can also assign a part of a frame for contention-based bandwidth requests [6]. This means that a group of subscriber stations share the same part of the uplink frame to send their resource requests. In this scenario, no zero byte resource requests have to be sent if no resources are required. In some cases, two or more subscriber stations might attempt to send their resource requests simultaneously. As the two transmissions interfere with each other, none of them will get the requested resources and the procedure has to be repeated.

A MAC PDU consists of three parts: the MAC header, the checksum at the end, and the payload part which is filled with user data of higher layers. The MAC header has a length of six bytes which is very small compared to an 802.11 WLAN header, which already requires the same number of bytes to encode only one of the three MAC addresses required for the delivery of the packet. The reduction of the header length is due to the centralized nature of the network which makes many parameters required in other systems unnecessary (e.g. destination address). Furthermore, many values have already been agreed during connection setup and are only renegotiated when necessary (e.g. modulation and coding schemes to use). Table 5.3 shows the fields of the MAC header, their lengths and their meanings.

Table 5.3 Parameters of a MAC header

Parameter name	Length	Description
Header type	1 bit	0 = Generic MAC header 1 = Bandwidth request header, which can be used by subscriber stations to request additional bandwidth during a scheduled uplink period instead of using the contention slot at the beginning of a frame
Encryption control	1 bit	0 = Packet is not encrypted 1 = Payload is encrypted
Type (extension header indicator)	6 bits	Each bit can be set individually to either 0 or 1 to indicate the presence/absence of special extension headers for functionalities like meshed networks, ARQ feedback, fragmentation, downlink fast feedback allocation, grant management, etc.
CI (CRC indicator)	1 bit	0 = No CRC checksum at end of PDU 1 = CRC checksum appended to payload
EKS	2 bits	Indexes which traffic encryption key and initial vector to use (see end of this chapter on authentication and encryption for details)
Length	11 bits	Total length of the MAC PDU including the header and the checksum. The maximum size of a MAC PDU can thus be 2.048 bytes. This is sufficient for most higher layer protocols like IP, which uses frame sizes between 500 and 1500 bytes
CID	16 bits	The connection identifier: identifies the subscriber station (used instead of the MAC address, see above)
HCS	8 bits	Header check sequence to protect misinterpretation of the header due to transmission errors

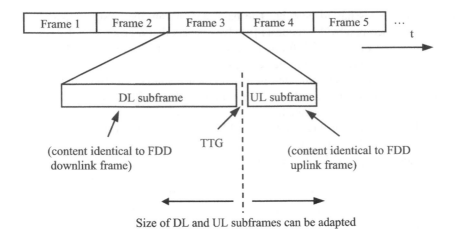

Figure 5.6   TDD frame structure

5.4.2 Frame Structure in TDD Mode for Point-to-Multipoint Networks

In TDD mode, downlink and uplink are sent on the same frequency band instead of using two independent bands as in the FDD mode. This is done by time multiplexing uplink and downlink transmission. TDD frames are split into a downlink subframe and an uplink subframe. The composition of the subframes is identical to the composition of the FDD uplink and downlink frames as described in Section 5.4.1. To allow a subscriber station to switch its transceiver from transmit to receive mode, a transmit/receive transition gap (TTG) has to be inserted between the downlink and the uplink subframe. To switch its transceiver from transmit mode back into receive mode for the downlink subframe of the next frame, a receive/transmit transition gap (RTG) has to be inserted between two frames. These gaps must have a length of at least 5 microseconds. Compared to the smallest frame duration of at least 2.5 milliseconds (2500 microseconds), these gaps are very short and thus do not waste a lot of bandwidth.

The lengths of downlink and uplink subframes in a frame are not fixed and can be changed by the network as shown in Figure 5.6. In many cases, the downlink subframe takes more space in a frame then the uplink subframe, as subscribers usually request more data than they send to the network. Exceptions to this rule are applications like voice or video over IP, which require the same amount of bandwidth in both directions.

5.5 Ensuring Quality of Service

While the PHY layer supplies the means of transferring data over the air interface, the management and control functionality of the network is part of the MAC layer. One of the primary goals of the 802.16 standard is to ensure a certain quality of service for a data stream. While some applications require constant bandwidth, low delay, and no jitter (variations in the delay over time), other applications are delay tolerant but have rapidly changing bandwidth requirements. To accommodate these different requirements, a connection-oriented model is used that transfers data over unidirectional connections. A connection is identified via its

CID, which is part of the MAC header of each packet as mentioned in Section 5.4. For an IP session between a user and the network, one CID is used in the downlink direction and another CID is used in the uplink direction. This allows the network to control the properties of the downlink and uplink independently. The maximum guaranteed bandwidth is one of the properties that might be different between the downlink and uplink directions. The 802.16 standard defines the following quality of service (QoS) classes to accommodate the requirements of different applications:

The unsolicited grant service (UGS): This service guarantees a fixed bandwidth and constant delay for a connection. The base station assigns sufficient uplink opportunities to a subscriber station for the allocated bandwidth and ensures that the subscriber station can use the uplink at the correct times in order to ensure a minimum jitter due to changing delay times. In the downlink direction, the UGS in the base station ensures that incoming packets from the network are forwarded to the subscriber station at the appropriate rate. Applications for this service are wireless bridges for T1 and E1 services (see Chapter 1) and voice transmission. As the requested bandwidth is ensured, the subscriber station does not need to send uplink bandwidth requests at the beginning of an uplink frame. In some cases, the buffer at the subscriber station might grow over time due to a slight misalignment of the data rate of the network and the subscriber station. In such cases, the subscriber station can set the slip indicator bit in the management grant subheader, which is used in MAC frames of UGS connections. The base station then schedules additional uplink resources to allow the subscriber station to empty its uplink buffer.

The real-time polling service (rtPS): This QoS class has been designed to fulfill the needs of streaming applications such as WebTV or other kinds of MPEG streams. To ensure the required bandwidth, the base station provides sufficient unicast request opportunities to the subscriber station. This means that instead of using the contention-based uplink resource request area at the beginning of a frame, the network schedules some time in the second field of an uplink subframe where only the particular device is allowed to send an uplink bandwidth request.

The non real-time polling service (nrtPS): Again, the network polls the subscriber station to find out if uplink bandwidth needs to be allocated. However, the frequency of such polling messages is in the order of one second or less. Thus, a certain bandwidth and delay time cannot be ensured by the network. In order to speed up the process, the subscriber station is also allowed to use the area dedicated for contention-based uplink bandwidth requests at the beginning of an uplink subframe. If there are too many subscribers in the cell for the base station to grant individual uplink request opportunities to each of them, the base station can group several devices into a multicast (uplink request) group. Then the base station informs the whole group when they can send their uplink bandwidth requests. As overlap of bandwidth requests may still occur, devices that do not get uplink bandwidth grants after their request have to repeat the procedure.

Best effort service (BS): This service does not ensure any bandwidth or delay and subscriber stations have to use the contention area at the beginning of an uplink subframe for their bandwidth requests. The number of collisions in the contention area and thus the extent of jitter depend on the number of users of the cell and the length of the contention area.

If a subscriber station is polled for bandwidth requests or if it has to use the contention-based region and the beginning of the uplink subframe to send a bandwidth request, an empty

MAC frame is sent with a special bandwidth request header. The main difference between a normal MAC header and a bandwidth request header is a parameter in which the subscriber station can request uplink bandwidth. The length of the parameter is 19 bits and the amount of required bandwidth is measured in bytes. Consequently, the maximum number of bytes that can be requested by a subscriber station is 2^{19} bytes $= 524.288$ bytes. Since bandwidth requests can also be cumulative, a subscriber station can request additional bandwidth later on if necessary. If a subscriber station is polled individually for reporting uplink bandwidth requirements and no resources are required, the subscriber stations sends a MAC bandwidth request header with the number of bytes requested set to 0. If the subscriber station is polled in a polling group (contention based) and no uplink resources are required, the subscriber station remains silent.

Uplink bandwidth can be granted by the network in two ways. An easy implementation from the network point of view is to grant uplink bandwidth per subscriber station (GPSS). If the subscriber station uses several connections to transfer data, it is the responsibility of the SS to distribute the granted uplink bandwidth among the connections based on their bandwidth requirements and QoS. The second implementation option is to grant uplink bandwidth per connection (GPC). Here, the base station is responsible for assigning sufficient bandwidth at the right time for each connection of an SS. This involves more signaling overhead but relieves the SS of this management task.

The bandwidth request schemes are lower layer mechanisms which by themselves only provide the means for ensuring quality of service. How these mechanisms are used to ensure quality of service attributes like a minimum sustained bandwidth is up to higher layers. Here, service flows are used to describe the QoS parameters of a connection. As will be shown in more detail in Section 5.6.1, service flows are created either after a subscriber station has joined the network or upon request of the subscriber station later on. The goal of establishing a service flow is for the network and the subscriber station to negotiate a number of QoS parameters that the network has to ensure throughout the lifetime of the service flow. In a second step, the base station then has to decide which of the bandwidth request schemes to use in order to grant the negotiated QoS profile. Table 5.4 shows the most important QoS attributes of a service flow. These parameters are exchanged during the service flow establishment so both the network and the subscriber station are aware which QoS parameters have been requested and granted.

A service flow is connected to a CID. As different CIDs are used in the downlink and uplink directions, QoS attributes can be set differently for each direction. This makes sense as in many cases, parameters like the maximum sustained traffic rate can and should be higher in the downlink direction than in the uplink direction, for example in applications like web browsing which require a higher downlink then uplink bandwidth. Furthermore, there is usually more bandwidth available in the downlink direction than in the uplink due to the higher transmission power of the base station and a better antenna system.

A single subscriber station can have several active connections/service flows established with the network at a time. This allows the subscriber to use several services simultaneously with different QoS requirements. Figure 5.7 shows how this could be used in practice. In the figure, a multi-function subscriber station offers two services to end users. On the one hand, it offers IP connectivity to the Internet via a number of Ethernet RJ-45 sockets and via a built-in wireless LAN card that acts as a WLAN access point as described in Chapter 4. For external devices like notebooks or PCs, it is transparent whether a wireless 802.16 backhaul

Table 5.4 Selection of service flow attributes

Service flow attribute	Description
Traffic priority	If a service flow has to compete for resources with other service flows with identical attributes but different traffic priority, the service flow with the highest traffic priority will be preferred
Service flow scheduling type	This attribute is used to negotiate the type of scheduling used for the service flow. Possible values are BS, nrtPS, rtPS, UGS and undefined (for a vendor-specific scheduling mechanism)
Maximum sustained traffic rate	This parameter contains the average transmission speed that the system will ensure for the service flow. The parameter has a length of four bits and the traffic rate is measured in bits per second
Maximum traffic burst	For services that do not require a constant bandwidth, this parameter can specify the maximum bandwidth that can be assigned to a subscriber station for a short time after a bandwidth request before the bandwidth is throttled down to meet the average transmission speed requirement set in the parameter above
Minimum reserved traffic rate	An optional parameter to specify the minimum bandwidth that is reserved for a device. If present, the network will ensure the minimum amount of resources for the connection as soon as possible. Some delay might occur as the aggregate bandwidth of the cell is allowed to be smaller then the sum of the minimum reserved traffic rates of all devices
Minimum tolerable traffic rate	This parameter is used by the base station to decide if the traffic rate that can be sustained over the air interface meets the requirements of the subscriber station. If the minimum tolerable traffic rate cannot be met due to radio conditions, the connection might be dropped. If polling is used for bandwidth allocation, the traffic rate parameters are used by the base station to decide the intervals at which to poll the subscriber stations and how fast the data is to be transferred if a bandwidth request is made by the subscriber station
Tolerated jitter	For real-time services like voice and video over IP the jitter of a connection should be as small as possible. The parameter contains the maximum jitter that is acceptable in milliseconds
Maximum latency	The base station will ensure the latency time contained in this parameter (in milliseconds) for the minimum tolerable traffic rate
ARQ parameters (window size, retry timeout, etc.)	On service flow creation, the use of the ARQ feature is also set. Its implementation is optional

(continued overleaf)

Table 5.4   (*continued*)

Service flow attribute	Description
Convergence sublayer types	In order for the base station and the subscriber station to be able to use the correct MAC convergence sublayer, this parameter is used to specify the higher layer protocol that is to be transported over the connection. The following protocols have been defined: 1: IPv4 2: IPv6 3: 802.3 Ethernet 4: 802.1Q VLAN 5, 6: IPv4 or IPv6 over 802.3/Ethernet 7, 8: IPv4 or IPv6 over 802.1Q VLAN 9: ATM For each convergence sublayer type, subparameters are defined to transport protocol-specific setup information. For IP connections, these can be used to configure the IP stack on both sides

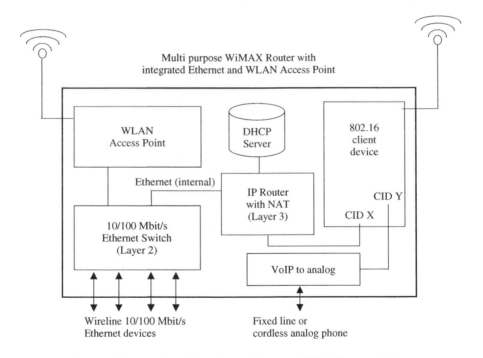

Figure 5.7   Functionalities of a multipurpose WiMAX router device

connection is used instead of a fixed-line connection such as ADSL or a cable modem. The second service offered to end users by the subscriber station is standard telephony service via one or several RJ-11 sockets for analog telephones. While the analog line is simulated for the telephone on the RJ-11 connector, an IP connection is used on the other side which uses its own CID with a QoS setting that ensures the required bandwidth and minimal delay. On the IP side, several protocols can be used for signaling and transporting the voice signal over IP. The most widely known voice over IP standard is the session initiation protocol (SIP). Other protocols could also be used for the service. One of the most popular proprietary protocols used for Internet telephony is Skype.

If only a single CID is used for the Internet and voice traffic, the same approach can be used as before. However, the subscriber station then has to ensure that IP packets generated by the embedded voice over IP converter are preferred over the IP packets that arrive from the fixed-line Ethernet port or via the WLAN. This method is inferior because a general Internet connection does not usually guarantee a certain bandwidth or delay behavior.

In many cases, users will decide to run their own voice over IP software on their notebooks and desktop PCs. In this case, it is difficult for the combination of WLAN access point, Ethernet hub and 802.16 subscriber station to distinguish IP packets of different services. Thus, bandwidth and delay can be larger issues and the voice quality degrades quickly if other services are used in parallel.

5.6 MAC Management Functions

The MAC common part sublayer is responsible for the management of the link between the subscriber station and the network. This includes the initial setup of the link between the subscriber station and the network as well as the maintenance of the communication session. The MAC layer also includes functionality to update the configuration and the software of the subscriber station, and methods to re-establish the link to the network in case the signal is lost.

5.6.1 Connecting to the Network

The first management task of the subscriber station after powering up is to find and connect to a network. This is done in several steps.

In the first step, the subscriber station retrieves the last known system parameters from non-volatile memory and listens on the last known frequency to check if a downlink channel from a base station can be detected. If unsuccessful, it will start to scan all possible channels in the bands it supports for a detectable signal. The subscriber station recognizes valid downlink signals if it is able to successfully decode the preamble at the beginning of the frames. Decoding the preamble is possible without further information as it contains a well-known bit pattern. The device has found a valid 802.16 channel if several preambles can be decoded. At this point the device is also aware of the length of the downlink frames. The device then decodes the beginning of the received downlink subframes to get the current system parameters and configuration from the downlink channel description (DCD), the DL-MAP, and the UL-MAP (see Section 5.4.1) as shown in Figure 5.8. This procedure is again executed if the client loses synchronization to the network and is unable to successfully decode DCD and DL-MAP messages for a configured amount of time, which has a maximum value of 600 milliseconds. Network synchronization can be a relatively slow process as the maximum time allowed between two DCD messages is 10 seconds.

Once all parameters for the initial network access are known, the subscriber station starts the initial ranging procedure by sending a ranging request message (RNG-REQ) with a

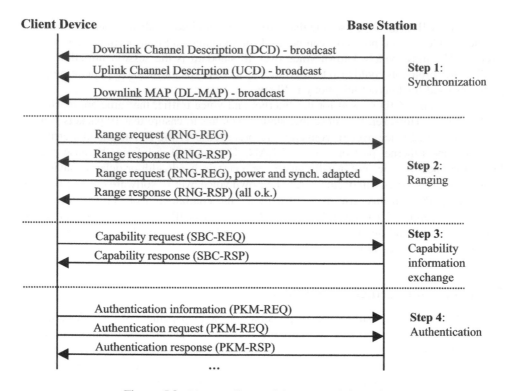

Figure 5.8   Message flow to join a network (part 1)

low transmission power in the contention-based ranging area at the beginning of an uplink subframe. The length of the contention-based ranging area and other parameters are broadcast via the DL-MAP message. If no answer is received, the message is repeated with a higher power level. The procedure is repeated until a response is received or the maximum number of retries has been reached. If no answer is received, the subscriber station goes back to step one and searches for a channel on a different frequency. The maximum time allowed between two ranging regions is 2 seconds. Thus, the ranging process could take several seconds if several RNG-REQ messages need to be sent. The RNG-REQ message also contains a parameter for the network containing the modulation and coding schemes that the subscriber station thinks are suitable to use in the downlink direction. The selection of these values on the client side is based on the downlink reception conditions that the subscriber station has experienced thus far.

When the network receives a ranging request message it analyzes the content, the signal power, and the timing of the message. The network then sends a ranging response message (RNG-RSP) to inform the subscriber station if timing adjustments are required to synchronize further uplink messages to the exact start time of a symbol (see 'timing advance' in Chapter 1). The network can also adjust the subscriber station's power level at this point to reduce interference with other subscribers in neighboring cells. If the subscriber station receives these instructions in the response message, it makes the requested changes and sends another RNG-REQ message. If the network is satisfied with the timing and power settings, it returns a RNG-RSP message which contains the MAC address of the subscriber station and CIDs for the uplink and downlink direction of the basic and primary management connection. In the RNG-RSP message, the network also informs the subscriber station if it has accepted the proposed modulation and coding schemes or if the network prefers to use more conservative values. Each subscriber station is assigned an individual set of CIDs for the primary management connection and the basic connection, which are used to exchange further management messages for the connection setup. These connections are also used later on to exchange management information between the subscriber station and the network to maintain the connection. The CIDs used for the management connection and the basic connection are not used to exchange user data packets. For this purpose, further CIDs are allocated as shown below.

In the next phase of the initial connection establishment procedure, network and subscriber station exchange basic capability information like supported modulation and coding schemes and their duplexing support (full duplex/half duplex). As the subscriber station has received CIDs for the exchange of management messages, it monitors UL- and DL-MAP messages for its management CID and uses allocated slots in an uplink frame to send its capability information via an SS basic capability request (SBC-REQ) message instead of using the contention-based part of an uplink frame. The network responds to the message with an SBC-RSP response message to confirm the proper reception. The SBC-RSP message also contains the basic capabilities that the network supports.

Once capabilities are exchanged, the subscriber station has to authenticate to the network to verify its identity and to start encryption of the air interface link. This is done by the subscriber station sending a privacy key management authentication information message and an authentication request (PKM-REQ) message to the base station. The network verifies the credentials and ends the authentication process with a privacy key management authentication response (PKM-RSP) message. Encryption of the data flow to and from the subscriber is enabled on a service flow basis. Therefore, activating encryption is part of the creation of the service flow which is described below. Further information about the methods used to authenticate the subscriber station and how encryption is performed can be found in Section 5.8.

After authentication, the subscriber station needs to register with the network by sending a registration request (REG-REQ) message as shown in Figure 5.9. The network acknowledges the message by returning a registration response (REG-RSP) message. At this point, the subscriber station is accepted into the network and service flows for the exchange of user data can be established. Optionally, the network registration process can also be used by the network to establish a secondary management connection to update the configuration of the subscriber station. As this step is optional, it is covered later on in this chapter.

The final mandatory step of the initial connection establishment procedure is to set up service flows and the corresponding CIDs used to transfer user data to and from the subscriber station. In most situations, service flows are pre-provisioned in the network. The network starts the process by sending a dynamic service addition request (DSA-REQ) message if the base station is able to fulfill the QoS requirements of the service flow. The message contains the CIDs for the user data service flow, the service flow id (SFID) and the QoS parameters of the service flow. The subscriber station then prepares to send user data over the service flow and replies to the message with a dynamic service addition response (DSA-RSP) message.

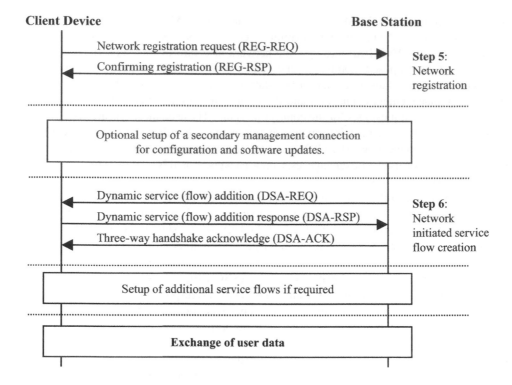

Figure 5.9 Message flow to join a network (part 2)

To finish the process, the base station sends a DSA-ACK (acknowledge) message and the three-way handshake is complete.

The subscriber station is also able to establish service flows on its own by sending a DSA-REQ message while its registration to the network is valid. Depending on the configuration of the network, the request may or may not be granted.

QoS parameters like the guaranteed bandwidth of the service flow can be changed by the network at any time by sending a dynamic service change request (DSC-REQ) to the subscriber station. The subscriber station verifies the validity of the request, makes the appropriate changes to the service flow, and acknowledges the change with a DSC-RSP message. Again, a confirmation is sent by the network to the subscriber station (DSC-ACK) to complete the three-way handshake.

The network or the subscriber station can delete a service flow at any time by sending a dynamic service deletion request (DSD-REQ) message. Both sides then stop using the service flow and the transaction is confirmed by the other side by sending a DSD-RSP message.

A unique feature not standardized in other systems is a configuration and software update capability during connection establishment. As shown in Figure 5.9, the network can optionally establish a secondary management connection for this purpose as part of the network registration procedure. The protocol used on the secondary management connection to transfer configuration information and software updates is the trivial file transfer protocol

(TFTP) over IP [7]. During the establishment of the secondary management connection, the network and subscriber station first negotiate the IP version to use (IPv4 or IPv6). Then, the subscriber station uses the dynamic host configuration protocol (DHCP) [8] to acquire an IP address and all other information required to find an updated configuration file, new software, etc. If present, the subscriber station downloads the information and changes its configuration or software as required. As many different subscriber stations are used in the network, information about the vendor, the hardware version of the device, and the current active software release of the subscriber station are used to determine if a software upgrade is necessary. If a software update was performed, the subscriber station might have to restart. Afterwards the network entry procedure has to be repeated and a secondary management connection might be established again. This time, however, the software version for the specific vendor and hardware is up to date and the network entry procedure commences.

The secondary management connection can also be used to update the date and time in the subscriber station via the time protocol specified in RFC 868 [9]. The time returned by the network is expressed in elapsed seconds since midnight on January 1, 1900. The universal coordinated time received from the time server can then be converted to local time by using the time offset received in the answer of the DHCP server, which was previously queried to get the IP address and other configuration information. RFC 868 also specifies that the time reference is only valid until the year 2036, when the seconds counter will overflow.

5.6.2 Power, Modulation, and Coding Control

Even though the WirelessMAN-OFDM specification has been designed only for stationary subscriber stations, there is still a need to monitor signal conditions and to adapt to them. Variable parameters are the power output of the subscriber station and the modulation and coding schemes for the uplink and downlink directions. A change of these parameters after the initial network entry procedure might be necessary due to changing weather conditions or if the subscriber station detects fading on the channel.

The current power and maximum possible transmission power is transmitted by a subscriber station to the network during the network attach procedure as part of the SBC-REQ message. If the network needs to increase or decrease the output power of the subscriber station at a later stage, it can do so by sending an unsolicited RNG-RSP message to the subscriber station with a new power value.

For the uplink direction, the base station can change the modulation and coding used by a subscriber station at any time by assigning a different burst of an uplink subframe which uses a higher or lower modulation scheme. The decision is based on the reception quality of previous MAC packets at the subscriber station.

For the downlink direction, the base station has no direct information about the change in reception quality over time for a subscriber station using a certain modulation and coding scheme. Thus, it is the client's device responsibility to request a change in the modulation or coding scheme if required. This can be done by the subscriber station by either sending an unsolicited RNG-REQ message or by sending a downlink profile burst change request (DBPC-REQ). The base station then decides to accept or reject the request and replies with a DBPC-RSP message.

5.6.3 Dynamic Frequency Selection

When operating in an unlicensed band, which is by definition open for the use of anyone and any technology, other systems like WLAN 802.11 might already use some of the frequencies in that band. Therefore, an 802.16 base station is required by regulatory bodies of some countries to ensure that the frequency band it wants to use is not in use by another system. This also includes base stations of its own network if the use of frequencies is not coordinated by the network, or base stations of other 802.16 network operators. This is done by the base station listening for some time on the channel it wants to use for interference caused by another system. If there is measurable interference on a channel, the base station selects another frequency band and repeats the procedure. Other systems described in this book like GSM and GPRS only used licensed bands and ensure by rigorous network planning that interference between cells is minimized. Only a few GSM base station products, mainly designed for indoor use, have implemented a similar scheme and can select their own operating frequency based on interference measurements and thus help the operator to save planning costs. Other systems like UMTS or CDMA are based on a different radio architecture and are thus able to use the same frequency band for all cells of the network (see Chapter 3).

As there exists no method to prevent the use of an allocated frequency band by other systems, the 802.16 standard also defines a number of mechanisms for a base station to check for interfering systems once it has selected a frequency band for operation. The easiest way to detect that a new interferer is now operating in the current frequency band of the cell is to stop transmitting and receiving periodically and listen on the empty channel for interference. If a new interferer has entered the area, the base station can then start to search for a new frequency band and continue its operation on a different frequency. Before it can do so, it has to inform the subscriber stations in the cell of the pending switch to a new frequency by announcing the new channel number in the DCD message.

In addition to listening for interferers, the base station can also ask one or several subscriber stations to do the same. This is done by including a report measurement information element in a DL-MAP entry of a subscriber station and then suspending activity of the cell while the subscriber stations measure their perceived interference of the channel. A measurement report can also be requested by a base station by sending a measurement report request (REP-REQ) message to a subscriber station which then has to send its report via a REP-RSP message. The base station can also instruct subscriber stations to measure interference in other frequency bands to test their suitability from the point of view of the subscriber stations. This method is quite helpful to ensure that subscriber stations do not experience strong interference if the base station itself is too far away from the source to experience the interference itself. If a subscriber station detects interference, it reports back to the system if it just experiences general interference or if it is caused by another 802.16 base station or subscriber.

5.7 MAC Management of User Data

Apart from managing the overall connection between the subscriber station and the network, the MAC layer also performs a number of tasks on the user data being delivered from higher layers of the protocol stack such as IP or 802.3. A number of these functions are described in this section.

5.7.1 Fragmentation and Packing

Similar to other systems that coordinate the use of the air interface from a central point, 802.16 subscriber stations are assigned uplink opportunities by the base station to send their data. The maximum size of MAC frames is 2048 bits. Therefore it is possible to put a complete higher layer frame into a single MAC packet. While the uplink is assigned to a subscriber station, it can send several consecutive and independent MAC frames. This is especially useful if IP is used as a higher layer protocol as packet sizes can vary from packet to packet and MAC frames that only contain a single higher layer frame may not fit exactly into the uplink opportunity allocated to the subscriber station. This situation is shown in Figure 5.10.

To facilitate efficient use of resources on the air interface, a sender can split a higher layer packet (e.g. an IP packet) into two (or more) parts and send them in two (or more) MAC frames. This eliminates transmission gaps as the MAC frame with the first part of a higher layer packet can be tailored to fit exactly into the end of the transmission opportunity. This is called fragmentation.

To decrease the overhead created by the MAC header, the standard also allows several small higher layer packets to be bundled together into a single MAC frame as shown in Figure 5.11. This is called segmentation. A single MAC frame could thus contain the end of a fragmented higher layer packet at the beginning, one or more non-fragmented higher layer packets in the middle, and the first fragment of the next higher layer packet at the end.

While packing and fragmentation are also used in GPRS and UMTS, wireless LAN does not use this functionality, as air interface management is decentralized. Thus, subscriber stations usually put a single higher layer packet into a MAC frame and use the air interface for exactly the time required to send the frame.

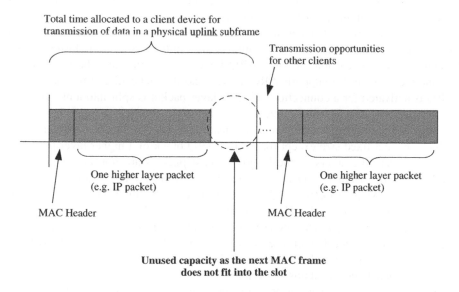

Figure 5.10 Inefficient use of the air interface without fragmentation and packing

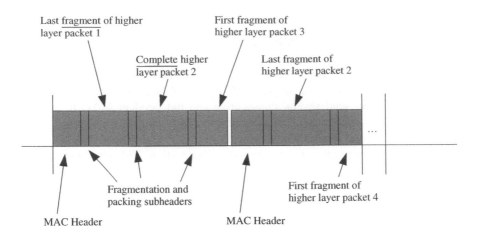

Figure 5.11 Efficient use of the air interface by using fragmentation and packing

5.7.2 Data Retransmission (ARQ)

While other technologies like HSDPA rely on sophisticated data retransmission techniques on lower layers to ensure fast retransmission of erroneous packets, implementation of the 802.16 ARQ scheme is purely optional. However, it might be very useful to implement this scheme in both network and subscriber stations as higher layer protocols like TCP interpret any missing data as congestion. This in turn reduces the throughput after an error has occurred. By correcting errors via retransmission on lower layers, which are aware that packets are not missing due to congestion but due to a transmission error over the wireless interface, high TCP throughput can be maintained.

The ARQ scheme presented below can be activated on a per connection (CID) basis during connection setup. The subscriber station signals to the base station during connection establishment that it is capable of using ARQ for the connection and the base station can activate the mechanism if it supports ARQ as well and if it is beneficial for the connection.

If ARQ is activated for a connection, a higher layer packet is split into a number of ARQ blocks. The maximum ARQ block size is 2040 bytes. In practice, a much lower value is used to allow for quicker detection and retransmission of missing or faulty information. ARQ blocks have a fixed size except for the last ARQ block of a higher layer packet which can be shorter than the others as the length of most packets are not an exact multiple of the ARQ block size. Higher layer packets are divided into several ARQ blocks, which are then mapped onto MAC packets. ARQ can be used in combination with fragmentation and packing as shown in Figure 5.12. Here, two higher layer packets are first split into several ARQ blocks. Smaller ARQ blocks are used for the remaining bytes at the end of the packets. Fragmentation and packing is then used to fill the first MAC packet with all ARQ blocks of higher layer packet one and the first two ARQ blocks of higher layer packet two. The second MAC packet then contains the remainder of the second higher layer packet. The small parts of the MAC packets in gray are fragmentation and packing subheaders which mark the beginning of a new higher layer frame and contain the (ARQ) block sequence

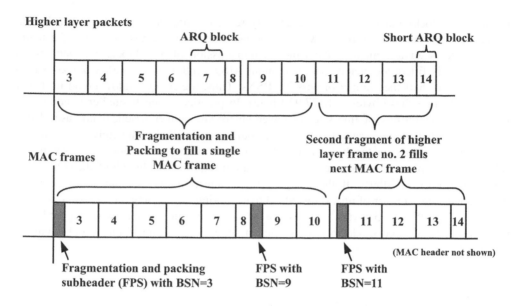

Figure 5.12 ARQ with fragmentation and segmentation

numbers (BSNs). Fragmentation and packing subheaders with BSNs are also inserted at the beginning of a new higher layer packet in a MAC frame to enable reassembly of the higher layer packets at the receiver.

There are two possibilities for the receiver to return ARQ feedback information to the sender: either, the ARQ feedback information is sent in an ARQ-FEEDBACK message over the basic management connection or directly on the user data service flow connection. In the latter case, an ARQ extension header is used which increases the length of the standard MAC header.

The standard offers several reporting formats to acknowledge correct blocks and to send a negative acknowledgment for bad (CRC checksum error) or missing blocks. The simplest format is a selective acknowledgment. This means, that the receiver acknowledges the reception of each ARQ block separately.

As single block acknowledgments are not very efficient for most scenarios, a cumulative acknowledgment format has been specified. This format is used to acknowledge all blocks up to a certain BSN by only specifying the highest BSN to be acknowledged.

If a number of good blocks have been received before a bad block can be reported, the receiver can use a third format, the selective acknowledgment. This is done by returning a selective ACK message that contains a bitmap in which each bit corresponds to one block. If the bit is set to one, the block is acknowledged. If the bit is set to zero, the block was not received correctly and the sender immediately reschedules the faulty blocks.

A refinement of the selective acknowledgment is the block sequence acknowledgment. Here, a bitmap is used in which each bit signals an acknowledgment or negative-acknowledgment for several blocks.

Once a faulty block has been received, the MAC layer cannot forward any data to higher layers until the block has been successfully retransmitted. Thus, the receiver has to store all blocks that have been successfully received after a faulty block in a buffer. To prevent the buffer from overflowing, a sliding window mechanism is used. If a faulty block prevents advancement of the window, transmission of new blocks will stop when the window is filled completely. The maximum ARQ window size is 2048 blocks, which corresponds to half the number of possible BSNs. Thus, the ARQ window could in theory be 2048 blocks × 2040 bytes = 4.177.920 bytes. In practice, a much smaller window size is used as data buffers of this size on lower layers are neither practical nor useful due to the delay that would occur if such a large buffer were filled completely. The window size mainly depends on the bandwidth delay product of the connection as shown in the following example. Let us assume that the maximum required bandwidth of the connection has been set to 1.5 Mbit/s. If the delay of the connection over the air interface is 50 milliseconds, the number of unacknowledged bytes during an ongoing data transfer is about 150 kbytes × 0.05 s = 7.5 kbytes. The ARQ window size should be somewhat bigger but in the same order of magnitude to have sufficient time to retransmit faulty blocks without stalling the overall data flow.

5.7.3 Header Compression

The MAC convergence sublayer also offers optional higher layer header compression schemes (packet header suppression, PHS) to decrease the signaling overhead. This can be quite beneficial for voice over IP connections, where small IP packets are used to minimize the latency introduced by the packetization of the data generated by the speech coder. In practice, such IP packets carry around 40 bytes of user data, an IP header of 20 bytes, and a UDP header of 8 bytes. As the headers usually contain the same content during an ongoing conversation, they can be easily compressed and a substantial amount of bandwidth on the air interface can be saved. Header compression can also be very beneficial to speed up call setup times. Typically, many interactions are necessary between the subscriber station, the telephony server (e.g. SIP server) and the device of the called subscriber to establish a voice or video call. As shorter messages can be transported faster over the air interface, call setup times are reduced.

The principle of the 820.16 header compression is to leave out the parts of the payload header (e.g. IP header) which are already present in a previous payload header. Among the parameters of an IP header, which never change for packets sent to the same destination device and application, are the IP addresses, the header length parameter, the time to live field, and the higher layer protocol indicator. The recipient of a packet with a compressed header can either be the subscriber station or the base station. To signal that packet header suppression is used for a packet, packet header suppression information is inserted at the beginning of the payload part of the MAC PDU. To enable the receiver to reconstruct the original header of a packet, sender and receiver have to exchange information about which parts of the header were removed before sending and on the content of the removed bytes. Rules and identifiers can be defined to construct a table at both sides for different headers, as a subscriber station usually communicates with several destination devices simultaneously, and header bytes and their contents vary per destination. It is also possible to change, add, and delete rules and entries in the compression table, as the destinations are usually not known at the setup of the connection and change throughout the lifetime of the service flow.

5.8 Security

As with all other technologies discussed in this book, security is of primary importance for 802.16 systems. Network operators require protection against fraudulent use and users need security measures to prevent third parties from using their subscription and generating costs for services they did not use. Furthermore, security mechanisms also have to ensure that the user's data cannot be intercepted and decoded by anyone but the network and the user by listening on the air interface. Similar to other wireless systems, security is achieved by user authentication during the network entry procedure and on a periodical basis for the duration of the connection. To protect transmitted data, encryption is used with an individual ciphering key per user. Methods to protect users against other users of the network who might try to hack into their systems is, as in other wireless systems, not specified and left to the network operator or the users.

5.8.1 Authentication

Contrary to other wireless systems, 802.16 uses public key authentication and cryptography to validate the credentials of a subscriber station. The method works as follows: each subscriber station is assigned a private and a public key by the manufacturer. The subscriber's public key is known to the network, while the private key remains secret and is never transmitted over the air interface. Data encrypted with the public key can only be decrypted with the private key. It is not possible to encrypt data with the public key and then to decrypt it again using the public key. The process also works in the other direction. Data encrypted with the private key can only be decrypted with the public key but not again with the private key. Unlike other systems, network and subscriber station do not have the same key for authentication but use a private/public key pair.

In addition to the private/public key pair, an X.509 certificate is used for authentication [10]. A certificate is issued by a certification authority (CA) and ties a well-known property of the subscriber station, the MAC address, to its public key. The CA adds a signature to the certificate, which is an encrypted copy of the X.509 certificate containing both the MAC address and the public key of the user. The encryption of the signature is performed with the CA's private key. The CA's public key is known to the base station of the network, which can thus decrypt the signature and compare the values with the MAC address and public key sent in the clear text part of the certificate. A potential hacker can only modify the clear text part of the certificate but not the signature, as only the private key (and not the public key) of the CA can be used to generate the signature. If the certificate is tampered with, i.e. an attacker tries to associate a different MAC address with a public key, the plain text and signature would not match and authentication would fail. As the public key of the CA is known, an attacker could also verify the validity of the certificate. However, the attacker would only be able to validate a subscriber. He could not use the information to get access to the network, as the private key of the user is not part of the certificate. Even if an attacker attempted to change the MAC address of his device to match the address in the intercepted certificate, the authentication would still be worthless, as the private key belonging to the certificate would still be unknown and further communication with the network would thus fail.

Note: X.509 certificates are also used for authenticating a web server to a web browser during establishment of a secure connection with the secure socket layer (SSL) protocol and secure http (https). A web-banking server for example sends an X.509 certificate to a web browser during https connection setup. The certificate, signed by a CA (e.g. Verisign), links the URL of the web-banking server (instead of the MAC address as in the example above) to the public key of the bank. The web browser knows and trusts the public key of the CA, as the public keys of most well-known CAs are usually preconfigured in web browsers. Thus, the web browser can easily verify whether the URL and the public key belong together, as it trusts the preconfigured public key of the CA. The bank on the other hand can only get a valid X.509 certificate from the CA during a registration process, which establishes the trust between the CA and the bank. The web browser trusts the CA and the CA trusts the bank due to the one-time registration process. For web applications, certificates signed by well-known CAs are usually not available for free as the CA charges for its services.

For WiMAX on the other hand, it is possible for the manufacturer of the subscriber station to be its own CA and sign the certificate itself. The public key of the manufacturer 'CA' is then given to the network operator, which establishes the trust relationship required for the certificate.

Figure 5.13 gives an overview of how the trust between the WiMAX subscriber station, the CA and the WiMAX network is established.

Having introduced the basic concepts of private keys, public keys and certificate, let us proceed with how the WiMAX authentication is performed. During the initial network entry procedure, the subscriber station sends its X.509 certificate to the base station. The base station checks the signature and returns an authentication key (AK) to the client. In order to prevent the AK from being compromised, it is encrypted with the public key of the subscriber station which was contained in the certificate. Decryption of the AK is only possible with the private key, which is safely embedded in the subscriber station. The AK is used to generate ciphering keys for all subsequent data exchange as described below.

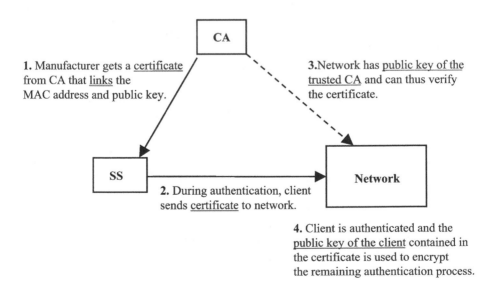

Figure 5.13   Establishment of trust relationships required for client authentication

5.8.2 Ciphering

Once an AK has been received from the network as part of the authentication process, the subscriber station prepares for the activation of encryption. In a first step, the network and the subscriber station derive a key encryption key (KEK) from the AK. In a second step, the subscriber station requests encryption activation. The network then generates a traffic encryption key (TEK) which is used to encrypt the connection to the user. The TEK is not returned to the user in clear text but is encrypted by using the KEK. As both network and client are aware of the KEK, the subscriber station can decrypt the TEK and use it for ciphering the user data in the future. User data traffic can only be exchanged once ciphering is in place. The ciphering key has a limited lifetime and it is the responsibility of the subscriber station to request a new TEK before the lifetime of the current ciphering key expires.

Figure 5.14 shows the message exchange required for authentication in a first step and negotiation of ciphering keys as a second step.

The TEK can be used as an input parameter for either a 56-bit DES (data encryption standard) or an AES (advanced encryption standard) encryption engine, both being capable of performing strong encryption of user data on the fly.

In addition to periodic ciphering key renewal, the subscriber station also has to perform periodic re-authentications as the lifetime of the AK is also limited. Both network and subscriber station use a state machine with timers to ensure that both authentication and ciphering are always valid and that new keys are obtained while the current ones are still valid.

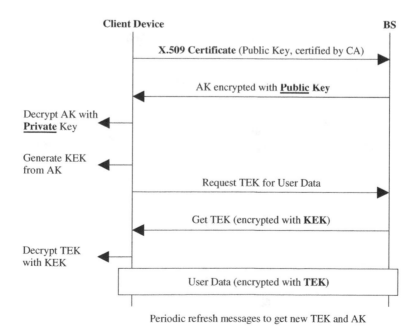

Figure 5.14 Authentication and exchange of ciphering keys

5.9 Advanced 802.16 Functionalities

While most of the functionality described in this chapter so far needs to be implemented in both base stations and subscriber stations to form a functional network, the 802.16 standard defines a number of optional and forward-looking functionalities that can extend the range and throughput of a network. It is unlikely that the functionalities described below will be implemented in the first generation of products. However, as the technology matures and the market for wireless high-speed Internet access increases, these functionalities offer possible options to grow the networks and the customer base.

5.9.1 Mesh Network Topology

One of the biggest problems of wireless networks is the limited coverage area of a single base station. As has been shown at the beginning of this chapter, the coverage area of 802.16 base stations is limited to a radius of two to five kilometers around the base station. This is similar to coverage areas of other systems like UMTS and HSDPA. The range of a base station in a conventional setup can only be increased by using directional outdoor antennas for subscriber stations, which is not always possible. A new approach to this problem are mesh networks. In the 802.16 mesh approach, not all subscriber stations communicate directly with the base station. Instead, distant subscriber stations communicate with neighboring subscriber stations instead of directly with the base station. These neighboring devices in turn forward their data packets to the base station or to another subscriber station if they are too far away from the base station themselves. This principle is shown in Figure 5.15. This approach

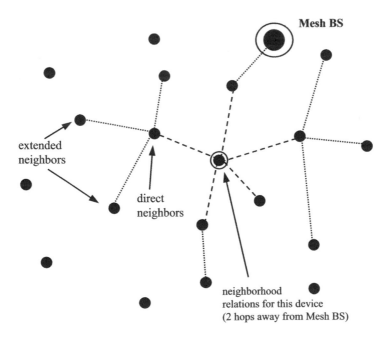

Figure 5.15   802.16 mesh network architecture

allows the coverage area of an 802.16 network to be extended into areas where it is not economically feasible to install a new base station. The mesh network is self-organized as it cannot be assured that all subscriber stations are switched on at all times. Thus, subscriber stations independently search for neighboring devices, which can be used as relays to the device that offers access to the backhaul network. In order to ensure that a neighboring client capable of forwarding packets to a subscriber station closer to a base station is always available, the number of neighbors in a network should either be high enough, or devices should be installed in a way to be active around the clock. This might be difficult to achieve, however, as the annual power consumption of a continuously operating device must not be underestimated and could make such an installation unpopular with end users.

Forwarding packets from one subscriber station to another until it finally reaches a base station reduces the overall throughput of the system, as a packet has to be sent over the air interface several times and only one subscriber station can be active at a time in an extended neighborhood, which includes the neighboring devices of a subscriber station and their neighboring devices. For throughput calculations, the standard assumes that 5% of the activity of a subscriber station is for its own purposes, which is based on current Internet usage patterns of private households. This means that a subscriber station can spend the remaining 95% of its operating time forwarding packets of other devices and remain silent while other devices in the area use the air interface.

Subscriber stations or base stations offering access to the backhaul network identify themselves as central mesh base station devices. At power up, each subscriber station searches for neighboring devices that broadcast their mesh network information, which includes information about the number of hops necessary to the closest mesh base station. A new subscriber station should start communicating either with the mesh base station directly if possible, or with the subscriber station that is closest to the mesh base station. Once the connection has been established, the new subscriber station should then start broadcasting mesh network information about itself, with the distance to the mesh base station increased by one compared to the mesh client it communicates with. Other devices, still further away from the base station, can then use the new client as a relay if required.

To avoid interference with other transmissions in an extended neighborhood, only one device is allowed to send a packet at a time. Due to this constraint, mesh networks require a different scheduling algorithm than centralized networks, where the base stations alone are responsible to schedule uplink and downlink opportunities. The standard offers the following scheduling algorithms:

- Distributed scheduling: scheduling is self-organized among the subscriber stations over a two-hop distance (extended neighborhood). Request grant schemes are used between neighbors to decrease the chance of simultaneous transmissions that would interfere with each other.
- Mesh base station (centralized) scheduling: here, the mesh base station device, which has access to a backhaul connection, is aware of all devices that want to use it to forward data to and from the backhaul network. Based on the bandwidth and delay requirements of each subscriber station, the mesh base station then calculates a scheduling pattern and distributes it via a broadcast to all neighbor stations, which in turn forward the scheduling pattern to their neighbors and so on.
- Combination: the standard allows a combination of distributed scheduling with centralized mesh base station scheduling.

As several subscriber stations might be necessary to forward packets from a distant device, the mesh base station requires the help of all subscriber stations between itself and the sender of a packet to ensure QoS attributes like a guaranteed bandwidth or latency. QoS thus has to be ensured on a packet-by-packet basis. Each packet contains QoS service parameters in the header from which a receiving subscriber station can deduct how to handle the packet, i.e. how quickly it has to forward the packet to the next hop. This ensures that packets with higher priority are sent first in case several packets are waiting to be transmitted.

Mesh network devices use a slightly different addressing scheme than devices in a standard 802.16 network as shown in Figure 5.15. Each subscriber station has a 16-bit node ID. To form the mesh, a connection is established to all subscriber stations, each using a unique 8-bit link ID. Afterwards, a broadcast message is sent over all links to inform the neighboring devices of the node ID and the number of hops that separate the subscriber station from the mesh base station.

5.9.2 Adaptive Antenna Systems

In order to minimize the costs of network deployment, the transmission capacity of a base station should be as high as possible to serve as many users as possible. In practice, the capacity of a base station is limited by factors such as the available bandwidth per base station, modulation and coding schemes, interference caused by neighboring base stations, as well as the distance of the wireless clients. Capacity can be increased if subscribers are not moving and directional antennas are installed on the rooftop pointing into the direction of the base station. In this case, lower power and better coding schemes can be used by the base station compared to moving subscribers with small omni-directional antennas. While systems like UMTS, HSDPA and CDMA1x allow subscribers to roam freely, the 802.16 profiles described in this chapter have been tailored specifically for non-moving subscribers with either rooftop antennas or omni-directional antennas in stationary subscriber stations. For these types of subscribers, it is relatively easy to increase the capacity and range of a base station by directing the signal energy towards specific devices. This concept is known as beam forming or as an adaptive antenna system (AAS). As shown in Figure 5.16, AAS can be used to limit the signal energy to a narrow beam which increases the range of the cell and lowers the interference with neighboring systems. Cell capacity can also be increased by transmitting data to different clients in parallel on the same frequency if they are located in different directions relative to the base station, as a single subscriber station only receives its own beam. It is more difficult to use AAS in systems that permit subscribers to roam freely and at high speed such as UMTS and HSDPA. Here, the bandwidth and processing power required to constantly adapt the direction of the beam towards a moving subscriber could easily outweigh the benefits.

Beam forming is achieved by sending the signal via several antennas, which are coupled with each other electrically. To form a beam, the signal is sent over each antenna with a calculated phase shift and amplitude relative to the other antennas. There are no moving parts required for directing the antennas in a certain direction, as the beam-forming effect is based on the phase and amplitude differences of the signal sent over the antennas. Usually, AAS is combined with sectorized antennas as described in the GSM and UMTS chapters to further increase the capacity of the system. In order to form the beam, at least two antennas are required that are separated by a multiple of the wavelength. At 2.5 GHz, the wavelength

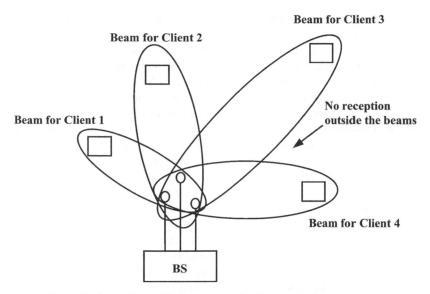

Base Station with multiple antennas for beam forming

Figure 5.16 Adaptive antenna systems and beam forming

is equal to $(1/2.5\,\text{GHz}) \times$ speed of light $= 12$ centimeters. In practice, antennas are typically separated about 1.5 meters.

To use AAS for 802.16, the standard has been designed in a backwards compatible way to allow the operation of both AAS capable and standard subscriber stations in the same cell. At network entry, the subscriber station is informed by the base station if it is capable of supporting AAS users. If AAS is supported by the base station, each uplink and downlink subframe has a special AAS area at the end which is preceded by an AAS preamble sequence. The AAS area has been put at the end of a frame because standard subscriber stations only listen to the beginning of a frame and their assigned downlink bursts and thus simply ignore AAS transmissions at the end of the frames. The AAS area is again split into two parts as shown in Figure 5.17. At the beginning of the AAS area, 'AAL alert' slots can be used by subscriber stations to join the network after power on as

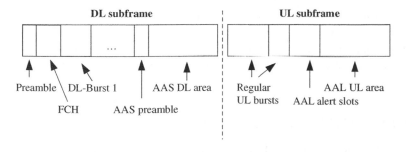

Figure 5.17 TDD uplink and downlink subframes with AAS areas

described in more detail below. The remaining part of the AAL area can then be used by the base station to send data simultaneously to several subscribers by forming individual beams to the subscribers. As for conventional transmissions, the DL-MAP and UL-MAP messages are used to broadcast information to all subscribers about when data will be sent to them and when they are allowed to send data. For AAS-capable devices, the DL-MAP message contains an extended concurrent transmission information element with the system parameters required for properly receiving the data transmitted in the AAL area.

There are two possibilities for a subscriber station to join the network: if a subscriber station is located close enough to the base station, it can use the standard network entry procedures as described in Section 5.6.1. If only a directed beam allows proper communication with the subscriber station, the subscriber station enters the network by sending a notification on all available alert slots of the AAS area. The base station receives the transmission and calculates the parameters required to form a beam towards the subscriber station whenever data is to be transferred.

In order to optimize the system, a number of MAC management messages are available to control the AAS parameters for each subscriber station. To keep a beam tuned correctly to a subscriber station, AAS feedback request and response (AAS-FBCK-REQ + RSP) messages and AAS beam request and response (AAS-BEAM-REQ + RSP) messages are used. Their purpose is to request channel measurements and to report their results to fine-tune the beams. Additionally, the AAS beam select (AAS-BEAM-SELECT) message has been defined to allow a subscriber station to indicate to the base station that it would like to use a different beam. Such a message might be used if a beam is directed to several subscribers instead of only one.

5.10 Mobile WiMAX: 802.16e

To improve the position of WiMAX in competition with UMTS and other 3G standards, the IEEE and the WiMAX forum have decided to enhance the standard with mobility functionality. As will be shown in the following section, the 802.16e standard introduces a number of enhancements on all layers of the protocol stack. On the physical layer, a new multiple access scheme is used. On the MAC layer, many additions were made to enable true mobility for wireless devices in and between networks. In addition, efficient power management functionalities for battery-driven devices have been defined. As client devices are enabled to roam through the network, they are now referred to as mobile stations. For national and international roaming, a network infrastructure has been standardized to support mobility management and subscriber authentication over network boundaries. These functions are outside the scope of the 802.16 standard, as it only describes the air interface. The WiMAX forum thus extended its work beyond promoting and certifying the technology and established a networking group to define and standardize how the network behind the base stations supports roaming and subscriber management. By specifying an end-to-end network topology, large and even nationwide networks can be built with components of different vendors.

5.10.1 OFDM Multiple Access for 802.16e Networks

For the 802.16e standard, the IEEE decided not to use the 256-OFDM physical layer used in first-generation networks. Instead, it was decided to evolve the OFDMA (orthogonal

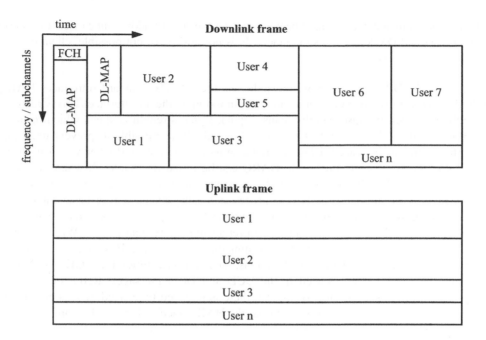

Figure 5.18 OFDMA subchannelization in the uplink and the downlink direction

frequency division multiple access) physical layer (PHY). This PHY was already specified in the previous version of the standard, and functionality has been added to address the requirements of mobile subscribers. In OFDM networks, subscribers transmit and receive their data packets one after another by using all available subchannels. OFDMA allows several subscribers to transmit and receive data simultaneously in different sets of subchannels. This principle is shown in Figure 5.18. Depending on the total channel bandwidth, 2048, 1024, 512, or 128 subchannels can be used compared to the fixed number of 256 subchannels of the OFDM PHY of first-generation networks. In an OFDMA system, the data rate of users cannot only be adapted by varying the length of their bursts as in OFDM, but also by varying the number of allocated subchannels.

The OFDMA physical layer is not backwards compatible with the 256-OFDM physical layer used by first-generation 802.16 networks. In practice, this creates a problem for operators of first-generation networks. Depending on the capabilities of their base stations and deployed stationary client devices, they have the following options to update their networks to support mobile devices:

- If base stations of a network operator support both OFDM and OFDMA via software upgrade, one carrier frequency is used for stationary devices while a second carrier frequency is used for mobile devices.
- If an operator has deployed stationary client devices that can be upgraded to support OFDMA, the network and the stationary client devices are updated. Afterwards, the same carrier frequencies are used to support stationary and mobile devices.

- If stationary client devices cannot be upgraded, and the use of additional carriers to support OFDM and OFDMA devices simultaneously is not desired or not possible, client devices have to be replaced.

Similar to HSDPA and other 3G technologies, the 802.16e standard introduces HARQ (hybrid ARQ) for fast error detection and retransmission on the air interface. This is required for mobile devices, because mobility causes quick signal strength changes which result in higher error rates. These have to be corrected as quickly as possible to prevent undesired side effects such as increased delay and retransmissions on the TCP layer which limit the overall throughput. An introduction to HARQ can be found in Chapter 3, where its use is discussed for HSDPA. In 802.16, HARQ can be activated per device or per service flow and the number of simultaneous HARQ processes are negotiated during basic capabilities exchange (SBC-REQ/RSP) and service activation (DSA-REQ/RSP). Both chase combining and incremental redundancy are supported to retransmit faulty data blocks. While HSDPA only uses HARQ to correct errors in the downlink direction, the 802.16 standard uses HARQ to secure data transmission in both directions. The response times for ACK and NACK messages are fixed and announced in the UCD and DCD messages. Retransmissions of faulty HARQ packets are asynchronous, i.e. there is no fixed time window in which faulty packets have to be retransmitted. In addition, the HARQ mechanism can be combined with adaptive modulation and coding techniques to quickly adapt to changing signal conditions. This reduces the number of retransmissions and increases throughput.

5.10.2 MIMO

To further increase transmission speeds, the 802.16e standard specifies MIMO (multiple input–multiple output) techniques for the network and the client devices. This is especially the case in urban environments, where a signal is often split into several transmission paths due to reflection and refraction caused by objects in the direct line of sight between the transmitter and the receiver. As the transmission paths have different lengths, each copy of the signal arrives at a slightly different time at the receiver as shown in Figure 5.19. For traditional GSM receivers, this phenomenon causes multipath fading due to the quickly changing paths and the resulting changes in interference of the different paths with each other. In systems such as UMTS, rake receivers are used to combine the signal energy received from different paths (see Chapter 3). Instead of trying to compensate for the effects of multipath transmissions at the receiver side, MIMO uses the effect by using multiple antennas at both the transmitter and receiver to send data on different paths but on the same frequency. If the same data stream is sent on all paths, robustness of the transmission is increased. If a different data stream is sent on each path, the data rate is increased. The MIMO variant used by 802.16 uses the second approach to increase the data rate.

MIMO requires a dedicated antenna for each transmission path both at the receiver (multiple input) and the transmitter (multiple output). Furthermore, each transmission path requires its own transmission and reception chain in the base station and the client device. A typical MIMO system makes use of two or four paths, which requires two or four antennas respectively. In current systems, antenna designs are used which already incorporate two antennas to pick up horizontally and vertically polarized signals created by reflection and refraction to counter the multipath fading effect (polarized diversity). An example of such

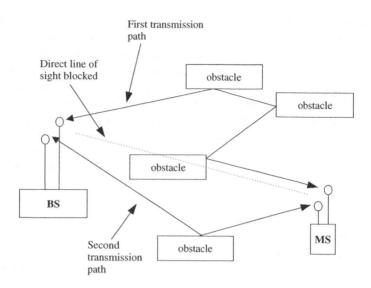

Figure 5.19 A signal is split into multiple paths by objects in the transmission path

an antenna is shown in Chapter 1, Figure 1.18. MIMO reuses this antenna design. Instead of combining the horizontally and vertically polarized signals for a single reception chain, the signals remain independent and are fed into independent reception chains. To send four individual data streams on the same frequency, two such antennas are required and must be separated in space by at least a quarter of a wavelength. Together with HARQ, AMC (adaptive modulation and coding), and AAS (adaptive antenna systems for beam forming), which were discussed above, MIMO techniques can multiply the overall bandwidth of a base station and the achievable data rates per client device [11]. It should be noted, that UMTS, HSDPA and HSUPA (see Chapter 3) do not make use of AAS and MIMO today, as those standards were developed earlier. Therefore, 802.16e networks using these enhancements will have a competitive advantage over enhanced UMTS networks. It is expected that the 3GPP will react to this and specify similar techniques in further evolutions of the UMTS standards.

5.10.3 Handover

The physical layer enhancements ensure a stable connection between the network and the user while roaming through a cell. To ensure connectivity beyond the user's serving cell, the MAC layer was enhanced to enable handovers between cells without dropping the client's context with the network. As handovers between cells also require routing changes in the network behind the base stations, the WiMAX radio and core network have to support the new mobility functionality. The required network functionalities are described in Section 5.11.

The 802.16e standard defines that both the mobile station and the network are allowed to initiate a handover. This is in contrast to systems like UMTS, where the network is always responsible for preparing and initiating a handover. For the handover decision, the mobile station and the network must be aware of neighboring cells and their reception levels at the current location relative to the current serving cell. The network can assist the mobile station in its search for neighboring cells by sending neighboring cell information in MOB_NBR-ADV messages. These messages contain the frequencies used by neighboring cells and the contents of their UCD and DCD messages. If this information is not available in the current serving cell, the mobile station is also allowed to search for neighboring cells on its own and retrieve the UCD and DCD messages itself. To synchronize with neighboring cells a mobile station can then perform an initial synchronization, ranging and association to ensure that a cell can be used as quickly as possible after a handover. This procedure is called cell reselection. It should be noted that cell reselection has a different meaning in GSM, GPRS and UMTS. Here, the term is used for the procedure that is performed by mobile stations in idle mode to move from one cell to another.

During the time required for the cell reselection procedure, the mobile station cannot receive data from the cell. To ensure that the cell buffers incoming data during this time, the network assigns scanning periods to the mobile station. The mobile station can also request them if required. Once the mobile station returns to the current serving cell, it sends a measurement report to the network. The network can then use this information to prepare a handover into a neighboring cell in a similar way as described in Chapters 1 and 3 for GSM and UMTS. If the mobile station finishes cell reselection early, it can exit this state by sending a MAC PDU to the serving cell.

A timer is used in the mobile station to renew its associations to neighboring cells frequently. This is required as signal conditions change when the subscriber changes its location and the parameters acquired during the association procedures become invalid. Associations have to be deleted if they cannot be renewed before the timer expires.

Handover times vary depending on how the handover is performed. Longer data transfer outages are to be expected if an uncoordinated handover is performed in which the mobile station initiates the handover on its own, is not synchronized to the new cell, and has not informed the network of the handover. In this case, most steps as described for normal network entry have to be performed before service can resumed. In order to restore service flow parameters like the IP addresses used by the mobile terminal, the new cell has to request information about the subscriber from the previous cell. For this purpose, the handover message of the mobile station includes the ID of the previous cell.

The interruption of an ongoing data transfer is much shorter if the handover is prepared and initiated by the network. Figure 5.20 shows the basic principle of the handover procedure, if the mobile is already associated with the target cell and the target cell is already prepared for the handover. If these conditions are met, contention-based initial ranging is not required. In addition, the network can prepare a target cell for a handover by forwarding all subscriber-related information like authentication information, encryption information, and parameters of active service flows. Once the mobile station establishes contact with the new cell, basic capability negotiation, PKM authentication, TEK establishment, and registration messaging can be skipped and service flows can be immediately reactivated. Figure 5.20 shows such an optimized handover procedure, which requires non-standardized messaging to exchange subscriber information between the current serving cell and the new cell.

As the CIDs of active service flows are cell specific, the REG-RSP message at the end of the handover procedure contains a list that maps the previous service flow identifiers to those of the new cell. The mobile station can thus keep its IP addresses.

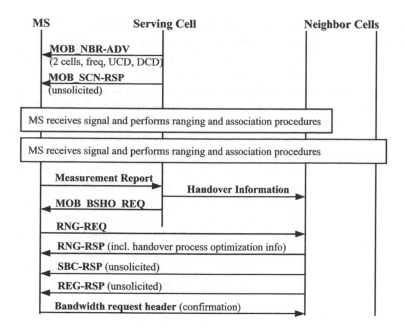

Figure 5.20 Optimized handover

How the traffic to and from the subscriber is rerouted to the new cell in the network is out of scope of the 802.16 standard and was defined separately by the WiMAX forum networking group. These mechanisms are described in Section 5.11.

Despite much optimization, the handover described above still requires the mobile device to disconnect from the current base station before starting communication with the new base station. As the resulting transmission gaps may have a negative impact on real-time applications such as voice and video over IP, additional enhancements are required to seamlessly handover such connections. For this purpose, two optional handover procedures have been specified which can be used if the network and the mobile device announce in registration request and response messages that they support them.

One optional handover procedure is fast base station switching (FBSS) [12]. If used, the mobile device frequently scans for neighboring base stations and reports measurement results to the network. Network and mobile device can then agree on using several base stations simultaneously by putting several base stations in a diversity set list which is kept in both the network and the client device. Adding and deleting cells in the diversity set is performed by the mobile sending MOB_MSHO_REQ messages. If the diversity list contains more than a single base station, the mobile station can dynamically inform the network from which base station it would like to receive data in the downlink direction via another MOB_MSHO_REQ message. The network is also allowed to trigger the handover process by sending a MOB_BSHO_REQ message. At any time, only a single base station is responsible for forwarding data to the mobile device in the downlink direction.

FBSS requires all base stations in the diversity set to be synchronized and to use a synchronized frame structure. This way, the mobile device must not resynchronize itself to a new base station in the downlink direction, which minimizes the interruption caused by the handover. In addition, all base stations included in the diversity set have to operate on the same frequency. As neighboring base stations transmitting on the same frequency interfere with each other, optional beam forming (AAS) and power adaptation functionality in the downlink direction help to reduce this unwanted side effect.

The base station that is responsible for sending data to the subscriber in the downlink direction is referred to as the anchor base station. Apart from data transfer, the anchor base station is also responsible for the administration of the subscriber context. When an FBSS handover is performed, the new base station assumes control of the context.

In the uplink direction, all base stations of the diversity set listen to transmissions of the mobile device. This requires a further logical synchronization in the radio network between the base stations in the diversity set, as all base stations have to schedule uplink opportunities for a mobile device at the same time. Each base station then forwards only correctly received frames to the core network. This requires functionality in the radio network to combine the different uplink data streams in order to forward only a single uplink data stream to the core network.

The macro diversity handover (MDHO) is an even smoother form of handover. Like the FBSS handover, it is also optional. When MDHO is activated for a connection, e.g. due to effects such as deteriorating signal conditions, all base stations of the diversity set synchronously transmit the same data frames in the downlink direction. As all base stations transmit on the same frequency, the mobile device can either use RF energy combining or soft data combining to benefit from the multiple simultaneous transmissions. If the reception of one of the base stations in the diversity set becomes too weak, it is removed from the diversity set. Additions and deletions in the diversity set are performed by the mobile using MOB_MSHO_REQ messages. As several base stations communicate with the client device simultaneously, anchor responsibilities only have to be transferred to another base station if the current anchor base station is removed from the diversity set. If only one base station remains in the diversity set, the MDHO state ends and the handover has been performed without any interruption of the ongoing data transfer.

In the uplink direction, the MDHO and FBSS handover behavior is identical.

The concept of an anchor base station cannot be found in other systems such as GSM, UMTS, or CDMA. In these systems, handovers are controlled from a central controlling element in the radio network such as a BSC or an RNC. In 802.16e radio networks on the other hand, the anchor base station concept has been introduced because the base stations organize themselves. The functionalities of the radio controller node between the base stations and the gateway to the core network (e.g. an SGSN) have been partly put into the base stations and party into the access service network gateway (ASN-GW) node, which is further described in Section 5.11.

5.10.4 Power-Saving Functionality

While a connection is active, a mobile terminal requires a considerable amount of energy to keep listening to the network for incoming data. To increase the battery operating time, the mobile can reduce its energy consumption in times of low activity by entering power-save mode. Several power-saving modes have been defined in the standard and each active service flow can use a different power-saving mode. As a consequence, the mobile can only deactivate its transceiver at times in which all active service flows have entered the power-saving state.

Power-saving class I is activated by the mobile station and confirmed by the base station. In this mode, active periods with a static length alternate with sleeping periods which increase over time. As the length of the sleeping periods increase over time up to a predefined value, activity of the mobile and energy consumption is automatically reduced over time. If data arrives for the mobile station while in this mode the network aborts the sleep mode by sending a MOB_TRF-IND message during an active period. The mobile station also automatically leaves the sleep mode if data has to be sent in the uplink direction. As no data can be sent or received in this mode, power-saving class I is most suitable for non-real-time and background service flows.

For real-time services, power-saving class II introduces fixed activity periods that alternate with predefined sleeping periods. In contrast to class I, data can be exchanged in active phases in both directions without leaving the overall power-save mode state. This is important for real-time services, as data with fixed or varying bandwidth requirements is constantly transmitted. By choosing appropriate activity and sleeping periods the system can ensure sufficient bandwidth for the connection and required delay times can be met. This is possible because real-time services do not require the full bandwidth offered by the air interface. Power-saving class II thus offers the system the possibility of limiting transmissions to certain frames, which helps to save battery power by deactivating the transceiver in a mobile station during frames which are sent in the sleeping periods.

Power-saving class III has been designed for management connections and broadcast services. When the mobile requests such a connection to be set into this sleep mode variant, the base station calculates a sleep window during which no broadcast data or management message needs to be sent in the downlink direction. The mobile station then enters sleep mode for the granted duration and becomes active again automatically once the sleeping period has expired.

5.10.5 Idle Mode

To further reduce power consumption during times of longer inactivity the 802.16e standard introduces an optional idle mode for mobile stations. Its basic functionality is similar to the concept of a UMTS UE in idle mode with an active PDP context (see Chapter 3). As in UMTS the mobile station retains its service flows, i.e. its IP addresses, while no active communication connection is maintained with the network. If new data is received by the core network for a mobile station in idle mode, a paging procedure has to be performed in all cells belonging to the same paging group. Paging a mobile in several cells requires a central paging controller in the network. As the 802.16 standard only defines the air interface part of the network, the implementation of this function is out of the scope of the standard and has been left for further standardization by the WiMAX forum networking group.

The concept of a paging group is similar to the UMTS concept of a location area. Unlike location areas, paging groups can overlap in the network and a cell can belong to several paging groups simultaneously. This is shown in Figure 5.21. This prevents frequent paging-group updates of mobiles in paging-group border areas.

While in idle mode, the mobile station can roam to cells belonging to the same paging group without performing a handover or notifying the network about the cell change. From

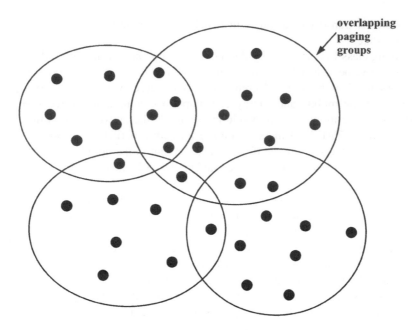

Figure 5.21 Overlapping paging groups

time to time, the mobile station has to send a location update to the network in order to keep the service flows active. For most of the time, the mobile station's transceiver is deactivated while in idle mode. In order to be able to react to incoming paging messages, the mobile has to periodically reactivate its transceiver to listen for incoming paging messages in which the mobile is identified via its MAC address. Furthermore, the mobile station has to periodically check the reception level of the current serving cell, search for neighboring cells, and perform and select a new serving cell if required.

If a mobile station receives a paging message, it has to perform ranging and network entry procedures. As the base station can retrieve the context of the mobile station from the mobility management controller in the network, most steps of the network entry procedure can be skipped in a similar way as described above for a handover. The mobility management controller function is typically implemented together with the paging controller function in a central element in the network.

5.11 WiMAX Network Infrastructure

Many features such as mobility management for handover and idle mode paging require coordination between different nodes of the network. How these features are implemented is beyond the scope of the 802.16 specification, as it only deals with the air interface between base stations and client devices. Other features, such as national and international roaming between networks, user authentication, administration, and billing, are also not part of the 802.16 specification. As many vendors are developing mobile devices and network infrastructure components, a standard is required that describes these functionalities. This ensures interoperability of networks and components of different manufacturers within the network.

The main benefits of standardized WiMAX networks for subscribers are standardized hardware and software that can be mass-produced and can thus be competitively priced and used in any network. For the operator, standardized components and functionalities ensure competition among vendors resulting in competitive pricing of network components. In addition, standardized interfaces for roaming enable operators to offer services to visiting subscribers.

The following sections describe the main aspects of the WiMAX network infrastructure, which is standardized by the networking group of the WiMAX forum. Members of this body are vendors such as Intel, Samsung, Motorola, Nortel, and many others who are involved in developing products for the WiMAX ecosystem ranging from chipsets, user devices, base stations, and other network infrastructure. As specification work for the network infrastructure started relatively late in the overall design process, many first-generation networks are proprietary and not interoperable with each other. In the course of the evolution of these networks it is expected that they will be upgraded to be compliant to the WiMAX forum network infrastructure standards to benefit from the advantages listed above.

5.11.1 Network Reference Architecture

The network reference architecture of the WiMAX forum networking group specifies a number of reference points between logical functions of the network. Therefore network vendors can choose between different alternatives of where to put a number of functionalities in the radio and core network. Figure 5.22 shows one of the possible architectures [13] that is likely to be implemented by vendors. Similar to other types of wireless wide area networks, a WiMAX network is split into radio access and core network parts. The radio access network part is referred to as the access service network (ASN). It is connected to the core network via the ASN-gateway (ASN-GW) and the R3 reference point. A large network can comprise more than a single ASN if several ASN-GWs are required for the management of the radio access network.

An ASN contains two logical entities, the ASN-GW and the base stations. Compared to GSM and UMTS, it should be noted that the architecture no longer contains an entity between the gateway to the core network and the cells such as an RNC. The functionalities of this node, such as radio channel management and mobility management, were moved partly to the base stations and partly to the ASN-GW.

Another difference from networks discussed in Chapters 1 to 3 is the use of the IP protocol on all interfaces (reference points) between all nodes of the network. This reduces complexity and cost as IP has become the dominant network protocol and can be used with almost any kind of underlying transport technology. For short distances between nodes, Ethernet over twisted pair copper cables can be used as it is a very cheap transport technology. For larger distances, optical technologies are most suitable, and IP is transported via ATM or via optical Ethernet. As IP is used with all technologies, only a single WiMAX-specific software stack is required for the different transmission technologies. This reduces cost and complexity. As WiMAX networks no longer use circuit-switched connections, using IP on all interfaces is easily possible.

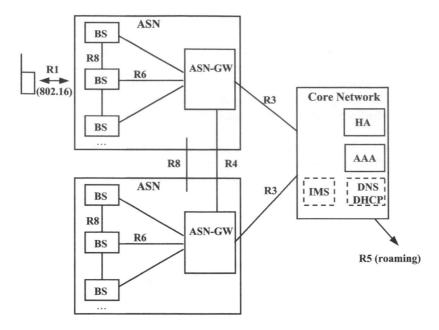

Figure 5.22 WiMAX network reference architecture

Fast base station switching (FBSS) and macro diversity handovers (MDHO) require a close synchronization between base stations. As the basic IP protocol does not ensure constant latency and bandwidth for a connection, IP QoS mechanisms have to be used over the R8 reference point. The reference points/interfaces inside the ASN (R6 and R8) have not been specified in the first version of the WiMAX network infrastructure standard and thus such solutions are proprietary.

Due to the use of IP on all interfaces, WiMAX network components can be directly connected with each. For longer distances, standard IP routers can be used to forward both user data and signaling traffic between the components. No special WiMAX software is required in the IP routers. This enables operators to use cheaply available IP hardware. In addition, operators can lease IP bandwidth from other companies, for example to connect base stations to the ASN-GW. To ensure security and confidentiality, encryption (e.g. IPSec) and tunneling mechanisms should be used on these interfaces.

Apart from offering direct Internet access, operators may also be interested in offering value-added services such as voice and video over IP, push to talk, voice and video mail, IP television, and other advanced multimedia services. It is likely that operators will host a variety of multimedia nodes in their core networks such as the IP multimedia subsystem (IMS, see Chapter 3).

Authentication, authorization, and accounting (AAA) is another functionality of the core network. It is used to flexibly bill services such as Internet access and IMS services used by the subscriber. To allow subscribers to roam between networks, AAA is another important functionality that has to be standardized in order to be interoperable. For this purpose, the R5 reference point has been defined to allow foreign networks to access the AAA server in the home network of a subscriber.

5.11.2 Micro Mobility Management

When establishing a connection to the network, an IP address is assigned to the subscriber device. When moving between base stations, the IP address has to remain the same to preserve communication connections established by higher layer applications. As routing decisions in the network are based on IP addresses and static routing tables, the mobility of the subscriber has to be hidden from most of the network. This is done in several ways.

While moving between base stations of a single ASN, the mobility of the subscriber is managed inside the ASN, and the ASN-GW hides the mobility of the subscriber from the core network and the Internet (R6 mobility). As long as the subscriber roams between base stations connected to the same ASN-GW, all IP packets flow through the same ASN-GW. Inside the ASN, IP tunnels are used to direct IP packets to the base station currently serving a subscriber. Three layers of tunnels are used as shown in Figures 5.23 and 5.24. On the first layer, each base station is connected to the ASN-GW via a secure and possibly encrypted IP tunnel to protect the data flowing between the two nodes. This allows the use of third-party networks to forward traffic between a base station and the ASN-GW.

Inside the base station IP tunnel, a further IP tunnel is established per subscriber. When a subscriber roams from one base station to another, the ASN-GW redirects this tunnel to another base station tunnel. By tunneling the IP packets through the IP network, only the routing table of the ASN-GW has to be modified when the subscriber roams to another cell. The routing tables of routers in between the ASN-GW and the base stations do not have to be altered, as the routing is based on the IP address of the base stations. The IP packets for a client device including its IP address is embedded in the payload part and is thus not used for the routing process inside an ASN.

This micro mobility management concept is similar to that of the GPRS tunneling protocol (GTP) which is used in GPRS and UMTS networks to tunnel user data between the GGSN and the SGSN (see Chapter 2). It should be noted, however, that in GPRS and UMTS

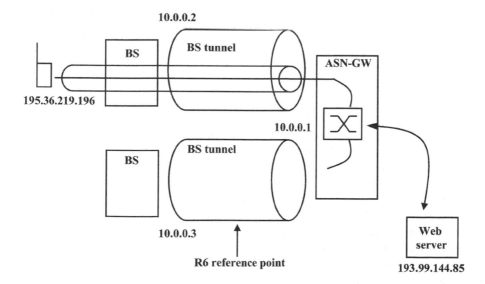

Figure 5.23 Micro mobility management inside an ASN

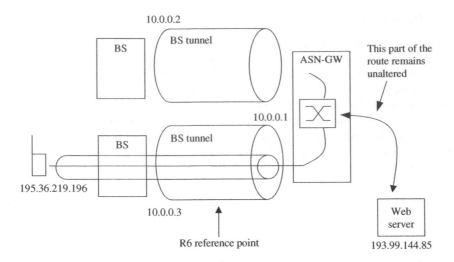

Figure 5.24 Subscriber tunnel after handover to new cell

networks IP tunneling is used in the core network while in WiMAX networks IP tunneling is used in the radio access network (ASN). In the WiMAX core network, mobile IP is used for subscriber mobility management, which is discussed in the next section.

A client device can have several active service flows, each with its own IP address. To separate these service flows, a third tunnel layer is used.

The 802.16 standard offers several convergence sublayers on the air interface to embed IP packets in a MAC frame. The WiMAX networking group has chosen the IP convergence sublayer (CS) as shown on the left side of Figure 5.1 for its network architecture [14]. This CS only generates a small overhead compared to other CS and reduces the complexity of developing dual-mode devices capable of seamlessly roaming between WiMAX and other networks types such as UMTS.

5.11.3 Macro Mobility Management

If a subscriber roams to a base station of another ASN, traffic needs to be redirected to the new ASN. This can be done in several ways. If the anchor ASN-GW is to be maintained, the traffic from and to the core network continues to flow through the ASN-GW of the subscriber's original ASN. The original ASN then forwards all user data frames and management messages to the new ASN via the R4 reference point shown in Figure 5.22.

While one of the cells of the old ASN is still part of the diversity set of the enhanced FBSS or MDHO handover variants, the A8 interface can be used if present to include cells of several ASNs in the diversity set.

To optimize the routing in the network, it might be beneficial at some point to change the route of the incoming and outgoing traffic of a user to flow only through the ASN-GW of the new ASN. For this dynamic rerouting, mobile IP (MIP) is used between the ASN-GW and

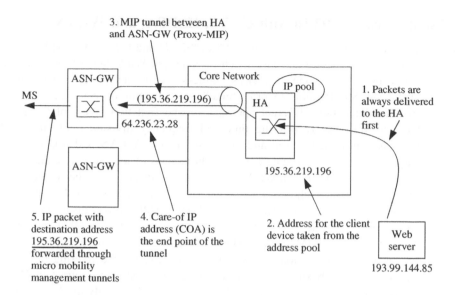

Figure 5.25 Principle of (proxy) mobile-IP in a WiMAX network

the subscriber's home network. The principle of MIP is shown in Figure 5.25. If a subscriber establishes an IP version 4 connection, the ASN-GW acts as a proxy and terminates the MIP connection instead of the mobile device (proxy-MIP). This allows the use of a standard IP version 4 stack on the client device without MIP capability. During the connection setup procedure the ASN-GW registers with the MIP home agent (HA) in the user's home network and sends its local IP address to the HA. This IP address is also known as the user's care-of IP address (COA) as it can change at any time during the lifetime of the connection. The HA then assigns an IP address for the user and returns it to the ASN-GW. The ASN-GW in turn forwards this IP address to the client device, which will use it for all incoming and outgoing data packets. The IP address assigned by the HA to the ASN-GW (and thus to the client device) belongs to a local pool of IP addresses and all data packets which use this IP address as the destination will always be routed to the HA. If an external host sends an IP packet to the mobile device, it is routed to the home agent first. There the packet is forwarded inside an MIP tunnel to the COA, i.e. the ASN-GW. The ASN-GW is the end of the MIP tunnel and in turn forwards the IP packet through the micro mobility management tunnels described in the previous section. Any change in the COA, i.e. a change to another ASN-GW, is transparent to external hosts and routers. From their point of view, the home agent remains the destination for the packet.

In the reverse direction, mobile devices use the IP address assigned by the HA as the originating IP address of a packet to an external host and not the COA (of which it is not even aware as the ASN-GW acts as MIP proxy). As routing decisions in an IP network are not based on the originating IP address but on the terminating IP address of a packet, it is routed directly to the external host instead of via the HA.

If a client device uses IPv6, no proxy MIP mechanisms are required in the ASN-GW, as IPv6 natively offers MIP functionality.

5.12 Comparison of 802.16 with UMTS, HSDPA, and WLAN

As has been shown in this chapter, wireless LANs (802.11) and the wireless MANs defined
in the IEEE 802.16 standard do not have much in common. While WLAN is designed for
home and office use to interconnect devices wirelessly with each other and the Internet
over short distances, 802.16 aims to offer broadband connections over larger distances.
This requires a fundamentally different approach on the first two layers of the protocol
stack compared to WLAN. By offering time and frequency division duplexing, 802.16
systems can be used in both licensed and unlicensed bands. WLAN and 802.16 are thus
complementary technologies, as some devices for home and office use may combine them
by offering wireless connectivity via WLAN to notebooks, PDAs, and other devices, while
using 802.16 as a backhaul technology to connect the local network to the Internet. By
providing fast Internet access with speeds between 1 and 10 Mbit/s over distances of several
kilometers in a real environment, 802.16 networks can compete with other metropolitan
network technologies such as UMTS and HSDPA, which have been discussed in Chapter 3.
Given similar bandwidths allocations, both systems are capable of delivering fast Internet
access to both private and business users at comparable speeds. In contrast to HSDPA,
which is a natural evolution for UMTS networks and will thus be mostly used by incumbent
wireless operators, 802.16 is an interesting technology for new network operators that want
to compete with other methods of broadband fixed and wireless Internet access. While
HSDPA is designed for both fixed and mobile use, the 802.16-2004 standard is limited to
stationary wireless clients with internal antennas if they are close enough to the base station,
or roof-mounted directional antennas for larger distances. This limitation greatly reduces
the complexity of the solution, which in turn helps to reduce network infrastructure costs.
While systems like UMTS and CDMA1x are end-to-end network systems with sophisticated
service architectures to allow national and international roaming of subscribers, the 802.16-
2004 standard only deals with the first two layers of the network protocol stack. Thus, such
networks are limited to regional coverage. The 802.16e extension to the standard aims to
improve the situation by adding mobility, and notebook component manufacturers such as
Intel have shown interest in delivering chipsets which support the mobility extension of the
802.16 the standard. In addition a network architecture has been defined that allows national
and international roaming. At the time of publication, the first 802.16 networks only support
stationary devices, with the first notebooks using 802.16e mobility chipsets expected in the
2007 timeframe. This will help to further increase the competition with UMTS and CDMA1x
networks, which should result in lower prices for end customers. 802.16 should prove to be
the technology of choice for offering fast Internet access in rural areas, where other forms
of broadband access such as DSL or cable are not economically viable. In countries where
networks such as UMTS and CDMA1x are available, this will increase competition and will
help to drive operators to evolve their networks in order to hold on to their market shares
and revenues. In developing countries, 802.16 allows operators to offer Internet access to
a broader market in the same way that GSM networks have allowed operators to deliver
telephony services to millions of people without access to a public fixed-line telephony
network. Like other technologies described in this book, 802.16 networks have not appeared
in the marketplace as quickly as predicted by analysts, sales managers, and the media.
However, if the long-term success of these systems can be taken as an example, networks
based on 802.16 should have an interesting and exciting future.

5.13 Questions

1. What are the theoretical and practical bandwidths offered by an 802.16 system when used for connecting end users to the Internet?
2. How does the coding scheme influence user throughput and overall throughput in a cell?
3. Why does the 802.16 support both FDD and TDD mode of operation?
4. What is a service flow?
5. Which difficulties are encountered when a license-free band is used for the operation of an 802.16 cell?
6. Why is fragmentation and packing used for transmitting IP packets over the 802.16 air interface?
7. What is the difference between the coordination scheme used in WLAN (802.11) and the one used in 802.16 systems?
8. Why is the MAC address of a device not used in the header of a MAC packet?
9. Which steps are required for a subscriber station to connect to the network?
10. How can a mesh network extend the range of a base station?
11. What is the advantage of using an adaptive antenna system?
12. What is the basic architectural difference between a WiMAX radio network and other radio networks described in this book?
13. What is fast base station switching?
14. How can MIMO improve transmission speeds?

Answers to these questions can be found on the companion website for this book at http://www.wirelessmoves.com.

References

[1] The Institute of Electrical and Electronics Engineers, Inc., '802.16-2004 IEEE Standard for Local and Metropolitan Area Networks – Part 16: Air Interface for Fixed Broadband Wireless Access Systems', IEEE standard, October 2004.
[2] The Worldwide Interoperability for Microwave Access Forum, 'IEEE 802.16a Standard and WiMAX Igniting Broadband Wireless Access', white paper, available at http://www.wimaxforum.org.
[3] David Johnston and Hassan Jaghoobi, 'Peering into the WiMAX Spec: Part 1', white paper, January 2004, available at http://www.commsdesign.com.
[4] WiMAX Forum, Eugene Crozier and Allen Klein, 'WiMAX's Technology for LOS and NLOS Environments', white paper, available at http://www.wimaxforum.org.
[5] Arunabha Ghosh, David R. Wolter, Jeffrey G. Andrews and Runhua Chen, 'Broadband Wireless Access with WiMax/802.16: Current Performance Benchmarks and Future Potential', February 2005, *IEEE Communications Magazine*, pp. 129–36.
[6] Govindan Nair *et al.*, 'IEEE 802.16 Medium Access Control and Service Provisioning', August 2004, *Intel Technology Journal*, **8**(3), 212–28.
[7] K. Sollins, 'RFC 1350 – The TFTP Protocol (Revision 2)', *Internet RFC Archives*, July 1992.
[8] R. Droms, 'RFC 2131 – Dynamic Host Configuration Protocol', *Internet RFC Archives*, March 1997.
[9] J. Postel and K. Harrenstien, 'RFC 868 – Time Protocol', *Internet RFC Archives*, May 1983.
[10] R. Housley *et al.*, 'RFC 2459 – Internet X.509 Public Key Infrastructure Certificate and CRL Profile', *Internet RFC Archives*, January 1999.

[11] A. Jeffries *et al.*, 'New Enabling Technologies: Building Blocks for Next-Generation Wireless Solutions', *Nortel Technical Journal*, 2, July 2005, available at http://www.nortel.com.

[12] Bill Cage *et al.*, 'WiMAX: Untethering the Internet User', *Nortel Technical Journal*, 2, July 2005, available at http://www.nortel.com.

[13] Parviz Yegani, 'WiMAX Overview', Presentation for the IETF-64 Conference, November 2005.

[14] Max Riegel, 'IEEE 802.16 Convergence Sublayer', Presentation for the IETF-64 Conference, November 2005.

6

Bluetooth

To connect devices such as computers, printers, mobile phones, PDAs and headsets with each other, a number of cable and infrared technologies have been developed over the years. Wired connections are mostly used for big or stationary devices, while infrared connections have advantages for small or mobile devices. In practice, however, the use of wired or infrared connections is often complicated and also not very practical in many situations. The Bluetooth technology offers an ideal solution to this problem. In order to show the possibilities of Bluetooth, this chapter provides an overview of the physical characteristics and the general functionality of the system, as well as the organization and functionality of the protocol stack. We then discuss the concept of Bluetooth profiles and demonstrate how they can be used in practice. While Bluetooth and wireless LAN are two very different systems, they also have many things in common. Thus, a comparison is made between the two technologies at the end of the chapter to show which technology is the best choice for which application.

6.1 Overview and Applications

Due to the ongoing miniaturization and integration, more and more small electronic devices are used nowadays in everyday life. Bluetooth enables these devices to wirelessly communicate with each other without a direct line-of-sight connection. This enables a wide range of new applications and possibilities. Some of them are described below.

The mobile phone is at the center of many new applications. In addition to normal voice telephony, mobile phones are also used today to connect to the Internet. Apart from the embedded WAP browser, external devices like notebooks or PDAs can also use the mobile phone as a gateway to the Internet. In order to establish a connection between the devices, the mobile phone simply has to be in range and does not even have to be taken out of a pocket or case. Thus, it is no longer necessary to connect devices with a cable or position them in a certain way for an infrared source. This is a big advantage especially when traveling in trains, buses, cars or the metro, where there is usually only limited space and freedom of movement.

The Bluetooth module embedded in a mobile phone can be used for many other things as well. Calendar entries, addresses, notes, etc., which are stored on a mobile phone, can be

Communication Systems for the Mobile Information Society   Martin Sauter
© 2006 John Wiley & Sons, Ltd

quickly exchanged with other personal devices like PDAs, notebooks, or devices of friends while they are at close range.

Many mobile phones are also equipped with a photo camera and file systems in order to take and store pictures. By using Bluetooth, these pictures can be sent quickly and for free to other mobile phones, PDAs, notebooks and PCs in the vicinity.

Mobile phone file systems are not only suitable for photos but can also be used for a great variety of other file types. Thus, it is also possible to send files from a PC or notebook to a mobile phone and retrieve the information at another location with another device. This application replaces a universal serial bus (USB) memory stick and the files can be copied to and from the mobile phone without attaching a cable or plugging the phone into a USB port of a PC.

Speech transmission between a mobile phone and a headset is another interesting application for Bluetooth. For an incoming call, the user simply accepts the call by pressing a button on the Bluetooth headset. Some Bluetooth headsets even have a small display in which the number or the name of the caller is displayed. For outgoing calls, the mobile phone's voice recognition feature can be used and thus a single button on the headset is enough to establish an outgoing connection. All this can be done while the phone remains in your pocket!

Bluetooth, however, is not limited to use with mobile phones. As great emphasis has been put on easy and fast configuration of a new connection, Bluetooth is also ideally suited for data transmission between PCs, notebooks and PDAs. With only minimal configuration of the devices, it is possible to exchange files, calendar entries and notes and to synchronize calendars and address books.

Furthermore, Bluetooth can be used to connect PCs with peripheral devices. Bluetooth-enabled printers, mouse devices, keyboards and modems are available in order to reduce the number of cables and clutter on the desktop.

Bluetooth technology can also be used for mobile game consoles, where the technology can be used to network the consoles of different players with each other.

As there are a great number of different Bluetooth devices from different vendors, reliable interoperability is of utmost importance for the success of Bluetooth. This is ensured by the Bluetooth standard and interoperability tests, which are performed during so-called 'unplug fests' and by certified Bluetooth qualification test facilities [1] for final products.

Table 6.1 lists the different Bluetooth protocol versions. Generally, a new version is always downward compatible to all previous versions. This means that a Bluetooth 1.1 device is still able to communicate with a Bluetooth 2.0 device. Functionality, however, that has been introduced with a newer version of the standard, can of course not be used with a device that supports only a previous version of the standard.

6.2 Physical Properties

Up to version 1.2 of the standard, the maximum data rate of a Bluetooth transmission channel is 780 kbit/s. All devices that communicate directly with each other have to share this data rate. The maximum data rate for a single user thus depends on the following factors:

- number of users that exchange data with each other at the same time;
- activity of the other users.

Table 6.1 Bluetooth versions

Version	Approved	Comment
1.0B	Dec. 1999	First Bluetooth version, which was only used by a few first-generation devices
1.1	Feb. 2001	This version corrects a number of errors and ambiguities of the previous version (errata list). This further increases the interoperability between devices of different vendors
1.2	Nov. 2003	Introduction of the following new features:
		• faster discovery of nearby Bluetooth devices. Devices can now also be sorted on the signal quality, as described in Section 6.4.2 • fast connection establishment, see Section 6.4.2 • adaptive frequency hopping (AFH), see Section 6.4.2 • improved speech transmission, e.g. for headsets (eSCO) as described in Sections 6.4.1 and 6.6.4 • improved error detection and flow control in the L2CAP protocol • new security functionality: anonymous connection establishments
2.0	2004	Enhanced data rates extends the Bluetooth 1.2 specification with faster data transmission modes. Further details can be found in Sections 6.2 and 6.4.1. The complete standard can be found in [2]

The highest transmission speed can be achieved if only two devices communicate with each other and only one of them has a large amount of data to transmit. In this case, the highest data rate that can be achieved is 723 kbit/s. After removing the overhead, the resulting data rate is about 650 kbit/s. The remaining bandwidth for the other device to send data in the reverse direction is about 57 kbit/s. This scenario occurs quite often, for example during web surfing or when transferring a file. In these cases, one of the two devices sends the bulk of the data while the other device only sends small amounts of data for requests or acknowledgment. Figure 6.1 shows the achievable speeds for this scenario on the left.

If both ends of the connection need to send data as quickly as possible, the speed that can be achieved at each side is about 390 kbit/s. Figure 6.1 shows this scenario in the middle section.

If more than two devices want to communicate with each other simultaneously, the maximum data rate per device is further reduced. This is shown on the right side of Figure 6.1.

In 2004, the Bluetooth 2.0 + EDR (enhanced data rate) standard [2] was released. This enables data rates of up to 2.178 Mbit/s by using additional modulation techniques. This is discussed in more detail in Section 6.4.1.

In order to reach these transmission speeds, Bluetooth uses a channel in the 2.4 GHz ISM (industrial, scientific, and medial) band with a bandwidth of 1 MHz. Gaussian frequency shift keying (GFSK) is used as modulation up to Bluetooth 1.2, while DQPSK and 8DPSK are used for EDR packets. Compared to a 22 MHz channel required for wireless LAN, the bandwidth requirements of Bluetooth are quite modest.

For bi-directional data transmission, the channel is divided into timeslots of 625 microseconds. All devices that exchange data with each other thus use the same channel and are

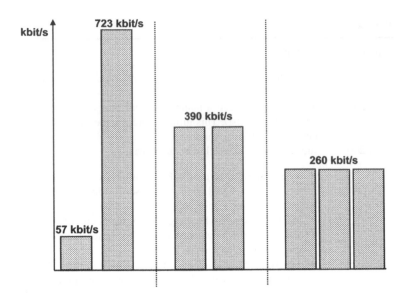

Figure 6.1 Three examples of achievable Bluetooth data rates depending on the number of users and their activity

assigned timeslots at different times. This is the reason for the variable data rates shown in Figure 6.1. If a device has a large amount of data to send, up to five consecutive timeslots can be used before the channel is given to another device. If a device has only a small amount of data to send, only a single timeslot is used. This way, all devices that exchange data with each other at the same time can dynamically adapt their use of the channel based on their data buffer occupancy.

As Bluetooth has to share the 2.4 GHz ISM frequency band with other wireless technologies like wireless LAN, the system does not use a fixed carrier frequency. Instead, the frequency is changed after each packet. A packet has a length of either one, three or five slots. This method is called frequency hopping spread spectrum (FHSS). This way, it is possible to minimize interference with other users of the ISM band. If some interference is encountered during the transmission of a packet despite FHSS, the packet is automatically retransmitted. For single slot packets (625 microseconds), the hopping frequency is thus 1600 Hz. If five slot packets are used, the hopping frequency is 320 Hz.

A Bluetooth network, in which several devices communicate with each other, is called a piconet. In order to allow several Bluetooth piconets to coexist in the same area, each piconet uses its own hopping sequence. In the ISM band, 79 channels are available. Thus, it is possible for several wireless LAN networks and many Bluetooth piconets to coexist in the same area as shown in [3].

The interference created by wireless LAN and Bluetooth remains low and hardly noticeable as long as the load in both the wireless LAN and the Bluetooth piconet(s) is low. As has been shown in Chapter 4, a wireless LAN network only sends short beacon frames while no user data is transmitted. If a wireless LAN network, however, is highly loaded, it blocks a 25 MHz frequency band for most of the time. Therefore, almost a third of the available channels for Bluetooth are constantly busy. In this case, the mutual interference of the two systems is high, which leads to a high number of corrupted packets. In order to prevent this, Bluetooth 1.2 introduces a method called adaptive frequency hopping (AFH). If all devices in a piconet are Bluetooth 1.2 compatible, the master device (see Section 6.3) performs a channel assessment to measure the interference encountered on each of the 79 channels. The link manager (see Section 6.4.3) uses this information to create a channel bitmap and marks each channel that is not to be used for the frequency-hopping sequence of the piconet. The channel bitmap is then sent to all devices of the piconet and thus, all members of the piconet are aware of how to adapt their hopping sequence. The standard does not specify a single method for channel assessment. Available choices are the received signal strength indication (RSSI) method or other methods that exclude a channel due to a high packet error rate. Bluetooth 1.2 also offers dual mode devices, which are equipped with both a wireless LAN and a Bluetooth chip, to inform the Bluetooth stack which channels are to be excluded from the hopping sequence. In practice, this is quite useful, as the device is aware which wireless LAN channel has been selected by the user, and it can then instruct the Bluetooth module to exclude 25 consecutive channels from the hopping sequence.

As Bluetooth has been designed for small, mobile and battery-driven devices, the standard defines three power classes. Devices like mobile phones usually implement power class 3 with a transmission power of up to one milliwatt. Class 2 devices send with a transmission power of up to 2.5 milliwatts. Class 1 devices use a transmission power of up to 100 milliwatts. Only devices such as some USB Bluetooth sticks for notebooks and PCs are usually equipped with a class 1 transmitter. This is due to the fact that the energy consumption compared to a class 3 transmitter is very high and should therefore only be used for devices where the energy consumption does not play a critical role. The distances that can be overcome with the different power classes are also quite different. While class 3 devices are usually designed to work reliably over a distance of 10 meters or through a single wall, class 1 devices can achieve distances of over 100 meters or penetrate several walls. The range of a piconet also depends on the reception qualities of the devices and the antenna design. In practice, newer Bluetooth devices have a much-improved antenna and receiver design, which increases the size of a piconet without increasing the transmission power of the devices. All Bluetooth devices can communicate with each other, independently of the power class. As all connections are bi-directional, however, it is always the device with the lowest transmission power that limits the range of a piconet.

Security plays an important role in the Bluetooth specifications. Thus, strong authentication mechanisms are used to ensure that connections can only be established if they have been authorized by the users of the devices that want to communicate. Furthermore, encryption is also a mandatory part of the standard and must be implemented in every device. Ciphering keys can have a length of up to 128 bits and thus offer good protection against eavesdropping and hostile takeover of a connection.

6.3 Piconets and the Master/Slave Concept

As previously described, all devices that communicate with each other for a certain time form what is called a piconet. As shown in Figure 6.2, the frequency hopping sequence of the channel is calculated from the hardware address of the first device that initiates a connection to another device and thus creates a new temporary piconet. Therefore devices can communicate with each other in different piconets in the same area without disturbing each other.

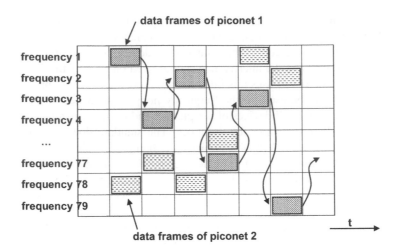

Figure 6.2 By using different hopping sequences, many piconets can coexist in the same area

A piconet consists of one master device that establishes the connection and up to seven slave devices. This seems to be a small number at first. However, as most Bluetooth applications only require point-to-point connections as described in Section 6.1, this limit is therefore sufficient for most applications. Even if Bluetooth is used with a PC to connect with a keyboard and a mouse, there are still five more devices that can join the PC's piconet at any time.

Each device can be a master or a slave of a piconet. Per definition, the device that initiates a new piconet becomes the master device as described in the following scenario.

Consider a user who has a Bluetooth-enabled mobile phone and headset. After initial pairing of the two devices (see Section 6.5.1), the two devices can establish contact with each other at any time and thus form a piconet for the duration of a phone call. At the end of a phone call the Bluetooth connection ends as well, and the piconet thus ceases to exist. In the case of an incoming call, the mobile phone establishes contact with the headset and thus becomes master of the connection. In the reverse case the user establishes an outgoing phone call by pressing a button on the headset and by using the voice-dialing feature of the mobile phone. In this case, it is the headset and not the mobile phone that establishes the connection and thus the headset becomes the master of the newly established piconet. If another person in the vicinity also uses a Bluetooth-enabled mobile phone and headset, the two piconets overlap. As each piconet uses a different hopping sequence, the two connections do not interfere with each other. Because of the initial pairing of the headset and the mobile phone it is ensured that each headset finds its own mobile phone and thus always establishes a connection for a new phone call with the correct mobile phone.

The master of a piconet controls the order and the duration of slave data transfers over the piconet channel. To grant the channel to a slave device for a period of time, the master sends a data packet to the slave. The slave is identified via a three-bit address in the header of the data packet, which has been assigned to the device at connection establishment. The data packet of the master can have a length of one to five slots depending on the amount of data that has to be sent to the slave. If no data needs to be sent to the slave, an empty

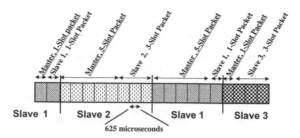

Figure 6.3 Data exchange between a master and three slave devices

one-slot packet is used. Sending a packet to a slave device implicitly assigns the next slot to the slave, independently of the packet containing user data. The slave can then use the next one to five slots of the channel in order to return a packet. With Bluetooth 1.1, slaves answer on the next hopping frequency of hopping sequence. The Bluetooth 1.2 specification slightly changes this behavior and thus Bluetooth 1.2 compliant devices answer on the same frequency the master has previously used. The slave sends an answer packet regardless if data is waiting in the buffer to be sent to the master. If no data is waiting in the slave's buffer, an empty packet is sent in order to acknowledge to the master that the device is still active and accessible. After a maximum number of five slots, the right to use the channel expires and is automatically returned to the master even if there is still data waiting in the slave's output buffer to be sent. Afterwards, the master device can decide if the channel is granted to the same or a different slave device. If the master did not receive any user data from the slave and the master's output buffer for the particular slave is also empty, it can pause the data transmission for up to 800 slots in order to save power. As the duration of a slot is 625 microseconds, 800 slots equal a transmission pause of 0.5 seconds. See Figure 6.3 for an example.

As a slave cannot anticipate when a new packet of a master arrives, a slave is thus not able to establish a connection to additional devices. In some cases, it is therefore necessary that master and slave change their roles during the lifetime of the piconet. This is necessary for example if a PDA has established a connection to a PC in order to synchronize data. As the PDA is the initiator of the connection, it is the master of the piconet. While the connection is still established, the user wants to use the PC to access a picture file on a mobile phone and thus has to include the mobile phone in the piconet. This is only possible if the PDA (master) and the PC (slave) change their roles in the piconet. This procedure is called a 'master–slave role switch'. After the role switch, the PC is the master of the piconet between itself and the PDA. Now, the PC is able to establish contact with the mobile phone while the connection to the PDA remains in place. By contacting the mobile phone and transferring the picture, however, the data rate between the PC and the PDA is reduced.

6.4 The Bluetooth Protocol Stack

Figure 6.4 shows the different layers of the Bluetooth protocol stack and will be used in the following sections as a reference. The different Bluetooth protocol layers can only be loosely coupled to the seven-layer OSI model, as some Bluetooth layers perform the tasks of several OSI layers.

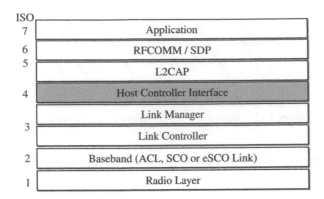

Figure 6.4 The Bluetooth protocol stack

6.4.1 The Baseband Layer

The properties of the physical layer, i.e. the radio transmission layer, have already been described. Based on the physical layer, the baseband layer performs the typical duties of a layer two protocol, such as the framing of data packets. For the data transfer, three different packet types have been defined in the baseband layer.

For packet data transmission, Bluetooth uses asynchronous connection-less (ACL) packets. As shown in Figure 6.5, an ACL packet consists of a 68–72-bit access code, an 18-bit header and a payload (user data) field of variable size between 0–2744 bits.

Before the 18 header bits are transmitted, they are coded into 54 bits by a forward error correction algorithm (1/3 FEC). This ensures that transmission errors can be corrected in most cases. Depending on the size of the payload field, an ACL packet requires one, three or five slots of 625 microseconds.

The access code at the beginning of the packet is used primarily for the identification of the piconet to which the current packet belongs to. Thus, the access code is derived from the device address of the piconet master. The actual header of an ACL packet consists of a number of bits for the following purposes: The first three bits of the header are the logical transfer address (LT_ADDR) of the slave, which the master assigns during connection establishment. As three bits are used, up to seven slaves can be addressed.

After the LT_ADDR, the four-bit packet type field indicates the structure of the remaining part of the packet. Table 6.2 shows the different ACL packet types. Apart from the number of slots used for a packet, another difference is the use of forward error correction (FEC)

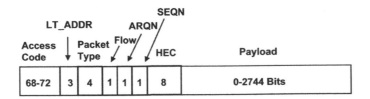

Figure 6.5 Composition of an ACL packet

Table 6.2 ACL packet types

Packet type	Number of slots	Link type	Payload (bytes)	FEC	CRC
0100	1	DH1	0–27	No	Yes
1010	3	DM3	0–121	2/3	Yes
1011	3	DH3	0–183	No	Yes
1110	5	DM5	0–224	2/3	Yes
1111	5	DH5	0–339	No	Yes

for the payload. If FEC is used, the receiver is able to correct transmission errors. The disadvantage of using FEC, though, is the reduction of the number of user data bits that can be carried in the payload field. If a 2/3 FEC is used, one error correction bit is added for two data bits. Instead of two bits, three bits will thus be transferred (2/3). Furthermore, ACL packets can be sent with a CRC checksum in order to detect transmission errors, which the receiver was unable to correct (see Figure 6.6).

To prevent a buffer overflow, a device can set the flow bit to indicate to the other end to stop data transmission for some time.

The ARQN bit informs the other end if the last packet has been received correctly. If the bit is not set, the packet has to be repeated.

The sequence bit (SEQN) is used to ensure that no packet is accidentally lost. This is done by toggling the bit in every packet. The following example shows how the bit is used in a scenario in which device-1 and device-2 exchange data packets: If device-2 receives two consecutive packets with the sequence bit set in the same way, it indicates that device-1 was unable to receive the previous packet and has thus repeated its data packet. The repetition is necessary as it is not clear to device-1 if only the return packet is missing or if its own packet was also lost. Device-2 then repeats its packet that includes the acknowledgment for the packet of device-1 and ignores all incoming packets as long as no packet with a correct SEQN bit is received by device-1. Even if multiple packets are lost, all data is eventually delivered.

The last field in the header is the header error check (HEC) field. It ensures that the packet is ignored if the receiver cannot calculate the checksum correctly.

The payload field follows the ACL header and is composed of the following fields. The first bits of the payload header field again contain some administrative information. The first field is called the logical channel (L_CH) field. It informs the receiver if the payload

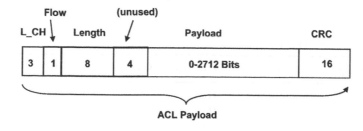

Figure 6.6 The ACL payload field including the ACL header and checksum

field contains user data (L2CAP packets, see Section 6.4.5) or link manager protocol (LMP) signaling messages (see Section 6.4.3) for the administration of the piconet.

The flow bit is used to indicate to the L2CAP layer above that the receiver buffer is full. Finally, the payload header includes a length field before the actual payload part is transmitted. After the actual payload, an ACL packet ends with a 16-bit CRC checksum.

As no bandwidth is guaranteed for an ACL connection, this type of data transmission is not well suited for the transmission of bi-directional real-time data such as a voice conversation. For this kind of application, the baseband layer offers a second transmission mode that uses synchronous connection-oriented (SCO) packets. The difference to ACL packets is the fact that SCO packets are exchanged between a master and a slave device in fixed intervals. The interval is chosen in a way that results in a total bandwidth of exactly 64 kbit/s.

When an SCO connection between a master and a slave device is established, the slave device is allowed to send its SCO packets autonomously even if no SCO packet is received from the master. This can be done very easily as the timing for the exchange of SCO packets between two devices is fixed. Therefore, the slave does not depend on a grant from the master and thus it is implicitly ensured that only this slave sends in the timeslot. This way, it is furthermore ensured that the slave device can send its packet containing voice data even if it has not received the voice packet of the master device.

The header of an SCO packet is equal to the header of an ACL packet with the exception that the flow, ARQN and SEQN fields are not used. The length of the payload field is always 30 bytes. Depending on the error correction mechanism used, this equals 10, 20 or 30 user data bytes. Table 6.3 gives an overview of the different SCO packet types.

The last line of the table shows a special packet type, which can contain both SCO and ACL data. This packet type can be used to send both voice data and signaling messages at the same time. As will be shown in Section 6.6.4 for the headset profile, an SCO connection between a mobile phone and a headset does not only require a speech channel but also a channel for signaling messages (e.g. to control the volume). The SCO voice data can then be embedded in the first 10 bytes of a 'DV' packet which are followed by up to nine bytes for the ACL channel. The forward error correction and the checksum are only applied to the ACL part of the payload.

It has to be noted that it is not mandatory to use DV packets if voice and data have to be transmitted simultaneously between two devices. Another possibility is to use independent ACL packets in slots that are not used by the SCO connection. Finally, a third possibility to send both ACL and SCO information between two devices is to drop the SCO information of a slot and to send an ACL packet instead.

As no CRC and FEC are used for SCO packets, it is not possible to detect if the user data in the payload field was received correctly. Thus, defective data is forwarded to higher layers

Table 6.3 SCO packet types

Packet type	Number of slots	Link type	Payload (bytes)	FEC	CRC
0101	1	HV1	10	1/3	No
0110	1	HV2	20	2/3	No
0111	1	HV3	30	None	No
1000	1	DV	10 (+0 − 9)	2/3	Yes

if a transmission error occurs. This produces audible errors in the reproduced voice signal. Furthermore, the bandwidth limit of 64 kbit/s of SCO connections prevents the use of this transmission mechanism for other types of interactive applications such as audio streaming in MP3 format that usually requires a higher data rate. Bluetooth 1.2 thus introduces a new packet type called enhanced-SCO (eSCO), which improves the SCO mechanism as follows.

The data rate of an eSCO channel can be chosen during channel establishment. Therefore a constant data rate of up to 288 kbit/s in full duplex mode (in both directions simultaneously) can be achieved.

eSCO packets use a checksum for the payload part of the packet. If a transmission error occurs, the packet can be retransmitted if there is still enough time before the next regular eSCO packet has to be transmitted. Figure 6.7 shows this scenario. Retransmitting a bad packet and still maintaining a certain bandwidth is possible, as an eSCO connection with a constant bandwidth of 64 kbit/s only uses a fraction of the total bandwidth available in the piconet. Thus, there is still some time to retransmit a bad packet in the transmission gap to the next packet. Despite transferring the packet several times, the data rate of the overall eSCO connection remains constant. If a packet cannot be transmitted by the time another regular packet has to be sent it is simply discarded. Thus, it is ensured that the data stream is not slowed down and the constant bandwidth and delay times required for audio transmissions are kept.

For some applications such as wireless printing or transmitting large pictures from a camera to a PC, the maximum transmission rate of Bluetooth up to version 1.2 is not sufficient. Furthermore, new technologies like HSDPA (see Chapter 3), which offer speeds of several megabits per second, also require higher transmission rates between a notebook and a mobile phone compared to what is offered by the standard Bluetooth transmission modes. Thus, the Bluetooth standard was enhanced with a high-speed data transfer mode called Bluetooth 2.0 + enhanced data rates (EDR). The core of EDR is the use of a new modulation technique for the payload part of an ACL or eSCO packet. While the header and the payload of the packet types described before are modulated using GFSK, the payload of an EDR ACL and eSCO packet is modulated using DQPSK (differential quadrature phase shift keying) or 8DPSK (eight phase differential phase shift keying). These modulation techniques allow the encoding of several bits per transmission step. Thus, it is possible to increase the data rate while the total channel bandwidth of 1 MHz and the slot time of 625 microseconds remain constant. In order to be backward compatible, the header of the new packets is still encoded using standard GFSK modulation. Thus, the system becomes backward compatible

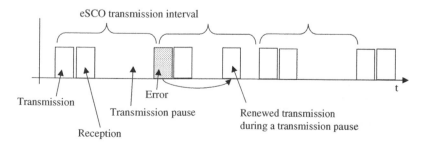

Figure 6.7   Retransmission of an eSCO packet caused by a transmission error

as legacy devices can at least decode the header of an EDR packet and thus be aware that they are not the recipient of the packet. The same approach is used by wireless LAN (see Chapter 4) to ensure backward compatibility of 802.11g networks with older 802.11b devices. Furthermore, a coding scheme for the packet type field was devised that enables non-EDR devices to recognize multislot EDR packets which are sent by a master to another slave device in order to be able to power down the receiver and thus save energy for the time the packet is sent. Table 6.4 gives an overview of all possible ACL packet types and the maximum data rate that can be achieved in an asynchronous connection. In this example, five slot packets are used in one direction and only one slot packet in the reverse direction. The first part of the table lists the basic ACL packet types which can be decoded by all Bluetooth devices. The second and third parts of the table contain an overview of the EDR ACL packet types. 2-DH1, 3 and 5 are modulated using DQPSK, while 3-DH1, 3 and 5 are modulated using 8DPSK. The numbers 1, 3 and 5 at the end of the packet type name describe the number of slots used by that packet type.

Due to the number of EDR packet types, it is no longer possible to identify all packet types using the four-bit packet type field of the ACL header (see Figure 6.5). Since it was not possible to extend the field in order to stay backward compatible, the Bluetooth specifications had to go a different way. EDR is always activated during connection establishment. If master and slave recognize that they are EDR capable, the link managers of both devices (see Section 6.4.3) can activate the EDR functionality, which implicitly changes the allocation of the packet type field bit combinations to point to 2-DHx and 3-DHx types instead of the standard packet types. While the DQPSK modulation is a mandatory feature of the Bluetooth 2.0 standard, 8DPSK has been declared an implementation option. Thus, it is not possible to derive the maximum possible speed of a device by merely looking at the Bluetooth 2.0 + EDR compliancy. As currently most Bluetooth 2.0 devices are all capable of sending 3-DH5 packets, however, it is most likely that there will only be few devices in the future that will not be able to use the higher order modulation packet types.

Apart from ACL, SCO and eSCO packets for transferring user data there are a number of additional packet types that are used for the establishment or the maintenance of a connection.

Table 6.4 ACL packet types

Type	Payload (bytes)	Uplink data rate (kbit/s)	Downlink data rate (kbit/s)
DM1	0–17	108.8	108.8
DH1	0–27	172.8	172.8
DM3	0–121	387.2	54.4
DH3	0–183	585.6	86.4
DM5	0–224	477.8	36.3
DH5	0–339	723.2	57.6
2-DH1	0–54	345.6	345.6
2-DH3	0–367	1174.4	172.8
2-DH5	0–679	1448.5	115.2
3-DH1	0–83	531.2	531.2
3-DH3	0–552	1766.4	265.6
3-DH5	0–1021	2178.1	177.1

ID packets are sent by a device before the actual connection establishment to find other devices in the area. As the timing and the hopping sequence of the other device are not known at this time, the packet is very short and only contains the access code.

Frequency hopping synchronization (FHS) packets are used for the connection establishment during the inquiry and paging phases, which are further described below. An FHS packet contains the 48-bit device address of the sending device and timing information in order to enable a remote device to predict its hopping sequence and thus to allow connection establishment.

NULL packets are used for the acknowledgment of a received packet if no user data is waiting in the output buffer of a device that could be used in the acknowledgment packet. NULL packets do not have to be acknowledged and thus interrupt the mutual acknowledgment cycle if no further data is to be sent.

An additional packet type is the POLL packet. It is used to verify if a slave device is still available in the piconet after a prolonged time of inactivity due to a lack of user data to be sent. Similarly to the NULL packet it does not contain any user data.

6.4.2 The Link Controller

The link control layer is located on top of the baseband layer which was discussed previously. As the name suggests, this protocol layer is responsible for the establishment, maintenance and correct release of connections. To administrate connections, a state model is used on this layer. The following states are defined for a device that wants to establish a connection to a remote device.

If a device wants to scan the vicinity for other devices the link controller is instructed by higher layer protocols to change into the inquiry state. In this state, the device starts to send two ID packets per slot on two different frequencies in order to request listening devices with unknown frequency hopping patterns to reply to the inquiry.

If a device is set by the user to be detectable by other devices, it has to change into the inquiry scan state periodically and scan for ID packets on alternating frequencies. The frequency a device listens to is changed every 1.28 seconds. In order to save power, or to be able to maintain already ongoing connections, it is not required to remain in the inquiry scan state continuously. The Bluetooth standard suggests a scan time of 11.25 milliseconds per 1.28 second interval. The combination of fast frequency change of the searching device on the one hand and a slow frequency change of the detectable device on the other hand results in a 90% probability that a device can be found within a scan period of 10 seconds.

In order to improve the time it takes to find devices, version 1.2 of the standard introduces the interlaced inquiry scan. Instead of only listening on one frequency per interval, the device has to search for ID packets on two frequencies. Furthermore, this version of the standard introduces the possibility to report the signal strength (RSSI, received signal strength indication) with which the ID fame was received to higher layers. Thus, it is possible to sort the list of detected devices by the signal strength and to present devices which are closer to the user at the top of the list. This is especially useful if many devices are in close proximity such as during an exhibition. In this environment, it is quite difficult to send an electronic business card to a nearby device, as the result of the scan reveals several dozen devices and it is necessary to scroll through a long list. If the list is ordered on the signal strength, however, it is very likely that the device that should receive the electronic business card is received with a high signal level and thus presented at the top of the list.

If a device receives an ID packet, it returns a frequency hopping synchronization (FHS) packet, which includes its device address, frequency hopping and synchronization information.

After receiving an FHS packet, the searching device can continue its search. Alternatively, the inquiry procedure can also be terminated in order to establish an ACL connection with the detected device by performing a paging procedure.

In order to be detectable, master devices can also enter the inquiry scan state from time to time. Thus, it is possible to detect and connect to them even if they are already engaged in a connection with another device. It has to be noted, however, that some devices like mobile phones do not support this optional functionality.

If a user wants its device to remain invisible, it is possible to deactivate the inquiry scan functionality. Thus, a device can only initiate a paging procedure and thus a connection with the user's device, if it already knows the device's hardware address. It is useful to activate this setting once a user has paired all devices (see Section 6.5.1) that are frequently used together. In this way the devices of the user remain invisible to the rest of the world but are still able to establish connections with each other. This drastically reduces the opportunity for malicious attacks on Bluetooth devices, which may try to take advantage of security holes of some Bluetooth implementations [4].

In order to establish an ACL connection by initiating a paging procedure, a device must be aware of the hardware address of the device to connect to from either a previous connection or as a result of an inquiry procedure. The paging procedure works in a similar way as the inquiry procedure, i.e. ID packets are sent in a rapid sequence on different frequencies. Instead of a generic address, the hardware address of the target device is included. The target device in turn replies with an ID packet and thus enables the requesting device to return an FHS packet which contains its hopping sequence. Figure 6.8 shows how the paging procedure is performed and how the devices enter the connected state upon success.

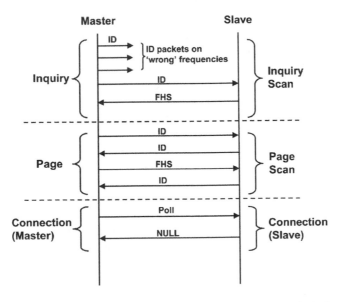

Figure 6.8   Connection establishment between two Bluetooth devices

The power consumption of a device that is not engaged in any connection and thus only performs inquiry and page scans in regular intervals is very low. Typically, the energy consumption in this state is less than one milliwatt. As mobile phones have a battery capacity of typically 2000–3000 milliwatt-hours, the Bluetooth functionality only has a small effect on the standby time of a mobile device.

After successful paging, both devices enter the connection active state and data transfer can start over the established ACL connection.

During connection establishment it can happen that the slave device is master of another connection at the same time. In such cases, the Bluetooth protocol stack enables the device to indicate during the connection establishment that a connection is only possible if a master–slave role change is performed after establishment of the connection. This is necessary, as it is not possible to be a master and a slave device at the same time. However, as a device needs to be a slave in order to be contacted, this feature allows a device to temporarily violate this rule in order to include another requesting device in its piconet.

During an active connection, the power consumption of a device mainly depends on its power class (see Section 6.2). Even while active, it is possible that for some time no data has to be transferred. Especially for devices like mobile phones or PDAs, it is very important to conserve power during those periods in order to maximize the operating time on a battery charge. The Bluetooth standard thus specifies three additional power-saving substates of the connected state.

The first substate is the connection-hold state. In order to change into this state, master and slave have to agree on the duration of the hold state. Afterwards, the transceiver can be deactivated for the agreed time. At the end of the hold period, master and slave implicitly change back into the connection-active state.

For applications that only transmit data very infrequently, the connection-hold state is too inflexible. Thus, the connection-sniff state might be used instead, which offers the following alternative power-saving scheme. When activating the sniff state, master and slave agree on an interval and the time during the interval in which the slave has to listen for incoming packets. In practice it can be observed that the sniff state is activated after a longer inactivity period (e.g. 15 seconds) and that an interval of several seconds (e.g. 2 seconds) is used. This reduces the power consumption of the complete Bluetooth chip to below one milliwatt. If renewed activity is detected, it can be noted that some devices immediately leave the sniff state even though this is not required by the standard. If the power consumption of a Bluetooth device during sniff state is compared to the wireless LAN 802.11 power-save mode, a big difference can be observed. Even while in power-save mode, a WLAN card still requires 200 to 500 milliwatts of power on average. The comparison to the power requirement of less than one milliwatt during the Bluetooth sniff state shows quite clearly that a special emphasis has been put on power-saving techniques during the development of the Bluetooth standard.

The connection-park state can be used in order to even further reduce the power consumption of the device. In this state, the slave device returns its piconet address (LT_ADDR) to the master and only checks very infrequently if the master would like to communicate.

6.4.3 The Link Manager

The next layer in the protocol stack (see Figure 6.4) is the link manager layer. While the previously discussed link controller layer is responsible for sending and receiving data packets depending on the state of the connection with the remote device, the link manager's task is to establish and maintain connections. This includes the following operations:

- Establishment of an ACL connection to a slave and assignment of a link address (LT_ADDR).
- Release of connections.
- Configuration of connections, e.g. negotiation of the maximum number of timeslots that can be used for ACL or eSCO packets.
- Activation of the EDR mode if both devices support this extension of the standard.
- Conducting a master–slave role switch.
- Performing a pairing operation as described in Section 6.5.1.
- Activating and controlling authentication and ciphering procedures if requested by higher layers.
- Control of AFH which was introduced with the Bluetooth 1.2 standard.
- Management (activation/deactivation) of power-save modes (hold, sniff and park).
- Establishment of an SCO or eSCO connection and the negotiation of parameters like error correction mechanisms, data rates (eSCO only), etc.

The link manager performs these operations either due to a request of a higher layer (see next section) or due to requests of the link manager of a remote device. Link managers of two Bluetooth devices communicate using the link manager protocol (LMP) over an ACL connection as shown in Figure 6.9. The link manager recognizes if an incoming ACL packet contains user data or an LMP message by looking at the logical channel (L_CH) field of the ACL header.

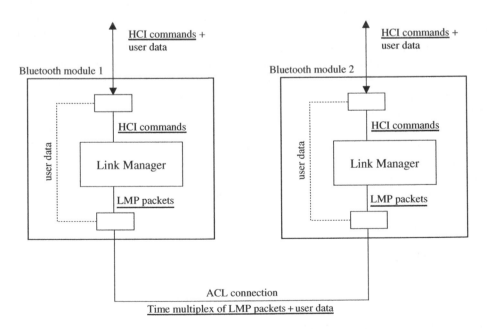

Figure 6.9 Communication between two link managers via the LMP

In order to establish a connection to higher layers of the protocol stack after a successful establishment of an ACL connection, the link manager of the initiating device (master) has to establish a connection to the link manager of the remote device (slave). This is done by sending an LMP_Host_Connection_Request message. Afterwards, optional configuration messages can be exchanged. The LMP connection establishment phase is completed by a mutual exchange of an LMP_Setup_Complete message. After this step, it is possible to transfer user data packets between the two devices. Furthermore, it is still possible at any time to exchange further LMP messages that are required for some of the operations that were described in the list at the beginning of this section.

6.4.4 The HCI Interface

The next layer of the Bluetooth protocol stack is the host controller interface (HCI). In most Bluetooth implementations, this interface is used as a physical interface between the Bluetooth chip and the host device. Exceptions are headsets for example, which implement all Bluetooth protocol layers in a single chip due to their physical size and the limitation of using Bluetooth only for a single application, i.e. voice transmission.

By using the HCI interface, the device (host) and the Bluetooth chip (controller) can exchange data and commands for the link manager with each other by using standardized message packets. Two physical interface types are specified for the HCI.

For devices like notebooks, the USB is used to connect to a Bluetooth chip. USB is the standard interface that is used in all PC architectures today to interconnect with printers, scanners, mice and other peripheral devices. The Bluetooth standard references the USB specifications and defines how HCI commands and data packets are to be transmitted over this interface.

The second interface for the HCI is a serial connection, the universal asynchronous receiver and transmitter (UART). Apart from power levels, this interface is identical to the RS-232 interface used in the PC architecture. While an RS-232 interface is limited to a maximum speed of 115 kbit/s, some Bluetooth designs use the UART interface to transfer data with a speed of up to 1.5 Mbit/s. This is necessary, as the maximum Bluetooth data rate far exceeds the ordinary speed of an RS-232 interface used with other peripheral devices. Which bandwidth is used on the UART interface is left to the developers of the host device. In practice, some devices are not able to use the maximum bandwidth offered by the Bluetooth air interface. In such cases, the processing power of the host processor is sometimes not sufficient to support the full Bluetooth speeds at the UART interface.

The following packet types can be sent over the HCI interface:

- Command packets, which the host sends to the link manager in the Bluetooth chip.
- Response packets, which the Bluetooth controller returns to the host. These packets are also called events, which are either generated as a response to a command or sent on their own, e.g. to report that another Bluetooth device would like to establish a connection.
- User data packets to and from the Bluetooth chip.

On the UART interface, the different packet types are identified by a header, which is inserted at the beginning of each packet. The first byte is used to indicate the packet type to the receiver. If USB is used as a physical interface for the HCI, the different packet types are sent to different USB endpoints. The USB polling rate of one millisecond ensures that the user data and event packets, which are transmitted from the Bluetooth chip to the host, are detected with only minimal delay.

Some Bluetooth USB adapters, which use the Bluetooth protocol stack of Widcomm (which has since been acquired by Broadcom), sometimes contain a program on the installation CD which is called 'BTserverSpyLite.exe'. This program is very well suited to monitor, record and decode HCI packets.

Figure 6.10 shows how a Bluetooth module is instructed via the HCI interface to establish a connection to another Bluetooth device. This is done by sending an HCI_Create_Connection command, which includes all necessary information for the Bluetooth controller to establish the connection to the remote device. The most important parameter of the message is the device address of the remote Bluetooth device. The controller confirms the proper reception of the command by returning an HCI_Command_Status event massage and then starts the search for the remote device. Figure 6.8 shows how this search is performed. If the Bluetooth device address is known, the inquiry phase can be skipped. If the controller was able to establish the connection, it returns an HCI_Connection_Complete event message to the host. The most important parameter of this message is the connection handle, which allows communication with several remote devices over the HCI interface at the same time. In the Bluetooth controller, the connection handle is directly mapped to the L_CH parameter of an ACL or SCO packet.

Furthermore, there are a number of additional HCI commands and events to control a connection and to configure the Bluetooth controller. A selection of those commands is presented in Table 6.5.

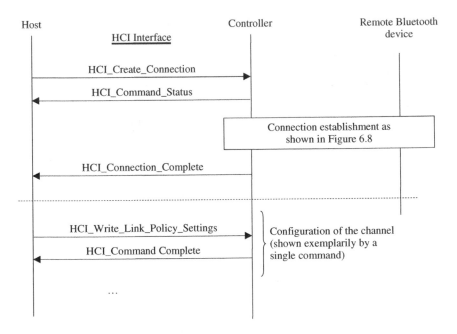

Figure 6.10 Establishment of a connection via the HCI command

Table 6.5 Selection of HCI commands

Command	Task
Setup_Synchronous_Connection	This command establishes an SCO or eSCO channel for voice applications (e.g. headset to mobile phone communication)
Accept_Connection_Request	The link manager informs the host device of incoming connections by sending a Connection_Request event message. If the host agrees to the connection request, it returns the Accept_Connection_Request command to the Bluetooth controller
Write_Link_Policy_Settings	This command can be used by the host devices to permit or restrict the hold, park and sniff states
Read_Remote_Supported_Features	If the host requires information about the supported Bluetooth functionality of a remote device, it can instruct the Bluetooth controller to request a feature list from the remote device by sending this message. Therefore the host is informed about the kind of multislot packets the remote device supports, which power-saving mechanisms it supports, if AFH is supported, etc.
Disconnect	This message releases a connection
Write_Scan_Enable	With this command, the host can control the inquiry and page scan behavior of the Bluetooth controller. If both are deactivated, only outgoing connections can be established and the device is invisible to other Bluetooth devices in the area
Write_Inquiry_Scan_Activity	This command is used to transfer inquiry scan parameters to the Bluetooth controller, e.g. the length of the inquiry scan window
Write_Local_Name	With this command, the host transfers a 'readable' device name to the Bluetooth module. The name is automatically given to remote Bluetooth devices searching for other Bluetooth devices. Thus, it is possible to assemble a list of device names instead of presenting a list of device addresses to the user as a result of a Bluetooth neighborhood search

6.4.5 The L2CAP Layer

In a next step of the overall connection establishment, an L2CAP (logical link control and adaptation protocol) connection is established over the existing ACL link. The L2CAP protocol layer is located above the HCI layer and allows the multiplexing of several logical connections to a single device via a single ACL connection. Thus, it is possible for example to open a second logical channel between a PC and a mobile phone to exchange an address book entry, while a Bluetooth dial-up connection is already established which connects the PC to the Internet via the mobile phone. If further ACL connections exist to other devices at the same time, L2CAP is also able to multiplex data to and from different devices. Such a scenario is shown in Figure 6.11. While a dial-up connection is established to slave 1, a file is transmitted simultaneously over the same connection, and an MP-3 data stream is received simultaneously from slave 2.

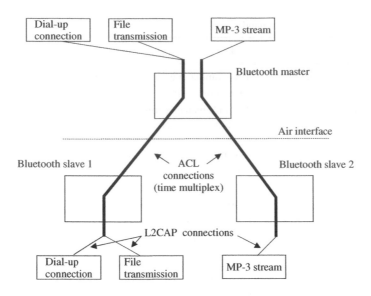

Figure 6.11 Multiplexing of several data streams

An L2CAP connection is established from the host device by sending an L2CAP_ Connection_Request message to the Bluetooth controller. The most important parameter of the message is the protocol service multiplexer (PSM). This parameter decides to which higher layer the user data packets are to be sent once the L2CAP layer is established. For most Bluetooth applications, PSM 0x0003 is used to establish a connection to the RFCOMM layer. This layer offers virtual serial connections to other devices for application layer programs and is described in more detail in Section 6.4.7. Furthermore, the L2CAP_Connection_Request message contains a connection-ID (CID) which is used to identify all packets of a particular L2CAP connection. The CID is necessary, as the RFCOMM layer can be used by several applications at the same time and thus the PSM is only unique during the connection establishment phase. If the remote device accepts the connection, it returns an L2CAP_Connection_Response message and also assigns a CID, which is used to identify the L2CAP packets in the reverse direction. Later, the connection is fully established and can be used by the application layer program. Optionally it is now also possible to configure further parameters for the connection by sending an L2CAP_Configuration_Request command. Such parameters are for example the maximum number of retransmission attempts of a faulty packet and the maximum packet size that is supported by the device.

Another important task of the L2CAP layer is the segmentation of higher layer data packets. This is necessary if higher layer packets exceed the size of ACL packets. A five-slot ACL packet for example has a maximum size of 339 bytes. If packets are delivered from higher layers that exceed this size, the packets are split into smaller pieces and sent in several ACL packets. Thus, the header of an ACL packet contains the information if the packet includes the beginning on an L2CAP packet or if the ACL packet contains a subsequent segment. At the other end of the connection, the L2CAP layer can then use this information to reassemble the L2CAP packet from several ACL packets, which is then forwarded to the application layer.

6.4.6 The Service Discovery Protocol

Theoretically it would be possible to begin the transfer of user data between two devices right after establishing an ACL and L2CAP connection. Bluetooth, however, can be used for many different applications, and many devices thus offer several different services to remote devices at the same time. A mobile phone for example offers services like wireless Internet connections (dial-up network), file transfers to and from the local file system, exchange of addresses and calendar entries, etc. In order for a device to detect which services are offered by a remote device and how they can be accessed, each Bluetooth device contains a service database that can be queried by other devices. The service database is accessed via the L2CAP PSM 0x0001 and the protocol to exchange information with the database is called the service discovery protocol (SDP). The database query can be skipped if a device already knows how a remote service can be accessed. As Bluetooth is very flexible, it offers services the option to change their connection parameters at runtime. One of those connection parameters is the RFCOMM channel number. More on this topic can be found in Section 6.4.7.

On the application layer, services are also called profiles. The headset service/headset profile ensures that a headset interoperates with all Bluetooth-enabled mobile phones that also support the headset profile. More about Bluetooth profiles can be found in Section 6.5.

Each Bluetooth service has its own universally unique ID (UUID) with which it can be identified in the SDP database. The dial-up server service for example has been assigned UUID 0x1103. In order for the Bluetooth stack of a PC to be able to connect to this service on a remote device like a mobile phone, the SDP database is queried at connection establishment and the required settings for the service are retrieved. For the dial-up server service, the database returns information to the requesting device that the L2CAP and RFCOMM layers (see next section) have to be used for the service and also informs the requestor of the correct parameters to use.

The service database of a Bluetooth device furthermore offers a general search functionality. This is required to enable a device to discover all services offered by a so far unknown device. The message sent to the database for a general search is called an SDP_Service_Search_Request. Instead of a specific UUID like in the example above, the UUID of the public browse group (0x1002) is used. The database then returns the UUIDs of all services it offers to other devices. The parameters of the individual services can then be retrieved from the database with SDP_Service_Search_Attribute_Request messages. For a service query, the database also returns the name for the requested service that can be set by the higher layers of the Bluetooth stack. Therefore it is possible to have country and language specific service names which are automatically assigned for example during the installation of the Bluetooth stack. The name, however, is just for presenting the service to the user. The Bluetooth stack itself always identifies a service by its UUID and never by using the service name. See Figure 6.12.

In practice, information that was initially retrieved from the service database of a remote device is usually stored on the application layer in order to access a remote device more quickly in subsequent communication sessions.

In order to finish the database request, the remote device releases the L2CAP connection by sending an L2CAP_Disconnection_Request message. If the device wants to establish a connection to one of the detected services right away, the ACL connection remains in place and another L2CAP_Connection_Request message is sent. This message, however, does not

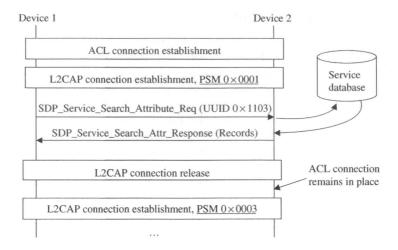

Figure 6.12   Connection establishment to a service

contain the PSM ID 0x0001 for the service database as before, but the PSM ID for the higher layer that needs to be contacted for the selected service. For most services this will be the RFCOMM layer, which offers a virtual serial connection. This service is accessed via PSM 0x0003. One of the few services that do not use the RFCOMM layer for data transfer are voice applications (e.g. headset profile), which use SCO connections for synchronous data transfer.

6.4.7 The RFCOMM Layer

As has been shown in Section 6.4.5, the L2CAP layer is used to multiplex several data streams over a single physical connection. The service database for example is a service which is accessed via the L2CAP PSM 0x0001. Other services can be accessed in a similar way by using other PSM IDs. In practice, many services also commonly use another layer, which is called RFCOMM and which is accessed with PSM 0x0003. RFCOMM offers a virtual serial interface to services and thus simplifies the data transfer.

How those serial interfaces are used depends on the higher layer service that makes use of the connection. The 'serial port' service for example uses the RFCOMM layer to offer a virtual serial interface to any non-Bluetooth application. From the application's point of view there is no difference between a virtual serial interface and a separate, physical serial interface. Usually, the operating system assigns COM port 3,4,5,6, etc. to the Bluetooth serial interfaces. Which COM port numbers are used is decided during the installation of the Bluetooth stack on the device. These serial interfaces can then be used for example during the installation of a new modem driver. When an application such as the Windows dial-up network uses the modem driver to establish a connection, the Bluetooth stack opens a connection to the remote device. This process can be performed automatically if the Bluetooth stack has previously assigned a certain COM port number to a specific remote device.

In order to simulate a complete serial interface, the RFCOMM layer does not only simulate the transmit and receive lines but also the status lines like the request to send (RTS), clear to send (CTS), data terminal ready (DTR), data set ready (DSR), data carrier detect (CD) and the ring indicator (RI) line. In a physical implementation of a serial interface, these lines are handled by a UART chip. Thus, the Bluetooth serial port service simulates a complete UART chip. A real UART chip translates the commands of the application layer into signal changes on physical lines. The virtual Bluetooth UART chip on the other hand translates higher layer commands into RFCOMM packets, which are then forwarded to the L2CAP layer.

RFCOMM is also used by other services, like the file transfer service (OBEX, Section 6.6.3) or the dial-up server service (Section 6.6.2). By using different RFCOMM channel numbers, it is possible to select during connection establishment which of the services to communicate with. The channel number is part of the service description in the SDP database. If a device asks the service database of a remote Bluetooth device for the parameters of the dial-up server service for example, the remote device will reply that the service uses the L2CAP and RFCOMM layer to provide its service. Thus, the device will establish a L2CAP connection by using PSM 0x0003 to establish a connection to the RFCOMM layer (L2CAP to RFCOMM). Furthermore, the database entry contains the RFCOMM channel number so the device can connect to the correct higher layer service. As the RFCOMM number can be assigned dynamically, the service database has to be queried during each new connection establishment to the service.

Figure 6.13 shows how different layers multiplex simultaneous data streams. While the HCI layer multiplexes the connections to several remote devices (connection handles), the L2CAP layer is responsible for multiplexing several data streams to different services per device (PSM and CID). This is used in practice to differentiate between requests to the

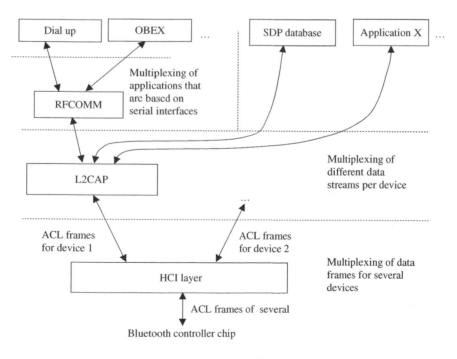

Figure 6.13 Multiplexing on different protocol layers

service database (PSM 0x0001) and the RFCOMM layer (PSM 0x0003). Apart from the service database, most Bluetooth services use the RFCOMM layer and thus can only be distinguished because they use different RFCOMM channel numbers.

The RFCOMM channel number also allows the use of up to 30 RFCOMM services between two devices simultaneously. Thus, it is possible during a dial-up connection to establish a second connection to transfer files via the object exchange (OBEX) service. As both services use different RFCOMM channel numbers, the data packets of the two services can be time multiplexed and can thus be delivered to the right services at the receiving end.

6.4.8 Bluetooth Connection Establishment Overview

Figure 6.14 gives an overview of how a Bluetooth connection is established through the different layers. In order to contact an application on a remote Bluetooth device, an ACL connection is initially established. Once the ACL link is configured, an L2CAP connection is established to the service database of the device by using the corresponding PSM number. Once the connection to the database is established, the record of the service to be used is retrieved. Then the L2CAP connection is released while the ACL connection between the two devices remains in place. In a next step, contact to the application is established over the still existing ACL connection. This is done by establishing another L2CAP connection. Most services use the RFCOMM layer for further communication, which provides virtual serial interfaces. By using the RFCOMM channel number, the Bluetooth stack can finally connect the remote device to the actual service, for example the dial-up server. How the two sides of the application communicate with each other depends on the application itself and is transparent for all layers described so far including the RFCOMM layer. In order to

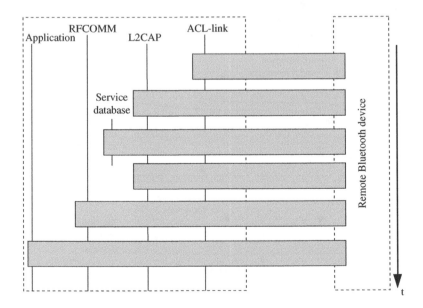

Figure 6.14 The different steps of a Bluetooth connection establishment

ensure interoperability on the application layer between devices of different manufacturers, Bluetooth defines so called 'profiles', which are described in more detail in Section 6.6.

6.5 Bluetooth Security

As Bluetooth radio waves do not stop at the doorstep, the Bluetooth standard specifies a number of security functions. All methods are optional and do not have to be used during connection establishment or for an established connection. The standard has been defined in this way as some services do not require security functionality. Which services are implemented without security is left to the discretion of the device manufacturer. A mobile phone manufacturer for example can decide to allow incoming file transfers without a prior authentication of the remote device. The incoming file can be held in a temporary location and the user can then decide to either save the file in a permanent location or to discard it. For services like dial-up data, such an approach is not advisable. Here, authentication should occur during every connection establishment attempt to prevent unknown devices from establishing an Internet connection without the user's knowledge.

Bluetooth uses the SAFER+ (secure and fast encryption routine) security algorithms, which have been developed by the ETH Zurich and are publicly available. So far, no methods have been found that compromise the keys generated by these algorithms and thus Bluetooth transmissions are considered to be safe. Reports about Bluetooth security problems as for example in [4] are specific implementation issues of different device manufacturers that have nothing to do with general Bluetooth security architecture.

6.5.1 Pairing

In order to automate security procedures during subsequent connection establishment attempts, a procedure called 'pairing' is usually performed during the first connection establishment between the two devices. From the user's point of view, pairing means to type in the same PIN number on both devices. The PIN number is then used to generate a link key on both sides. The link key is then saved in the Bluetooth device database of both devices and can be used in the future for authentication and activation of ciphering. The different steps of the paring procedure are shown in Figure 6.15 and are performed as follows.

To invoke the pairing procedure, an LMP_IN_RAND message is sent by the initiating device over an established ACL connection to the remote device. The message contains a random number. The random number is used together with the PIN and the device address to generate an initialization key, which is called K_{init}. As the PIN is not exchanged between the two devices, a third device is not able to calculate K_{init} with an intercepted LMP_IN_RAND message.

By using K_{init}, which is identical in both devices, each side then creates a different part of a combination key. The combination key is based on K_{init}, the device address of one of the devices and an additional random number, which is not exchanged over the air interface. Then the two combination key halves are XOR combined with K_{init} and are exchanged over the air interface by sending LMP_COMB_KEY messages. The XOR combination is necessary in order not to exchange the two combination key halves in clear text over the still unencrypted connection.

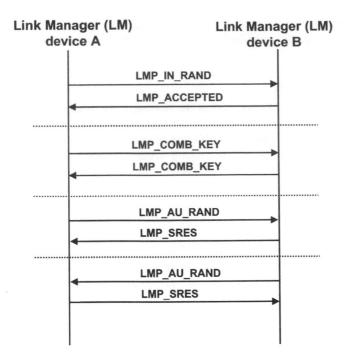

Figure 6.15   Pairing procedure between two Bluetooth devices

As K_{init} is known to both sides, the XOR combination can be reversed and thus the complete combination key is then available on both devices to form the final link key. The link key finally forms the basis for the authentication and ciphering of future connections between the two devices.

As the link key is saved in both devices, a pairing procedure and the input of a PIN by the user is only necessary during the first connection attempt. By saving the link key together with the device address of the remote device, the link key can be retrieved automatically from the database during the next connection establishment procedure. Authentication is then performed without requiring interaction with the user.

To verify that the link key was created correctly by both sides, a mutual authentication procedure is performed after the pairing. How authentication is performed is described in more detail in the next section. Figure 6.15 also shows how the complete pairing is performed by the link manager layers of the Bluetooth chips of the two devices. The only input needed from higher layers via the HCI interface is the PIN number to generate the keys.

6.5.2 Authentication

Once the initial pairing of the two devices has been performed successfully, the link key can be used for mutual authentication during every connection request. Authentication is performed using a challenge/response procedure, which is similar to procedures of systems

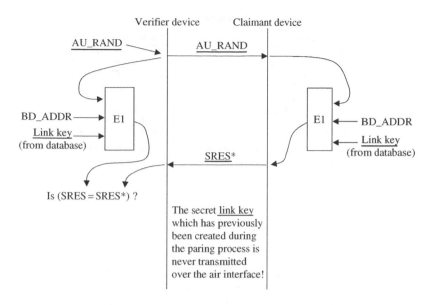

Figure 6.16   Authentication of a Bluetooth remote device

such as GSM, GPRS and UMTS. For the Bluetooth authentication procedure, three parameters are necessary:

- a random number;
- the Bluetooth address of the device initiating the authentication procedure;
- the 128-bit link key, which has been created during the pairing procedure.

Figure 6.16 shows how the initiating device (verifier) sends a random number to start the authentication procedure to the remote device (claimant). The link manager of the claimant then uses the BD_ADDR of the verifier device to request the link key for the connection from the host via the HCI interface.

With the random number, the BD_ADDR and the link key, the link manager of the claimant then calculates an answer, which is called the signed response* (SRES*), which is returned to the link manager of the verifier device. In the mean time, the verifier device has calculated its own SRES. The numbers can only be identical if the same link key was used to calculate the SRES on both sides. As the link key is never transmitted over the air interface, an intruder can thus never successfully perform this procedure.

6.5.3 Encryption

After successful authentication, both devices can activate or deactivate ciphering at any time. The key used for ciphering is not the link key that has been generated during the pairing process. Instead a ciphering key is used which is created on both sides during the activation of ciphering. The most important parameters for the calculation of the ciphering key are the link key of the connection and a random number, which is exchanged between the two

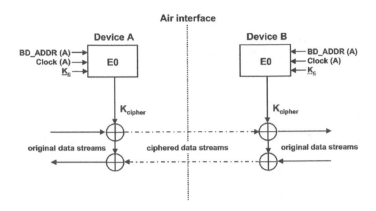

Figure 6.17 Bluetooth encryption by using a ciphering sequence

devices when ciphering is activated. Since ciphering is reactivated for every connection, a new ciphering key is also calculated for each connection. See Figure 6.17.

The length of the ciphering key is usually 128 bits. Shorter keys can be used as well if Bluetooth chips are exported to countries for which export restrictions apply for strong encryption keys.

Together with the device address of the master and the lower 26 bits of the master's real-time clock, the ciphering key is used as input value for the SAFTER+ E0 algorithm, which produces a constant bit stream. As the current value of the master's real-time clock is known to the slave as well, both sides of the connection can generate the same bit stream. The bit stream is then modulo-2 combined with the clear text data stream. Encryption is applied to the complete ACL packet including the CRC checksum before the addition of optional FEC bits.

6.5.4 Authorization

Another important concept of the Bluetooth security architecture is the 'authorization service' for the configuration of the behavior of different services for different remote users. This additional step is required in order to open services to some but not all remote devices. Thus, it is possible for example to grant access rights to a remote user for a certain directory on the local PC in order to send or receive files. This is done by activating the object exchange (OBEX) service for the particular user and his Bluetooth device. Other services, like the dial-up network service are handled in a more restricted manner and thus cannot be used by this particular remote user.

With the authorization service, it is possible to configure certain access rights for individual external devices for each service offered by the local device. It is left to the manufacturer of a Bluetooth device how this functionality is used. Some mobile phone manufacturers for example allow all external devices, which have previously performed a pairing procedure successfully, to use the dial-up service. Other mobile phone manufacturers have added another security barrier and ask the user for permission before proceeding with the connection establishment to the service.

Bluetooth stacks for PCs usually offer very flexible authentication functionality for the service offered by the device. These include:

- A service may be used by an external device without prior authentication or authorization by the user.
- A service may be used by all authenticated devices without prior authorization by the user. This requires a one-time pairing.
- A service may be used once or for a certain duration after authentication and authorization by the user.
- A service may be used by a certain device after authentication and one-time authorization by the user.

Furthermore, some Bluetooth stacks offer the display of short notices on the screen if a service is accessed by a remote device. The notice is displayed for informational purposes only, as access is automatically granted.

6.5.5 Security Modes

At which point during the establishment of an authenticated connection ciphering and authorization is performed depends on the implementation of the Bluetooth stack and the configuration of the user. The Bluetooth standard describes three possible configurations:

- If security mode 1 is used for a service, no authentication is required and the connection is not encrypted. This mode is most suitable for the transmission of address book and calendar entries between two devices. In many cases, the devices used for this purpose have previously not been paired. The electronic business card will in such a configuration therefore be first copied to a special temporary directory and only be included in the address book upon confirmation of the user.
- For security mode 2, the user decides if authentication, ciphering and authorization are necessary when a service is used. Many Bluetooth PC stacks allow individual configuration for each service. Security mode 1 therefore corresponds to using security mode 2 for a service without authentication and ciphering.
- If a service uses security mode 3, authentication and ciphering of the connection are automatically ensured by the Bluetooth chip. Both procedures are performed during the first communication between the two link managers, i.e. even before an L2CAP connection is established. For incoming communication requests, the Bluetooth controller thus has to ask the Bluetooth device database for the link key via the HCI interface. If no pairing has previously been performed with the remote device, the host cannot return a link key to the Bluetooth controller and thus the connection will fail. Security mode 3 is best suited for devices that only need to communicate with previously paired remote devices. Thus, this mode is not suitable for devices like mobile phones, which allow non-authenticated connections for the transfer of an electronic business card.

6.6 Bluetooth Profiles

As shown at the beginning of this chapter, Bluetooth can be used for a great variety of applications. Most applications have a server and a client side. A client usually establishes the

Bluetooth connection to the master and requests the transfer of some kind of data. Thus, the master and the client side of a Bluetooth service are different. For the transfer of a calendar entry from one device to another for example, the client side establishes a connection to the server. The client then transfers the calendar entry as the sending component. The server on the other hand receives the calendar entry as the receiving component. To ensure that the client can communicate with servers implemented by different manufacturers, the standard defines a number of Bluetooth profiles. For each application (dial-up services, calendar entry transmission, serial interface, etc.) an individual Bluetooth profile has been defined which describes how the server side and the client side communicate with each other. If both sides support the same profile, interoperability is ensured.

Note: The client/server principle of the Bluetooth profile should not be confused with the master/slave concept of the lower Bluetooth protocol layers. The master/slave concept is used to control the piconet, i.e. who is allowed to send and at which time, while the client/server principle describes a service and the user of a service. If the Bluetooth device, which is used as a server for a certain service, is the master or the slave in the piconet is thus irrelevant.

Table 6.6 gives an overview over a number of different Bluetooth profiles for a wide range of services. Some of them are described in more detail in the following sections.

Table 6.6 Bluetooth profiles for different applications

Profile name	Application
Dial-up networking (DUN) profile	Bluetooth connection between a modem or a mobile phone and a remote device like a PDA, PC or notebook
FAX profile	Profile for FAX transmissions
Common ISDN access profile	Profile for interconnecting an ISDN adapter with a remote device like a PDA, PC or notebook
LAN access profile	IP connection between a PDA, PC or notebook to a LAN and the Internet
Personal area network (PAN) profile	Same as the LAN access profile. However, the PAN profile does not simulate an Ethernet network card, but uses a number of newly created Bluetooth protocol layers for this purpose
File transfer profile	This profile can be used to exchange files between two Bluetooth devices
Object push profile	Simple exchange of calendar entries, address book entries, etc.; used for ad hoc transfers
Synchronization profile	Synchronization of personal information manager (PIM) applications for calendar and address book entries, notes, etc.
Basic imaging profile	Transfer of pictures from and to digital cameras
Hard copy cable replacement profile	Cable replacement between printers and a remote device like a PC
Basic printing profile	Printing profile for mobile devices like PDAs or mobile phones to enable them to print information without a printer driver

Table 6.6 (*Continued*)

Profile name	Application
Advanced audio distribution profile	Profile for the transmission of high-quality audio, e.g. music between an MP-3 player and a Bluetooth headset
Audio/video remote control profile	Profile to control audio/video devices remotely. This profile can be used for example with the advanced audio distribution profile to remotely control the audio player from the headset or an independent remote control
Headset profile	Profile for wireless headsets used with mobile phones. Voice quality transmissions only, not suitable for music
Hands-free profile	This profile is used to connect mobile phones with hands-free sets in cars
SIM-access profile	Provides access for hands-free equipment in cars to the data stored on the SIM card of a mobile phone
Human interface device (HID) profile	Connects mice, keyboards and joysticks to PCs, notebooks, PDAs and organizer phones

6.6.1 Basic Profiles: GAP, SDP, and the Serial Profile

The Bluetooth standard specifies two profiles, which the user is unable to directly see on the application level. The generic access profile (GAP) [2] defines how two devices can connect with each other in different situations and how to perform the connection establishment. The profile describes among other things:

- the presentation of Bluetooth specific parameters to the user like the device address (BD_ADDR) or the PIN;
- security aspects (security mode 1–3);
- idle mode behavior (e.g. inquiry, device discovery);
- connection establishment.

The GAP protocol thus ensures that the user interfaces for the configuration of the Bluetooth stack are similar on all devices. Furthermore, the GAP profile specifically defines which messages are sent during connection establishment, their order and which actions are taken when different options are discovered.

As shown in Section 6.4.6, each Bluetooth device has its own service database, in which each local service can store important data for the connection establishment to a remote device. The service discovery application profile [5] defines how the database is accessed and how it is structured for each profile.

The serial port profile (SPP) [6] is also a basic profile which many other profiles are based on. As the name implies, this profile simulates a serial interface for any kind of application. The profile uses the RFCOMM layer, which already offers all necessary functionalities on a lower layer. If a device has implemented this profile, any higher layer application that is able to transfer data over a serial interface is able to communicate with remote Bluetooth devices. A special adaptation of the application to the Bluetooth protocol stack is not necessary,

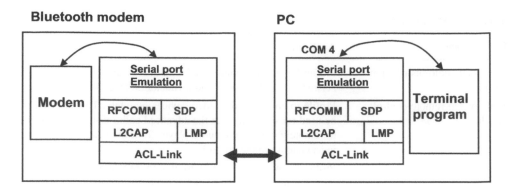

Figure 6.18 Protocol stack for the SPP

because on the application layer the simulated Bluetooth serial interface behaves like a physically present serial interface. Figure 6.18 shows the protocol stack of the SPP.

A practical example: The SPP can be used by a terminal program like Hyperterm to access a remote modem with a built-in Bluetooth interface. Before the Bluetooth connection can be used, the PC has to be paired with the modem. The Bluetooth configuration program is then used on the PC to assign a certain COM port number (e.g. COM 4) to the modem. The Bluetooth connection to the modem is automatically established whenever the terminal program is launched and the serial interface is accessed. All of this is transparent to the terminal program as it only sees the COM port which it treats as if it were a physically present interface.

6.6.2 The Network Profiles: DUN, LAP, and PAN

For network access, Bluetooth defines three different profiles.

The dial-up network (DUN) profile [7] replaces a cable connection between an end-user device like PCs or PDAs and a modem. As has been shown in Chapter 2, a modem emulation in a mobile phone can thus be used to establish an Internet connection via GPRS or UMTS. The mobile phone is accessed over the serial interface just like a modem that can be controlled via AT commands. These are specified in 3GPP TS 27.007 [8]. It is then possible to establish the GPRS or UMTS connection. The overall procedure is controlled by the dial-up networking software of the PC which only requires a serial interface and a modem. Thus, the DUN profile is based on the SPP. The most important difference is the fact that instead of a transparent connection between the both sides, a modem is simulated on the profile's server side that expects commands from the client side, i.e. from the PC's dial-up networking protocol stack.

Figure 6.19 shows how a modem is simulated by the DUN server side of the profile. In order to ensure interoperability, the DUN profile also specifies the AT commands that are to be supported. The modem side of the profile is also called the gateway. As the virtual modem for the DUN profile is already implemented in the mobile phone for other purposes, it is usually no additional effort for a mobile phone manufacturer to support not only the SPP, but the DUN profile as well.

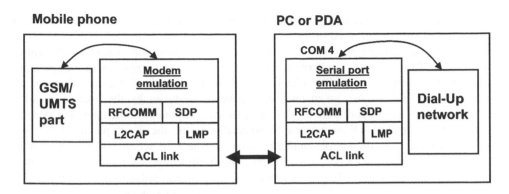

Figure 6.19 Protocol layers used by the DUN profile

On the client side of the profile, i.e. on the PC or the PDA, which is referred to as the 'data terminal' in the profile description, the profile provides a serial interface for the dial-up network of the operating system.

Once the connection to the Internet is established, the PC and the mobile phone start their PPP stacks and the network connection is established (see Chapter 2). The PPP part of the connection establishment, however, is not part of the DUN profile, as it can also be used to establish a general circuit-switched modem connection which might not require the establishment of a PPP connection over the link.

The second Bluetooth network profile is called LAN access profile (LAP) [9], and is used to interconnect a Bluetooth device with a local area network (LAN) and the Internet.

The LAN access profile is structured in a similar way as the DUN profile. On the server side of the profile, however, no modem is simulated on the highest layer of the protocol stack. Instead, the profile defines a LAN access point component. This component consists of a PPP server, which is responsible for establishing an IP connection over a serial interface.

The LAN access point can be a self-contained device with an Ethernet interface. Thus, this Bluetooth profile competes with wireless LAN, but the highest available data rates of 723 kbit/s, or about 2 Mbit/s with EDR, limits the application to pure Internet access. If it is compared with wireless LAN access points that are able to support speeds of up to 54 Mbit/s, it is obvious that this technology is better suited to support services like the exchange of large volumes of data between computers in a network.

The role of a Bluetooth LAN access point can also be taken by a PC, which is connected via Ethernet, DSL, ISDN, modem, etc. with the local area network and the Internet. The PC can then be used together with the LAP profile to connect mobile devices like PDAs quickly and cheaply to the LAN and to the Internet while being at home or in the office.

From the client's point of view, the establishment of an Internet connection using the LAN access profile is only slightly different from the establishment of an Internet connection via GPRS or UMTS and the DUN profile, as the mobile phone also implements a PPP server on top of the specified DUN functionality. Compared to the DUN profile, however, no modem commands are necessary to access the PPP server (see Chapter 2). If the DUN profile is used, the simulated or real modem at the server side can be used to establish either an Internet connection or a normal dial-up modem connection. The LAN access profile on the other hand is only used for the establishment of an IP connection over PPP. As no modem commands are necessary, the operating system's dial-up network and a so-called 'null-modem' driver can be used on the client side.

If a PC is used on the server side of the profile, the Bluetooth stack usually either implements a PPP server itself or alternatively uses the PPP server component of the operating system. On the client side, the PPP client of the dial-up network of the operating system is usually used. In this case, no manual steps have to be taken by the user to install a null-modem driver and to configure the dial-up network component of the operating system.

An advantage of the LAN access profile especially during the development of the first Bluetooth products was the simple implementation, as the PPP server and the PPP client were already part of the PC's operating system. The disadvantage for the user on the other hand was the cumbersome configuration process. This issue was solved with the emergence of the personal area network (PAN) [10] profile. Since the emergence of the PAN profile, the Bluetooth Special Interest Group (SIG) discourages further development and use of the LAN access profile. PAN is not based on a serial connection. Instead it emulates an Ethernet card from the point of view of the operating system. As shown in Figure 6.20, the RFCOMM and PPP layers are thus not used in the protocol stack for the PAN profile. Instead, the Bluetooth network encapsulation protocol (BNEP) protocol has been specified which can be accessed via L2CAP PSM 0x0015. The job of this protocol is the transmission of Ethernet packets via Bluetooth to and from virtual Ethernet networking adapters.

The PAN profile specifies three different roles for this task. The users of a PAN network are called PAN users (PANU). The users communicate with a network access point (NAP) in order to forward packets between each other or the Internet. Similar to the wireless LAN approach, the devices are not able to communicate directly with each other. The NAP is always the central point of the network and can thus forward Ethernet packets which are encapsulated by the BNEP protocol between the different PANU devices. This task is also called bridging.

If the PC of a user is used as a NAP of the network to act as a bridge between the Bluetooth users of the PAN, the network forms a so-called group ad-hoc network (GN). The PC, which takes over the responsibility of the access point does not act as a normal

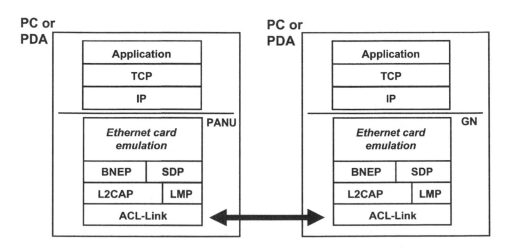

Figure 6.20 Protocol stack of the PAN profile

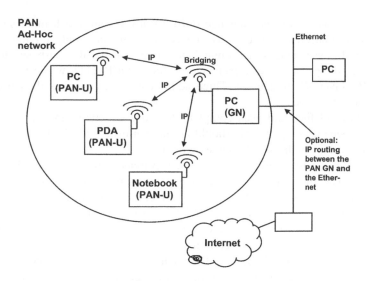

Figure 6.21   Roles in a personal area network (PAN)

PAN user device but is referred to as a GN device. GN devices fulfill the same tasks as a NAP, but are not able to bridge Ethernet packets to other non-Bluetooth network devices (no external bridging function). See Figure 6.21.

In practice this is not a big disadvantage, as instead of layer 2 bridging, most PAN devices are able to perform IP routing on layer 3 between the Bluetooth PAN users and other Ethernet devices as well as the Internet. This functionality, however, is not part of the PAN profile.

Apart from forwarding packets, NAPs and GNs usually also act as a DHCP server in order to automatically configure the IP stacks of the remote PAN users during connection establishment. This simplifies the connection establishment with the PAN network from the user's point of view to a single double click on the PAN icon of the Bluetooth user interface. Compared to the LAN access profile, which might require the installation of a null-modem and a dial-up connection profile in the dial-up network of the operating system, this is a huge advantage.

Thus the PAN profile is not only suitable to connect a PDA via the home or office PC to the Internet, but also for the quick installation of an ad-hoc IP network.

A Bluetooth PAN ad-hoc IP network has a number of clear advantages over the wireless LAN ad-hoc IP network configuration (see Section 4.3.1). While the wireless LAN ad-hoc IP network requires the configuration of a number of parameters like the selection of the appropriate ad-hoc profile, the configuration settings for the ciphering, selection of a channel and a service set ID, configuration of the IP stack, etc., a PAN ad-hoc IP network is configured very easily by pairing the devices and establishing a connection by double clicking on the PAN icon.

6.6.3 Object Exchange Profiles: FTP, Object Push, and Synchronize

The protocols that were described above for the establishment of IP connections are not suitable to quickly exchange files, electronic business cards, calendar and address book entries. For this task, the object exchange (OBEX) profiles are much more suitable. With these profiles, a connection between two devices is established only for the data transfer of one or several objects and is then immediately released. For this purpose, the Bluetooth standard defines the general object exchange profile (GOEP) [11], which uses the L2CAP and RFCOMM layers (Figure 6.22). Three additional OBEX profiles then use this basic profile for specific services.

For the transfer of files and even complete directory structures, the file transfer profile (FTP) [12] has been developed. This should not be confused with the file transfer protocol of the TCP/IP world, which uses the same acronym.

The OBEX file transfer profile is mostly used to transfer files between PCs, PDAs and mobile phones. The files can be located at any position in the file system. The GOEP defines the following commands for this task, which are sent in a binary coding over an established RFCOMM connection: DISCONNECT, PUT, GET, SETPATH and ABORT. Some PC Bluetooth stacks insert the directory tree of a remote Bluetooth device into the overall directory tree of the local device in a similar way as a remote file system on a local network. If the user clicks on the remote Bluetooth device in the directory tree, the general OBEX GET command is used to request the root directory of the remote Bluetooth device which is then presented to the user in the local file manager. The user can then select one or several files for the transfer to the local PC. For this purpose, the GOEP GET command is used. It is also possible to copy files or directories to the remote Bluetooth device. For this purpose, the general OBEX PUT command is used.

If the user changes into a subdirectory on the remote device, the OBEX SETPATH command is used in combination with another OBEX GET command to request the directory listing. Figure 6.23 shows how the content of a directory is XML encoded in a human readable format and sent to the requesting device.

In the OBEX protocol layer, the CONNECT, DISCONNECT, PUT, GET, SETPATH and ABORT commands and the corresponding answers are processed as packets. The first byte of a packet identifies the command. The command field is followed by two-byte length field and the parameters of the command. A parameter can be a directory name, a directory listing, or the contents of a requested file. The standard uses the term 'header' for a parameter, which is somewhat confusing. In order to be able to recognize the type of a parameter, each parameter contains a type information in the first byte. The type of a parameter can for example be 'filename' or 'body' (the content of the file).

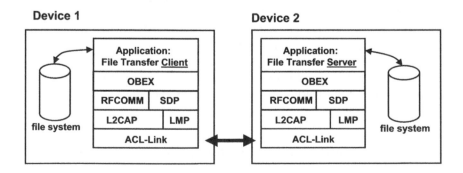

Figure 6.22 Protocol stack of the OBEX file transfer profile

```
<xml version="1.0">
<!DOCTYPE folder-listing SYSTEM „obex-folder listing.dtd">
<folder-listing-version="1.0">
    <folder name="Camera" modified="2004117T100840"
    user perm="RWD" group perm"W" />
    <folder name="other pics" modified="2004117T13321"
    user perm="RWD" group perm"W" />
</folder-listing>
```

Figure 6.23   XML-encoded directory structure

The maximum size of a packet is 64 kbytes. In order to transfer bigger files, i.e. 'header' of type 'body', a file is automatically split into several packets by the OBEX layer.

A somewhat simpler application for the GOEP is the object push profile [13] (Figure 6.24). This profile is used if the user wants to transmit a single calendar entry, address book entry or a single file via Bluetooth to another device. The profile works in the same way as the file transfer profile, as it also uses general OBEX commands like PUT and GET. The object push profile, however, does not support directory operations or the deletion of files. This simplification accelerates the process for single objects, as only a few decisions have to be made by the user before the object is transmitted.

Many devices allow an incoming object push transfer without prior authentication and ciphering. The object is then stored in a buffer and only inserted into the calendar, the address book or copied to the file system once the user has authorized the transfer.

For the transmission of calendar and address book entries, the Bluetooth standard requires the use of vCalendar and the vCard format, which are standardized in [14]–[18]. This is a precondition in order to exchange address book and calendar entries between any program and any end-user device. For other objects such as pictures, the file name extension (e.g. .gif, .jpg, etc.) can be used by the receiver to make a decision on how to treat the received object.

Even though the profile is called 'object push', it also defines an optional business card pull functionality which can be used to send a predefined business card to a remote device upon its request. The business card exchange feature extends the functionality to automatically

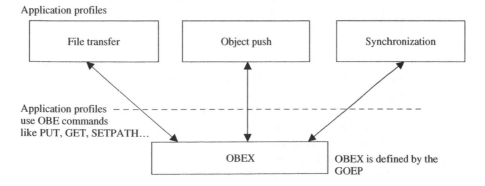

Figure 6.24   The FTP, object push and synchronization profiles are based on GOEP

send the business card stored in the retrieving device during a request for a business card to the remote device.

The third profile, which is based on GOEP, is the synchronization profile [19]. It allows automated synchronization of objects like calendar, address book entries, notes, etc. Again, general OBEX commands like GET and PUT are used. While the object push profile can only transfer a single address book entry to a remote device, the synchronization profile describes how to synchronize all records of a database. During the first synchronization attempt, all entries of the database on both devices are exchanged with each other. During all subsequent synchronizations, only objects that have changed since the last synchronization session are then updated on both sides. This is achieved by recording every change of an object in a journal. To allow the exchange database records of products of different vendors, the synchronization profiles also use the standardized vCard and vCalendar formats.

The Bluetooth standard does not itself define how the synchronization is performed, but uses the synchronization system defined in the IrMC standard [20] of the Infrared Data Association.

6.6.4 Headset, Hands-Free, and SIM Access Profile

Wireless headsets for mobile phones were the first Bluetooth devices on the market. To establish a voice channel between the mobile phone and the headset, the headset profile [21] is used. This profile is special, as it is one of the few profiles which uses SCO packets (see Section 6.4.1) for a connection. The SCO connection has a bandwidth of 64 kbit/s and carries the bi-directional audio stream between the headset and the mobile phone. If both devices are Bluetooth 1.2 compatible, eSCO packets are used for the voice path to add error correction and AFH. These features, which have been introduced with Bluetooth 1.2, particularly increase the speech quality if the error rate on the Bluetooth link increases due to an increased distance between the two devices or if there are obstacles in the transmission path which decrease the channel quality. If one of the two devices is not yet compatible with Bluetooth 1.2, the link manager layer automatically ensures that only SCO packets are sent and that AFH remains deactivated.

In order to use a headset with a mobile phone, the two devices have to be paired initially. Afterwards, the mobile phone tries to establish a connection to the Bluetooth headset for every incoming call. For the signaling between the headset and the mobile phone, referred to as the audio gateway (AG) in the Bluetooth headset standard, an ACL connection is used. The signaling connection uses the L2CAP and RFCOMM layers for communication as shown in Figure 6.25.

In order to exchange commands and the corresponding responses between the audio gateway and the headset, the AT command language is used, which was initially designed for the communication between a data terminal and a modem. The headset profile not only reuses some of the well-known AT commands, but also defines a number of extra commands to account for the special nature of the application. Figure 6.26 shows, how the audio gateway establishes a signal channel based on an ACL connection in order to send an unsolicited 'RING' response to the headset. The headset then informs the user about the incoming call by generating a 'ringing tone'. The user can then answer the call by pressing a key on the headset. When the user presses the accept button, the headset sends the following command to the audio gateway to open the speech path: 'at+ckpd=200'. The mobile phone

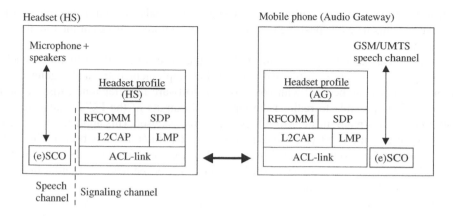

Figure 6.25 The headset profile protocol stack

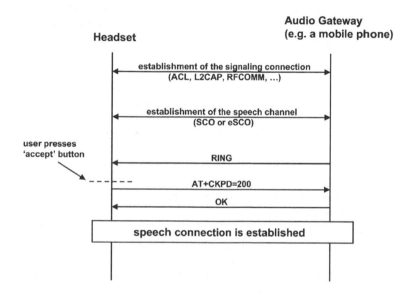

Figure 6.26 Establishment of the signaling and the speech channel

then accepts the call and starts acting as an audio gateway between the mobile network and the headset.

In order to conduct an outgoing call, the headset is also able to establish a new connection to the audio gateway. Together with the speech dialing function, which is usually part of the mobile phone, it is possible to initiate outgoing calls via the headset without any interaction with the mobile phone.

Due to the small size of the headset, only few functionalities of the remote device can be controlled via the headset. Thus, the only additional functionality of the headset profile is to control the volume of an ongoing conversation. This is done via +vgm AT commands, to control the volume of the microphone, and +vgs commands to control the volume of the loudspeaker. In this way it is possible to also control the volume settings of the mobile phone via the headset.

The headset can also be paired with a PC in the case where the Bluetooth stack on the PC supports the headset profile and where it has implemented the audio gateway role. Therefore a headset can be used with voice over IP software for telephone calls via the PC. Furthermore, it is possible to redirect the inputs and outputs of the PC's soundcard to the headset, in order to stream music, MP-3 files, etc. to the headset. This application is not very useful, however, as the (e)SCO channel is limited to 64 kbit/s and has been optimized for the transmission of mono audio signals only. Furthermore, the frequency band is limited to 300–3400 Hz. A much more suitable profile for this task is the advanced audio distributed profile, which is described in more detail in Section 6.6.5.

Closely related to the headset profile is the hands-free profile [22], which addresses the special needs of hands-free devices in cars that cannot be fulfilled by the headset profile. The most important feature of this profile is to replace the wired connection between the hands-free car kit and the mobile phone. By using this profile, the mobile phone need not be installed in the car and can thus remain in the pocket or bag of the user. Despite the similar purpose of a headset and a hands-free set, an additional profile was necessary, as a hands-free set today typically offers much more possibilities to interact with the mobile phone then a headset.

The basic mode of operation of the hands-free profile is identical to the headset profile. Commands and replies are exchanged between the hands-free unit and the mobile phone (audio gateway) via AT commands. Furthermore, the headset profile also uses SCO or eSCO connections for the voice path. In addition to the functionality of the headset profile, the hands-free profile offers the following functionalities:

- The transmission of the caller's number to the hands-free kit (CLIP).
- The hands-free set can reject incoming calls.
- The hands-free set can send a phone number to the AG, which the user has typed in via the keypad of the hands-free set.
- Call hold and multiparty calls.
- Transmission of status information such as remaining battery capacity and mobile network reception conditions of the mobile phone.
- Transmission of a roaming indicator to allow the hands-free set to indicate to the user that the phone is registered in a foreign network.
- Deactivation of the optional echo canceller of a mobile phone if the hands-free kit uses an integrated echo canceller.

If a mobile phone manufacturer would like to support headsets and hands-free kits, both headset and hands-free profiles should be implemented. Some headsets can also act as hands-free kits if they have implemented this profile as well.

Unfortunately the Bluetooth standard has not defined any interoperability between the two profiles. If the user of a mobile phone has a headset for phone calls outside the car, and a hands-free kit for calls while driving, it is not standardized which device a mobile phone should contact for an incoming call if both devices are in range. Simple implementations only communicate with either a headset or with a hands-free device, depending on which device the mobile phone was last paired with. As an alternative, it can be imagined that the mobile phone has a manual mechanism that allows the user to select the device to contact for an incoming call. At the beginning and the end of a trip, however, the user would have to manually configure the mobile phone, which greatly reduces usability. A possible solution would be a hands-free kit with a removable Bluetooth headset. If the Bluetooth headset is in the hands-free kit, the Bluetooth functionality is used in conjunction with the hands-free kit in the car. When the user leaves the car, the headset can be carried along. In both cases, the mobile phone communicates with the same device and thus the conflict is resolved.

Another possibility to use a headset and a hands-free car kit is the SIM access profile [23]. Contrary to the headset and hands-free profile, the mobile phone is not used as an audio gateway, i.e. as a bridge to the mobile phone network, but only offers access to its SIM card to an external device. Figure 6.27 shows this scenario. The external device, which will in most cases be a hands-free car kit, contains its own GSM/UMTS mobile phone except for the SIM card. When the hands-free kit is activated at the beginning of a trip, it establishes a Bluetooth connection to the mobile phone it has previously been paired with. Activating the SIM access server in the mobile phone deactivates the mobile phone's radio module. This is necessary as the radio module in the hands-free kit is used for the communication with the mobile network. Another big advantage of this method is the fact that the hands-free kit is usually connected to the power system of the car and an external antenna, which is not possible with the headset and the hands-free profile.

Figure 6.27 also shows the protocol stack that is used by the SIM access profile. Based on an L2CAP connection, the RFCOMM layer is used for a serial transmission between the hands-free kit (SIM access client) and the mobile phone (SIM access server). Apart from SIM access profile commands for activating, deactivating and resetting the SIM card, the Bluetooth connection is also used to send SIM card commands and responses. The commands and responses are sent as application protocol data units (APDUs) (see Section 1.10, Figures 1.49 and 1.50). Instead of exchanging APDUs between the radio part of the hands-free kit and the SIM card in the mobile phone via an electrical interface, the APDUs are exchanged via the Bluetooth channel. For the higher layers of the software of the hands-free kit, it is

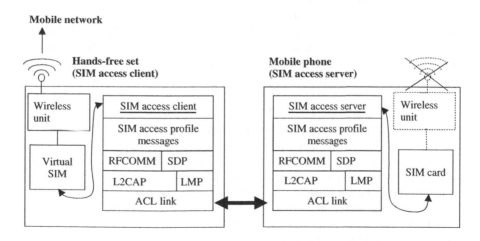

Figure 6.27 Structure of the SIM access profile

completely transparent that the SIM card is not embedded in the device but queried via a Bluetooth connection.

By using APDUs, it is not only possible to read and write files on the SIM card, but also to invoke the GSM or UMTS security mechanism embedded in the SIM card. This is done by sending an authentication command to the SIM card, including a random number (RAND) as described in Section 1.6.4. Furthermore, the SIM application toolkit can be used over the Bluetooth connection as these messages are also embedded into APDUs as described in Section 1.10.

6.6.5 High-Quality Audio Streaming

Both the headset and the hands-free profile have been designed to carry telephony grade (mono) voice channels with a limited bandwidth. For high-quality audio streaming, a much higher quality is required. Therefore, the advanced audio distribution profile (A2DP) [24] has been designed to carry audio data streams with bandwidths ranging from 127 kbit/s to 345 kbit/s depending on the audio stream type. As these data rates cannot be achieved by using SCO links, ACL links were selected to carry the audio stream. The first version of this profile has been available since 2003, but a number of devices that support this profile have only recently appeared on the market. Among them are mobile phones with built-in MP-3 players that are able to wirelessly stream MP-3 files to a high-quality audio headset. Some headsets even support the A2DP profile as well as the standard headset profile and can be used for both audio streaming and telephony.

One of the big problems of digital audio transmission is content protection and digital rights management on digital links. Therefore the A2DP profile offers methods to allow developers to use content protection over the Bluetooth link by encrypting the data stream. Therefore it is not possible to intercept the incoming audio stream on the link or at the receiving side by using a program other than the one intended by the manufacturer. However, the actual content protection method is not part of the standard. This proves to be problematic, as interoperability cannot be ensured with devices that, if implemented, use different content protection schemes. This might be one of the reasons why Bluetooth audio streaming products are only slowly appearing on the market.

Figure 6.28 shows the protocol stack used by the A2DP profile. The profile is based on GAP, which allows remote devices to query the supported features of the profile in the SDP database. Above the L2CAP layer, the audio video distribution transfer protocol (AVDTP) [25] is used to carry the audio data stream. As the protocol name implies, it can be used to carry both audio and video streams, and can thus be considered to be a generic transfer protocol for multimedia streams. The A2DP profile simply uses the protocol to transfer audio streams. Apart from the actual data stream, the protocol is also used to exchange control information between the two devices that are required for codec negotiation and to configure parameters like the bandwidth to be used for the stream. Higher layer control functionality like switching to the next music track or to pause the transmission from a remote device are not part of AVDTP and are handled by the audio/video control transport protocol (AVCTP), which is described further below.

The standard allows devices such as mobile phones to handle several Bluetooth applications simultaneously and to communicate with several remote devices at the same time. If this is supported by a device, it is possible to use an A2DP session between a notebook and

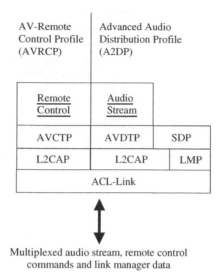

Multiplexed audio stream, remote control
commands and link manager data

Figure 6.28 The protocol stack used for A2DP and remote control

a headset together with a dial-up profile session to a mobile phone for Internet connectivity simultaneously. It has to be noted, however, that the A2DP session requires a substantial percentage of the overall capacity of the piconet, so Internet transfer speeds might be slower. If all devices support the Bluetooth version 2.0 + EDR standard, this is less of a problem due to the fact that the total bandwidth of EDR piconets is about 2 Mbit/s. Remember that Bluetooth version 1.2 only supports 723 kbit/s for standard devices, of which about 345 kbit/s are used for the highest quality audio codec.

The A2DP profile specifies two roles for a connection: the audio source is typically an MP-3 player, a multimedia mobile phone, or a microphone. The audio sink role is typically implemented in a headset or a Bluetooth-enabled loudspeaker set.

To ensure that A2DP compliant devices share at least a single common codec for audio transmissions, the profile contains the description of a proprietary audio stream format, called sub-band codec (SBC), which is mandatory for implementation in all A2DP compliant devices. A short description of this codec can be found below. Furthermore, the standard defines how audio streams encoded with MPEG 1-2 audio, MPEG-2,4 AAC and ATRAC shall be transported via the AVDTP. The implementation of these codecs is optional. The standard also offers the possibility to transport other codecs over AVDTP. To ensure inter-operability, it is defined that a device supporting additional codecs must always be able to recode the audio stream into SBC if the remote device does not support the codec.

On a high level, the SBC codec works as follows: At the input, the SBC coder expects a PCM-coded audio signal at a certain sampling frequency. For high audio quality, the standard suggests using either 44.1 or 48 kHz. The codec then separates the frequency range of the input signal into several frequency slices, which are also referred to as sub-bands. The standard suggests splitting the signal into either four or eight sub-bands, each dealing with a certain frequency range of the input signal. Afterwards, a scaling factor is calculated for each sub-band, which gives an indication of the loudness of the signal in the sub-band.

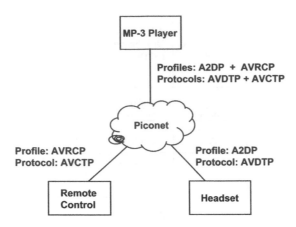

Figure 6.29   Simultaneous audio streaming and control connections to different devices

The scaling factors are then compared with each other in order to encode more important sub-bands with a higher number of bits. The recommendation for the number of bits to use for this purpose ranges from 19 for middle quality mono audio channels to up to 55 for high-quality joint stereo channels. The results of the different sub-bands are then compressed with a variable compression factor. Using the lowest compression factor in order to achieve the highest audio quality finally results in a bit stream of about 345 kbit/s.

To transfer user commands from the audio sink device (e.g. headset) like volume control, next/previous track, pause, etc. back to the audio source device (e.g. MP-3 player), the audio/video remote control profile (AVRCP) [26] is used. The profile uses the AVCTP [27] as shown in Figure 6.29 to send the commands from controller devices and to receive responses from target devices. To achieve interoperability between controller and target devices, the remote control profile specifies the following target device categories:

- Category 1: Player/Recorder.
- Category 2: Monitor/Amplifier.
- Category 3: Tuner.
- Category 4: Menu.

Depending on the device category, the standard then defines a number of control commands (operation IDs) and indicates for each device category if the support is mandatory or optional. Here are some examples of standardized control commands: 'select', 'up', 'right', 'root menu', 'setup menu', 'channel up', 'channel down', 'volume up', 'volume down', 'play', 'stop', 'pause', 'eject', 'forward', 'backward'. Vendor-specific control commands can be added to the list of commands which, however, reduce the interoperability between devices and should therefore only be added with care.

It has to be noted, that there is no interaction between the audio streaming session that uses the A2DP profile and a control session that uses the remote control profile. Thus, it is possible to form a piconet where an MP-3 player streams audio to a headset while it receives commands (e.g. volume control commands) from a third device such as a remote control.

6.7 Comparison between Bluetooth and Wireless LAN

Referring to the last sections, which discussed the different Bluetooth profiles, it has become apparent that wireless LAN and Bluetooth are not competing but rather complementing technologies.

The big strength of wireless LAN is communication in a local area network, as it replaces an Ethernet cable with a wireless connection. The focus of the evolution of wireless LAN is speed increase beyond the 11 and 54 Mbit/s, which are standardized today. Wireless LAN integrates perfectly into the network area, as from the point of view of a network application there is no difference between a wired and a wireless Ethernet. Despite power-saving mechanisms, the power consumption of a wireless LAN chip must not be underestimated, as it leads to a clear reduction in the operating times of small battery-driven devices like PDAs. Nevertheless, PCs, notebooks as well as PDAs are equipped with wireless LAN functionality today, as many homes, companies and public buildings such as hotels or cafés offer Internet access via a wireless LAN access point.

Bluetooth can also be used for networking purposes between notebooks, PCs and PDAs. Due to the achievable speeds of 723 kbit/s, i.e. 0.7 Mbit/s, or about 2 Mbit/s with EDR, Bluetooth only plays a minor role in this area due to its limited bandwidth in comparison to wireless LAN.

As has been shown in the last section, the big advantages of Bluetooth can be found outside of the local area network. These range from fast and simple transmission of files, business cards, calendar and address book entries, over cable replacement for Internet connections via mobile phones, to cable replacements for the desktop to interconnect PCs, printer, mice, keyboards, etc. For all of those applications, the achievable speeds are by far sufficient, and the integrated power-saving mechanisms are an indispensable precondition for many applications.

For these reasons, it can be foreseen that in the future, more and more PCs, notebooks and PDAs will use both Bluetooth and wireless LAN in order to benefit from the possibilities of both radio technologies.

6.8 Questions

1. What are the maximum speeds that can be achieved by Bluetooth and what do they depend on?
2. What is the frequency hopping spread spectrum (FHSS) and which enhanced functionalities are available with Bluetooth 1.2 in this regard?
3. What is the difference between inquiry and paging?
4. What kinds of power-saving mechanisms exist for Bluetooth devices?
5. What are the tasks of the link manager?
6. How can several data streams for different applications be transferred simultaneously by the L2CAP protocol?
7. What are the tasks of the service discovery database?
8. How can several services use the RFCOMM layer simultaneously?
9. What is the difference between authentication and authorization?
10. Why are such a high number of different Bluetooth profiles required?
11. Describe the necessary steps to establish an Internet connection via the dial-up networking (DUN) profile.

12. Which profiles can be used to quickly transfer files and objects between two Bluetooth devices?
13. What are the differences between the hands-free profile and the SIM access profile?
14. What are the advantages of integrating both wireless LAN and Bluetooth in a single device?

Answers to these questions can be found on the companion website for this book at http://www.wirelessmoves.com.

References

[1] Bluetooth Qualification Program website, http://qualweb.bluetooth.org.
[2] Bluetooth Special Interest Group, Bluetooth Specification Version 2.0 + EDR [vol. 0], November 2004, http://www.bluetooth.org.
[3] Jung-Hyuck Jo and Nikil Jyand, 'Performance Evaluation of Multiple IEEE 802.11b WLAN Stations in the Presence of Bluetooth Radio Interference', *IEEE International Conference on Communications*, vol. 26, May 2003.
[4] Adam Laurie and Ben Laurie, 'Serious Flaws in Bluetooth Security Lead to Disclosure of Personal Data', http://www.thebunker.net/security/bluetooth.htm, 2003.
[5] Bluetooth Special Interest Group, 'Bluetooth Specification Version 1.1 Part K:2 – Service Discovery Application Profile', February 2001, http://www.bluetooth.org.
[6] Bluetooth Special Interest Group, 'Bluetooth Specification Version 1.1 Part K:5 – Serial Port Profile', February 2001, http://www.bluetooth.org.
[7] Bluetooth Special Interest Group, 'Bluetooth Specification Version 1.1 Part K:7 – Dial-Up Networking Profile', February 2001, http://www.bluetooth.org.
[8] 3GPP, 'AT Command Set for User Equipment (UE)', TS 27.007.
[9] Bluetooth Special Interest Group, 'Bluetooth Specification Version 1.1 Part K:9 – The LAN Access Profile', February 2001, http://www.bluetooth.org.
[10] Bluetooth Special Interest Group, 'Personal Area Network Profile Version 1.0', February 2003, http://www.bluetooth.org.
[11] Bluetooth Special Interest Group, 'Bluetooth Specification Version 1.1 Part K:10 – Generic Object Exchange Profile', February 2001, http://www.bluetooth.org.
[12] Bluetooth Special Interest Group, 'Bluetooth Specification Version 1.1 Part K:12 – File Transfer Profile', February 2001, http://www.bluetooth.org.
[13] Bluetooth Special Interest Group, 'Bluetooth Specification Version 1.1 Part K:11 – Object Push Profile', February 2001, http://www.bluetooth.org.
[14] T. Howes, M. Smith and F. Dawson, 'MIME Content Type for Directory Information', RFC 2425, September 1998.
[15] F. Dawson, T. Howes and M. Smith, 'vCard MIME Directory Profile', RFC 2426, September 1998.
[16] F. Dawson and D. Stenerson, 'Internet Calendaring and Scheduling Core Object Specification (iCalendar)', RFC 2445, September 1998.
[17] S. Silverberg, S. Mansour, F. Dawson and R. Hopson, 'iCalendar Transport Independent Interoperability Protocol (iTIP) Scheduling Events, BusyTime, To-dos and Journal Entries', RFC 2446, September 1998.
[18] F. Dawson, S. Silverberg and S. Mansour, 'iCalendar Message-Based Interoperability Protocol (iMIP)', RFC 2447, November 1998.
[19] Bluetooth Special Interest Group, 'Bluetooth Specification Version 1.1 Part K:13 – Synchronization Profile', February 2001, http://www.bluetooth.org.
[20] Infrared Data Association, 'Specifications for Ir Mobile Communications (IrMC) V1.1', March 1999, http://www.irda.org.
[21] Bluetooth Special Interest Group, 'Bluetooth Specification Version 1.1 Part K:6 – Headset Profile', February 2001, http://www.bluetooth.org.
[22] Bluetooth Special Interest Group, 'Hands-Free Profile, Version 1.0', April 2003, http://www.bluetooth.org.

[23] Bluetooth Special Interest Group, 'SIM Access Profile Interoperability Specification, Revision V10r00', May 2005, http://www.bluetooth.org.

[24] Bluetooth Special Interest Group, 'Advanced Audio Distribution Profile Specification', Version 1.0, May 2003, http://www.bluetooth.org.

[25] Bluetooth Special Interest Group, 'Audio/Video Distribution Transfer Protocol Specification, Version 1.0', May 2003, http://www.bluetooth.org.

[26] Bluetooth Special Interest Group, 'Audio/Video Remote Control Profile, Version 1.0', May 2003, http://www.bluetooth.org.

[27] Bluetooth Special Interest Group, 'Audio/Video Control Transport Protocol Specification, Version 1.0', May 2003, http://www.bluetooth.org.

Index

Printed and bound in the UK by
CPI Antony Rowe, Eastbourne

Printed and bound by CPI Group (UK) Ltd, Croydon, CR0 4YY

27/10/2024

14580150-0003